D1426572

SPECTROMETRIC TECHNIQUES

VOLUME II

Contributors

JAMES B. BRECKINRIDGE
ISAIAH COLEMAN
GUY GUELACHVILI
LEON HEROUX
P. L. KELLEY
WILLIAM G. MANKIN
RUDOLF A. SCHINDLER
H. R. SCHLOSSBERG
ALEXANDER S. ZACHOR

SPECTROMETRIC TECHNIQUES

Edited by GEORGE A. VANASSE

Optical Physics Division
Air Force Geophysics Laboratory (AFGL)
Hanscom Air Force Base
Bedford, Massachusetts

VOLUME II

1981

ACADEMIC PRESS
A Subsidiary of Harcourt Brace Jovanovich, Publishers
New York London Toronto Sydney San Francisco

ACADEMIC PRESS, INC.
111 Fifth Avenue, New York, New York 10003

United Kingdom Edition published by
ACADEMIC PRESS, INC. (LONDON) LTD.
24/28 Oval Road, London NW1 7DX

Library of Congress Cataloging in Publication Data
Main entry under title:

Spectrometric techniques.

Includes bibliographies and index.
1. Spectrum analysis. 2. Spectrum analysis--Instruments.
I. Vanasse, George A.
QC451.S619 535.8'4 76-13949
ISBN 0-12-710402-X (v. 2) AACR2

PRINTED IN THE UNITED STATES OF AMERICA

81 82 83 84 9 8 7 6 5 4 3 2 1

Contents

List of Contributors

Numbers in parentheses indicate the pages on which the authors' contributions begin.

James B. Breckinridge (63), Jet Propulsion Laboratory, California Institute of Technology, Pasadena, California 91103

Isaiah Coleman (127), Bartlett Systems, Inc., Woody Creek, Colorado 81656

Guy Guelachvili (1), Laboratoire d'Infrarouge, Associé au C.N.R.S., Université de Paris-Sud, 91405 Orsay-Cedex, France

Leon Heroux (239), Aeronomy Division, Air Force Geophysics Laboratory, Hanscom AFB, Bedford, Massachusetts 01731

P. L. Kelley (161), Lincoln Laboratory, Massachusetts Institute of Technology, Lexington, Massachusetts 02173

William G. Mankin (127), National Center for Atmospheric Research, Boulder, Colorado 80307

Rudolf A. Schindler (63), Jet Propulsion Laboratory, California Institute of Technology, Pasadena, California 91103

H. R. Schlossberg (161), Air Force Office of Scientific Research, Bolling Air Force Base, Washington, D.C. 20332

Alexander S. Zachor (127), Atmospheric Radiation Consultants, Inc., Acton, Massachusetts 01720

Preface

This volume of Spectrometric Techniques contains articles that cover a broader scope than those of Volume I as far as spectral region and techniques are concerned. Specific aspects of the technique of Fourier transform spectroscopy are treated, referring the reader to Volume I for its general description. It was also decided to present the latest work in infrared spectroscopy with tunable lasers. A chapter on vacuum ultraviolet techniques is included to complete the demonstration of the diversity of techniques available to the spectroscopist interested in the ultraviolet visible and infrared spectral regions.

Contents of Volume I

Chapter 1

Distortions in Fourier Spectra and Diagnosis

GUY GUELACHVILI

LABORATOIRE D'INFRAROUGE, ASSOCIÉ AU C.N.R.S.
UNIVERSITÉ DE PARIS-SUD
ORSAY, FRANCE

1.1. Introduction

The main difficulty associated with Fourier spectroscopy has long been the computation of the Fourier transform itself, drastically limited by the limited power of the available computers. The general and tremendous

ISBN 0-12-710402-X

progress made in computer technology has overcome this (still recently) impassable barrier and allowed the method of Fourier spectroscopy to blossom. There is currently no serious technical reason preventing the numerical transformation of a great number of points, the computation time being expressed in seconds even for a 10^6 sample transformation (P. Connes, 1971; Delouis, 1971). Indeed, it is considerably more time-consuming to draw the computed spectrum.

Nowadays the main difficulty in performing Fourier spectroscopy rests in the correct realization of the interferogram, the recording of which has strong similarities to the ruling of a grating, which has long been known to be a delicate operation. Also, even if Fourier spectroscopy has exponentially expanded its influence in the past decade, it still remains a new spectroscopic approach. The interpretation of the effects observed in the spectrum due to features occurring in the interferogram often needs unusual mental gymnastics, which are facilitated neither by the general complexity of the recording procedure nor by the time interval (generally considerable) between acquiring the interferogram and obtaining the computed spectrum. This naturally led Fourier people to concentrate their efforts also on the achievement of real-time Fourier computers (Strong and Vanasse, 1958; P. Connes and Michel, 1971), which are now widely used. It is then relatively easy to know in real-time whether the spectrum is good or not. However if it is not, the same unavoidable mental gymnastics must be practiced to detect and correct, in the interferogram or its recording process, the origin of the defects in the spectrum, the correction of which is often not possible. Actually Fourier spectrometry has the specific features that it yields either good results or no results at all. This places even more importance on the examination of all possible kinds of errors affecting the interferogram, how they appear in the spectrum, and how they may be corrected.

The purpose of this chapter is to attempt to review a large number of systematic effects in the recording of an interferogram. The general principles and advantages of Fourier spectroscopy are assumed to be known (see, for instance, J. Connes, 1961; Loewenstein, 1966; Vanasse, 1973; Sakai, 1977). As much as possible, theoretical formulations will only be used if necessary, and replaced by a qualitative approximate representation of the phenomenon. Illustrations of the systematic effects will be given, starting usually from the particular version of interferometers developed by the French group (J. Connes *et al.*, 1970; Guelachvili, 1972). Although this technical choice may seem restrictive, especially concerning the driving mode, no difficulty should be encountered in extending these particular illustrations to the general diagnosis of systematic distortions.

After a brief review of Fourier transform spectrometry, we shall make general remarks on the recording of an interferogram, and give a brief illustration encompassing the particular French choice of approach to Fourier spectroscopy. The systematic errors are then classified in three different specific groups. The review in Section 1.6 of some systematic effects follows a more practical path. We concentrate first on the instrument itself, then on the control of the path difference, and finally on the measurement of the intensity of the interferogram. This section is partly taken from Guelachvili (1973). It is not exhaustive. The contrary would be an impossible task. The old-timers know very well how much the "unexpected" new distortions appearing in a familiar hardware must be expected. Some minor effects have been examined. This is justified by the high accuracy of the results obtained by the Fourier method, together with its extension to other fields (Murphy *et al.*, 1975; Fink and Larson, 1979; Durana and Mantz, 1979).

1.2. Brief Review of Fourier Transform Spectrometry

A. Principle

Suppose a Michelson interferometer is irradiated with a monochromatic source. If the path difference Δ varies linearly with time, the output signal is sinusoidally modulated. The frequency of this signal depends first on the variation of Δ versus time $\Delta(t)$, and second, on the wavenumber σ_0 of the monochromatic analyzed light. If one knows

$$\Delta = vt$$

(v constant), one can easily determine the value of σ_0 from the output signal. This simple way of determining σ_0 remains if, instead of one, the source consists of two monochromatic lines having respectively as wavenumber and intensity σ_1, I_1 and σ_2, I_2. The output signal is then a beat signal due to the combination of two different sinusoidal signals. Given $\Delta(t)$, one obtains without effort σ_1, I_1 and σ_2, I_2. This situation becomes more difficult when the source is complex. Then a complex harmonic analysis of the output signal of the interferometer must be done. The well-known mathematical tool indispensable for the harmonic decomposition of the interferogram is the Fourier transform, which will reconstruct the analyzed spectrum.

The Michelson interferometer is in that case acting as a coding system marking each spectral element (σ_i, I_i) of the optical source by a special and single seal. It works as a frequency divider, transforming each optical frequency σ to a reduced electrical frequency $\sigma \times (v/c)$ (c is the velocity of light) strictly depending on the wavenumber σ. This coding allows obtaining the multiplex advantage. As a matter of fact, each spectral component may be identified even if it has contributed simultaneously with a large number of others to build up the interferogram. No need exists to look at each spectral element without looking at the others which are all detected during all the time of the experiment. Consequently, a gain in the signal-to-noise ratio may be obtained depending on the detector used to record the interferogram.

The fundamental advantage, known as the *étendue* advantage due to the use of an interferometric device, is well known (Fellgett, 1958; Jacquinot, 1958; P. Connes, 1970). Also, the Michelson interferometer is not indispensable either for performing Fourier spectrometry or for getting the multiplex gain (Fellgett, 1967). It plays, however, a particular rôle in the development of this method. It is worth noting that this instrument is now as young as when Michelson conceived it and that it remains curiously the best apparatus for determining the unit of length. All metrological measurements leading to spectacular results have taken advantage of this interferometer. Almost one century after the constancy of the velocity of light was demonstrated by Michelson and Morley, it is needed again in the recent new measurement of c (Evenson and Petersen, 1976), and the instrument built by Michelson himself is still in use at the Bureau International des Poids et Mesures in Paris. This is particularly remarkable at the time when the obsolescence of an apparatus in physics is usually a very rapid phenomenon.

B. Theoretical Apparatus Function

Because it will be useful further in this paper, the formulation of the ideal response of a Fourier interferometer is given here.† It is well known that it is a sinc function due to the limitation of the path difference.

It is supposed that the interferogram starts at the path difference $\Delta_0 \simeq 0$ and stops at the maximum path difference Δ_M. The monochromatic source has a wavenumber σ_0 (wavelength $\lambda_0 = 1/\sigma_0$). The modulated part of the interferogram in this case is given by a sine function

$$I(\Delta) \simeq \sin 2\pi\sigma_0\Delta, \tag{1.1}$$

† No development is done here on the general technique of the Fourier method (sampling theorem, apodization, etc.), which can be found, for example, in J. Connes (1961) and Sakai (1977). Only the formulations necessary for treating the points considered in this paper are retained.

which is then sine transformed from $\Delta = 0$ to $\Delta = \Delta_{max}$ to recover the spectrum. Three cases are considered. First, the interferogram is actually and normally starting from $\Delta = 0$. Second, a slight misadjustment exists, and the interferogram actually starting from $\Delta = \lambda_0/4$ is believed to start from $\Delta = 0$. Third, the interferogram starts from $\Delta = l$, which can take any value.

1. $\Delta_0 = 0$

The spectrum $B(\sigma)$ gives the correct instrumental line shape, which is

$$f(\sigma) = \int_0^{\Delta_M} \sin 2\pi\sigma_0\Delta \sin 2\pi\sigma\Delta \, d\Delta, \tag{1.2}$$

$$= \frac{1}{2}\int_0^{\Delta_M} [\cos 2\pi(\sigma_0 - \sigma)\Delta - \cos 2\pi(\sigma_0 + \sigma)\Delta] \, d\Delta$$

$$= \frac{1}{2}\left[\Delta \frac{\sin 2\pi(\sigma_0 - \sigma)\Delta}{2\pi(\sigma_0 - \sigma)\Delta}\right]_0^{\Delta_M} - \frac{1}{2}\left[\Delta \frac{\sin 2\pi(\sigma_0 + \sigma)\Delta}{2\pi(\sigma_0 + \sigma)\Delta}\right]_0^{\Delta_M}$$

$$= \frac{\Delta_M}{2}\left[\frac{\sin 2\pi(\sigma_0 - \sigma)\Delta_M}{2\pi(\sigma_0 - \sigma)\Delta_M}\right] - \frac{\Delta_M}{2}\left[\frac{\sin 2\pi(\sigma_0 + \sigma)\Delta_M}{2\pi(\sigma_0 + \sigma)\Delta_M}\right]. \tag{1.3}$$

We obtain then two sinc functions of opposite signs centered, respectively, at $+\sigma_0$ and $-\sigma_0$. These are drawn in Fig. 1.1. For the sake of simplicity, only the profile centered at σ_0 will be considered in the rest of this paper.

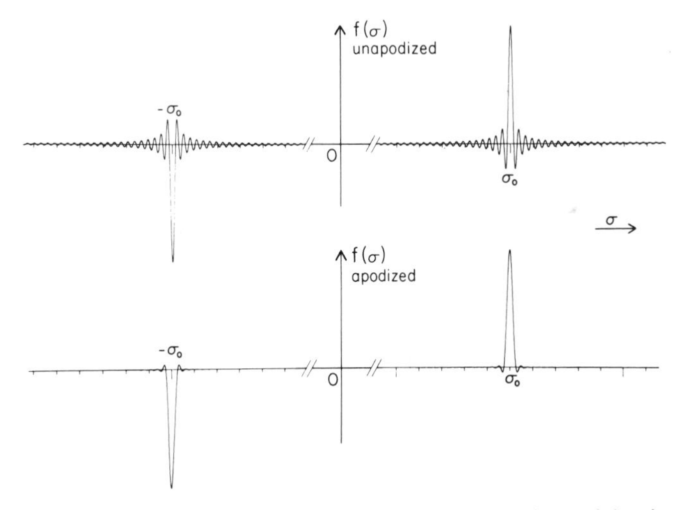

FIG. 1.1. Instrumental line shape $f(\sigma)$, given by the sine transform of the sine interferogram of a monochromatic line at σ_0.

2. $\Delta = \lambda_0/4$

This corresponds to an error in the recording procedure. Then $B(\sigma)$ is

$$
\begin{aligned}
h(\sigma) &= \int_0^{\Delta_M} \sin 2\pi\sigma_0\left(\Delta + \frac{\lambda_0}{4}\right) \sin 2\pi\sigma\Delta \, d\Delta \\
&= \int_0^{\Delta_M} \cos 2\pi\sigma_0\Delta \sin 2\pi\sigma\Delta \, d\Delta \\
&= \frac{1}{2}\int_0^{\Delta_M} [\sin 2\pi(\sigma_0 + \sigma)\Delta - \sin 2\pi(\sigma_0 - \sigma)\Delta] \, d\Delta \\
&= \frac{1}{2}\left[-\Delta \frac{\cos 2\pi(\sigma_0 + \sigma)\Delta}{2\pi(\sigma_0 + \sigma)\Delta}\right]^{\Delta_M} - \frac{1}{2}\left[-\Delta \frac{\cos 2\pi(\sigma_0 - \sigma)}{2\pi(\sigma_0 - \sigma)\Delta}\right]_0^{\Delta_M} \\
&= -\frac{\Delta_M}{2}\left[\frac{\cos 2\pi(\sigma_0 + \sigma)\Delta_M - 1}{2\pi(\sigma_0 + \sigma)\Delta_M}\right] \\
&\quad + \frac{\Delta_M}{2}\left[\frac{\cos 2\pi(\sigma_0 - \sigma)\Delta_M - 1}{2\pi(\sigma_0 - \sigma)\Delta_M}\right], \qquad (1.4)
\end{aligned}
$$

which corresponds to two functions of opposite signs, centered respectively at $-\sigma_0$ and σ_0. Figure 1.2 represents $h(\sigma)$ with and without apodization.

3. $\Delta = l$

The resulting incorrect apparatus function is

$$
\begin{aligned}
q(\sigma) &= \int_0^{\Delta_M} \sin 2\pi\sigma_0(\Delta + l) \sin 2\pi\sigma\Delta \, d\Delta \\
&= \cos 2\pi\sigma_0 l \int_0^{\Delta_M} \sin 2\pi\sigma_0 \sin 2\pi\sigma\Delta \, d\Delta \\
&\quad + \sin 2\pi\sigma_0 l \int_0^{\Delta_M} \cos 2\pi\sigma_0\Delta \sin 2\pi\sigma\Delta \, d\Delta, \qquad (1.5)
\end{aligned}
$$

which is a combination of the shapes $f(\sigma)$ and $h(\sigma)$ weighted by cosine and sine factors depending on $\sigma_0 l$. Figure 1.3 gives an example of such a function with $l = \pi/4$, i.e., an equal contribution of $f(\sigma)$ and $h(\sigma)$ for Eq. (1.5) to give $q(\sigma)$. For example, if

$$l = k\lambda_0,$$

$q(\sigma)$ becomes the correct instrumental line shape $f(\sigma)$ as expected since

$$I(\Delta) = I(\Delta + k\lambda_0) \qquad (k \text{ integer}).$$

As a consequence, any zero of the interferogram of a monochromatic line is suitable for starting the recording, provided one takes care of which side of the sine this zero is located.

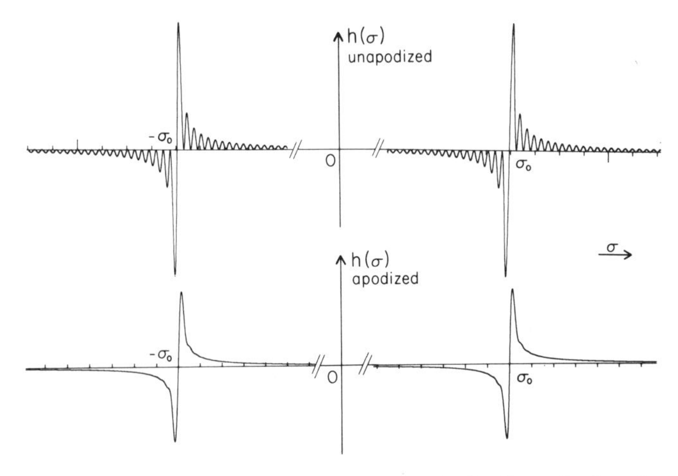

FIG. 1.2. Same as in Fig. 1.1, but a phase error of $\pi/2$ affects the interferogram.

1.3. Recording an Interferogram

A. General Remarks

Essentially two "orthogonal" determinations are necessary for obtaining an interferogram. This is generally expressed by two different simultaneous measurements in the course of an experiment.

As suggested in Section 1.1,A, the first one is connected to the recip-

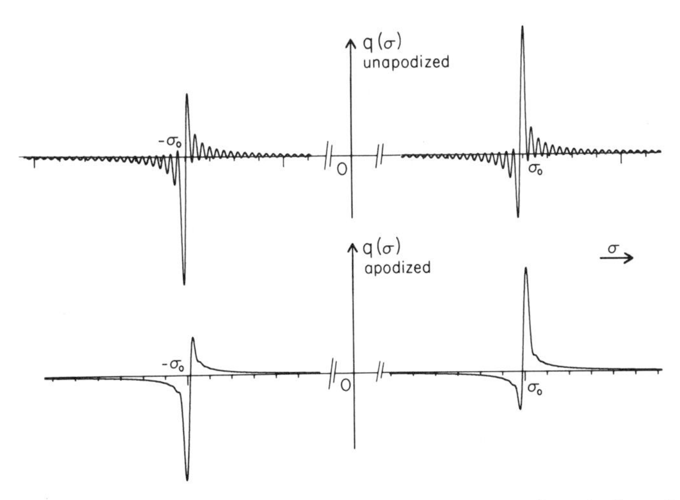

FIG. 1.3. Same as in Figs. 1.1 and 1.2, but the phase error has a value of $\pi/4$.

rocal mathematical relationship existing between Δ and σ in the Fourier transform. Determination of the wavenumber axis in the spectrum depends drastically on the correct measurement of the path difference during the recording of the interferogram. The more accurately Δ is known, the more accurately the spectrum is calibrated. The first care must then be the *rigorous control of* Δ. This is enough to ensure a good wavenumber scale. Generally, the measurement of Δ is obtained using a simultaneously recorded additional interferogram of a monochromatic source whose wavenumber is already well known.

The second determination of importance is that of *the intensity of the interferogram* $I(\Delta)$ as a function of Δ. It is obviously necessary to know Δ very accurately. In particular, its determination using the additional monochromatic interferogram corresponds to slightly different experimental conditions from which the path difference of the recorded interferogram is obtained. The intensity $I(\Delta)$ is also dependent on various parameters starting from the source going to the recorder through the detector, as in any other spectrometric method.

Consequently, at least two types of distortions affecting an interferogram may be defined: the *phase error* and the *intensity error* which, respectively, correspond to distortions on the path difference and on the intensity.

B. Calibration

1. *Wavenumber*

As stated above, only one standard is enough to give a complete calibration of the wavenumber axis. This is at the origin of the consistency of the Fourier results. However, slight misadjustments always exist between the two different paths whereby the monochromatic and the analyzed sources are recorded. When one desires highly accurate results, one must take these into account.

If Δ is the assumed path difference given by the calibration interferogram and if actually the path difference of the recorded interferogram is $(1 + \alpha)\Delta$ (α constant), then an error $d\sigma_0$ exists in the location of the line, normally at σ_0 in the spectrum. Instead of σ_0, one obtains $\sigma_0(1 + \alpha)$. [In Eq. (1.2) replace $\sin 2\pi\sigma_0$ by $\sin 2\pi\sigma_0(1 + \alpha)\Delta$.] Then each wavenumber σ must be corrected by $d\sigma$, with

$$d\sigma/\sigma = \alpha. \tag{1.6}$$

Due to systematic effects, such as the finite angle of the beam, the constant factor α is practically never equal to zero and must be taken into account.

The best way to get a correct estimate of α is to use the *internal standard method* (Guelachvili, 1973). It consists in recording under the same conditions in space and if possible in time the interferograms of the analyzed source and of a complementary standard source, which may not be monochromatic.† In that case the rigorous control of Δ (discussed in Section 1.2,A) by a monochromatic parallel interferogram is no more a measurement of Δ, but only a scaling system which is known to be proportional to Δ. After the computation of the spectrum, the N internal standard lines are measured and their wavenumbers σ_i compared to what they should be, i.e., σ_{0i}. The correct wavenumber scale is then obtained using

$$\alpha = \frac{1}{N} \sum_{i=1}^{N} \frac{\sigma_{0i} - \sigma_i}{\sigma_{0i}}$$

in Eq. (1.6). This calibration method has another important advantage in that it takes into account the slight asymmetries often affecting actual profiles which have, in the whole spectral range, practically the same distorted shape. Since no symmetry exists in these profiles, they should not in principle be exactly measurable. The internal standard method removes this difficulty. Indeed, in that case the standard and the measured lines have the same shape. Provided that they are measured in the same manner, no error remains in the final wavenumber scale that has been determined for the whole spectral range recorded.

2. *Intensity*

The accurate calibration of the intensity axis is strongly dependent on the measurement of the intensity of the interferogram. One finds here all the factors that one has to account for in classical spectrometric methods, the main distortion being nonlinearities. However, here these factors perturb the interferogram and appear Fourier transformed in the spectrum. In Fourier spectrometry their effects are pronounced and specific to the technique (see Section 1.6,C.2b). Distinction must also be made between emission and absorption spectra since the interferograms behave differently, even if the emission spectrum consists of numerous lines spread over a large spectral range.

In emission the modulated part of the interferogram has generally a peak-to-peak amplitude, which slowly varies from $\Delta = 0$ to the maximum path difference. This is not true when a broadband absorption spectrum is looked at. Then the very strong modulation near $\Delta = 0$ rapidly decreases

† For example, in absorption spectra the white-light beam is sent successively through the cell containing the sample of interest and through another additional cell with a standard gas having several well-measured absorption profiles such as the rotational fine structure of a vibrational transition.

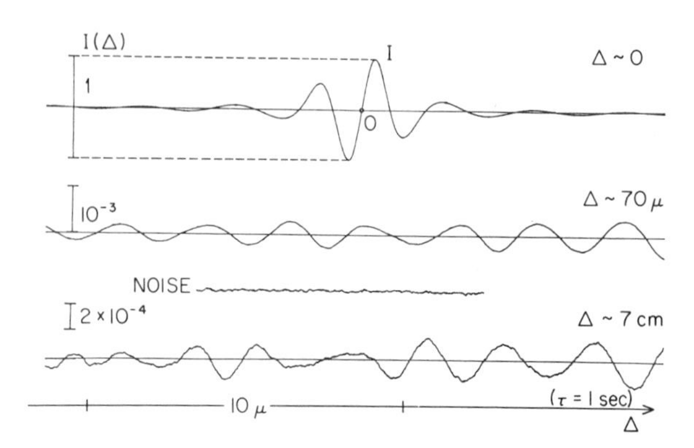

FIG. 1.4. Interferogram of a broadband spectrum (7000 cm^{-1}). With internal modulation the mean level remains zero. Only the small area around $\Delta = 0$ has a high contrast. The sampling interval is too small to be seen on the figure.

for higher values of Δ (see Fig. 1.4). Consequently, detection may easily be linear in emission, whereas in absorption care must be taken because of the great dynamic range of the measurements. A nonlinear detection introduces zero level distortion in the spectrum and harmonic spectral components. The measurement of intensities requires a careful investigation of all the sources of nonlinearities which are, of course, not the only possible sources of error. For example, thermal emission may also introduce parasitic spectral components in the final results.

As in the wavenumber measurement, the *internal standard method* is the best way to get a correct estimate of the intensity. It allows the determination of the transmission of the whole instrument as well as the calibration of the spectral intensities. Also, simple checks in the spectrum may be done to detect the existence of remaining nonlinearities. In case of an emission spectrum, no harmonic lines should appear at two and three times the wavenumber of each line or at their corresponding "images." For absorption spectra, a convenient procedure is to record a slightly larger free spectral range than the one of interest to look at a portion of the spectrum which should be zero; then any displacement of the level in that region indicates remaining nonlinearities.

1.4. One Practical Example of Fourier Transform Spectrometry

The aim of this section is to give an actual illustration of the general previous considerations. It also serves as an introduction to the detailed review, in Section 1.6, of the systematic effects that may disturb a Fourier spectrum. The illustration is concerned mainly with the French-type high-information Fourier spectrometer, which represents one specific technical choice of instruments among others (see A.F.C.R.L. Report No. 114.71.0019, Vanasse, ed., 1971; and *Applied Optics,* May 1978) already reviewed by Sakai (1977). Consequently, only the general outline will be taken into consideration. For more detail, see Pinard (1969), J. Connes *et al.* (1970), Guelachvili (1972), P. Connes and Michel (1975).

A. Recording Procedure

The basic choices of instrument implementation and techniques by the group led by P. Connes have been the *internal modulation* and the *stepping mode* recording, which essentially are concerned, respectively, with the measurement of the intensity of the interferogram and the control of the path difference. They took their final shape in the so-called Connes' type interferometer,† which is an adaptation of the initial Michelson apparatus for Fourier spectrometry. Its principal features are rigorous symmetry and the use of retroreflector cat's-eye systems instead of plane classical mirrors.

1. *Control of the Path Difference*

With an interferometer perfectly symmetric, no chromatic effect may occur due to chromatic behavior of the various coatings. Then only a geometrical determination of the path difference is necessary. The interferogram of a monochromatic line (3.50 μ superradiant line of Xe or more recently 0.63 μ or 3.39 μ laser line of Ne) is used for that purpose. It yields the knowledge of Δ which has to follow a particular law of variation versus time according to the analyzed spectral range, the desired resolution, and the time available for the recording. This law presupposes a servo control, the role of which is to detect and correct all the perturbations on Δ. The shorter the time between the occurrence of the perturbation and the end of return to initial position, the more efficient must be this servo control.

Another quality the servo system must have is the lack of hysteresis; which means that it must actually restore the previous unperturbed Δ. These two features, i.e., the accuracy and the speed of the error correction, depend on several factors (mechanical, optical, electrical). For example, it is important to avoid static friction in the moving system used to vary Δ in order to be free of hysteresis. The moving cat's-eye is supported either on oil or on air, or even on an "airless" air system. The force that

† This expression could also legitimately be applied to the Spherical Fabry Perot Interferometer and to the SISAM (P. Connes, 1958a,b).

causes its movement is obtained from a linear motor. Quickness of response implies that only a light optical component has to be displaced over a small distance. This is one of the advantages of the cat's-eye, consisting of a big mirror and a small mirror that may be mounted on a piezoelectric ceramic.

To determine whether the interferometer has the right path difference, the servo system reacts to an electrical signal equal to zero if Δ is correct and proportional to the sign and the amplitude of the perturbation, when one exists. This may be obtained, for instance, from the zero of the fringes of the monochromatic monitoring source, the continuous part of which has been removed. An appropriate electronic perturbation needs to be added in the circuit so that the servo reacts and Δ varies following a predetermined stepping program. It is obtained by a phase-modulation system. A polarimetric method allows choosing a stepping interval which is a multiple of $\lambda/100$. Then the free spectral range is flexibly adapted to the actual analyzed spectrum. In other words, there is no oversampling, and the number of samples in the interferogram is then equal to the number of spectral elements in the spectrum. Also, since stepping consists in defining fixed values of Δ, it has the obvious advantage of yielding accurate positioning necessary for obtaining accurate spectral profiles.

2. *Interferogram Measurement*

From the beginning to the end of the recording, all the stages of the procedure are performed and controlled by the path difference servo control system. The experiment starts at path difference zero, which may be easily found thanks to the odd (antisymmetric) shape of the interferogram, which is equal to zero at $\Delta = 0$. The odd shape of the interferogram is due to internal modulation, which consists in modulating the path difference in the recording mode (Mertz, 1958; J. Connes *et al.*, 1967). More precisely, after one sample is obtained, Δ is increased by a constant factor corresponding to the step, and around this new position a periodic modulation is performed which modulates the signal of the interferogram. This signal is synchronously detected and integrated. The integrated signal is the value of the new sample. Again Δ is increased and another sampled value obtained, etc.

Internal modulation essentially acts as a mathematical derivation of the classical interferogram. Instead of performing a cosine transform, one has to perform a sine transform of the interferogram to obtain the spectrum. Another effect of internal modulation depends on the modulation amplitude. If one considers a monochromatic interferogram, it is obvious that with an amplitude equal to the wavelength of the source one will always have zero as a result of the modulation. On the other hand, the maximum efficiency will be obtained for a peak-to-peak amplitude l equal to half the wavelength. Consequently, this suggests that the spectrum is not exactly reproduced. Actually it is multiplied by a function equal to zero at $\lambda = 2l$ and maximum at $\lambda = l$. This function depends on the form of the modulation. For instance, a square internal modulation of the interferogram multiplies the wavenumber spectrum by a sine function.

The main advantages of internal modulation are well known. Essentially it allows the filtering of the noise as in any synchronous detection, and also increases the dynamic range of the measurement, since it eliminates the continuoús background of the interferogram. This important feature is illustrated by Fig. 1.4, which represents an actual interferogram of a broadband absorption spectrum (3000–10,000 cm^{-1}) continuously recorded at three different path differences ($\Delta \sim 0$, $\Delta \sim 70\ \mu$, $\Delta \sim 7$ cm). The ratio of the amplitude of the modulation at $\Delta \sim 0$ and the noise are about 5×10^4, which is the dynamic range one needs in this particular example. Without internal modulation in a stepping-mode recording, the background would have been externally chopped. consequently a large useless modulated signal would have been added everywhere to the interesting part of the interferogram, which is the small modulation seen in Fig. 1.4. Noting that the background on which this modulation rides is approximately half of the central peak around $\Delta \sim 0$, the smallest detail of interest, corresponding to the noise, would be more than 10^4 times smaller than the parasitic signal. This may easily give rise to a decrease of sensitivity and a loss of information.

With a Connes' type interferometer, two outputs are available because of the lateral shift of the beam due to the cat's-eye. The two opposite signals are sent to a differential preamplifier, the output of which may be multiplied by an adjustable gain varying from 1 to 10^3. One is able then to use the dynamic range of a digital voltmeter even with very different input signals. The stepping allows spending the majority of the measurement time on the detection itself and consequently under the best S/N ratio conditions. This remains true even for high stepping rates. At 2000 steps/sec, 80% of the recording time is still devoted to the integration. Another advantage is the possible choice of a variable time during the integration of the signal to compensate for source intensity fluctuations. This compensation is also applicable through an additional adjustment of the electronic gain. The two outputs of the interferometer are used also to reduce the intensity fluctuations which appear in phase on both detectors. For that purpose, slight adjustments in amplitude and phases between the two signals are performed before each experiment in such a way that the externally chopped signal of the source at the internal modulation frequency gives zero after the differential preamplifier.

B. Typical Results

Among the main features of these high information content results one should emphasize, first, that they contain a great number of bits of information. It is usual to have interferograms with 10^6 samples recorded in a few hours, with transformed spectra having 10^6 spectral elements. Second, the spectra are well resolved. The maximum path difference may be 2 or 6 m. Consequently the so-called third generation interferometry delivers Doppler-limited spectra. Third, the results obtained are accurately measured. More details on the general characteristics of the high-information results may be found in Guelachvili (1978a).

Figure 1.5 gives an illustration of a typical emission spectrum (Guelachvili, 1978b). The whole spectral range is shown on trace (a) with a resolution equal to 0.20 cm^{-1}. It extends from 1800 to 2500 cm^{-1}. A region around 2096 cm^{-1} is drawn in (b) with 0.030 cm^{-1} resolution. It corresponds to about 230 cm^{-1}. Again an expansion of the wavenumber scale is performed on trace (c), which covers 20 cm^{-1}, with an apodized resolution of 0.005 cm^{-1}, corresponding to a maximum path difference of 2 m. On the trace (d), both the intensity and wavenumber scales have been expanded in order to be able to see over a 2-cm^{-1} range the noise level and the shape of the lines. The faintest line marked with an arrow is 300 times smaller than the brightest lines of the spectrum, which have a S/N ratio of about 5000. The half width of the profiles is 8×10^{-3} cm^{-1}. The instrumental resolution is consequently narrower than the linewidth.

Another illustration is given in Fig. 1.6. It concerns now an absorption spectrum of CH_3D (Chackerian *et al.*, 1978) with a relatively low density of lines, so that the background and consequently the measurement of the intensity of the lines may be easily determined. The same presentation as for Fig. 1.5 is used. In fact, the spectrum shown on trace (a) has absorbing conditions different from those of the spectra shown on traces (b) and (c). It has, however, the same apodized resolution (0.005 cm^{-1}). The CO spectrum, whose position is accurately known, is obtained from a different absorption cell and is used to calibrate the spectrum according to the internal standard method discussed in Section 1.3,B.

In Fig. 1.7, estimation of the noise in the crowded C_3O_2 absorption spectrum (Fusina *et al.*, 1980) becomes difficult because no place in the spectral range of interest is definitely displaying the background. Again the same presentation as for Fig. 1.6 has been adopted. The upper part of

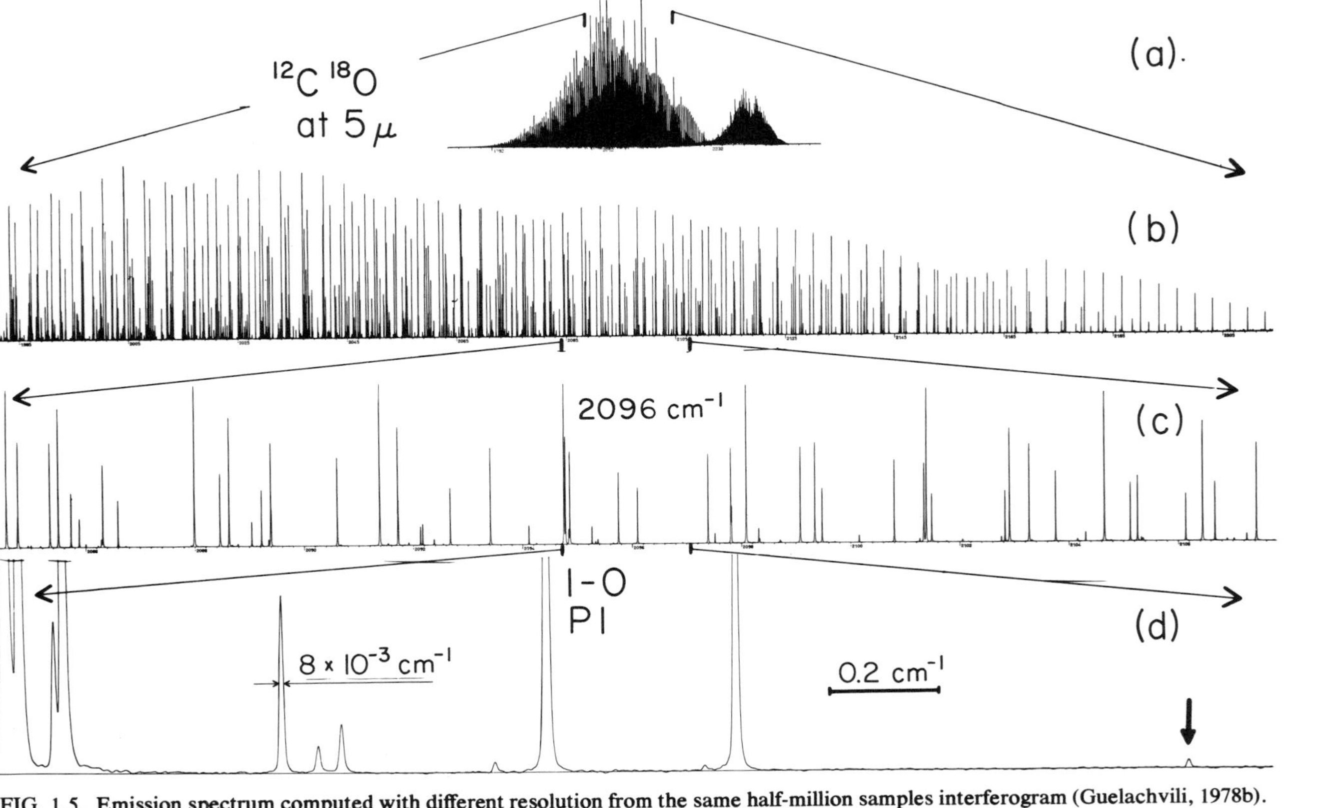

FIG. 1.5. Emission spectrum computed with different resolution from the same half-million samples interferogram (Guelachvili, 1978b).

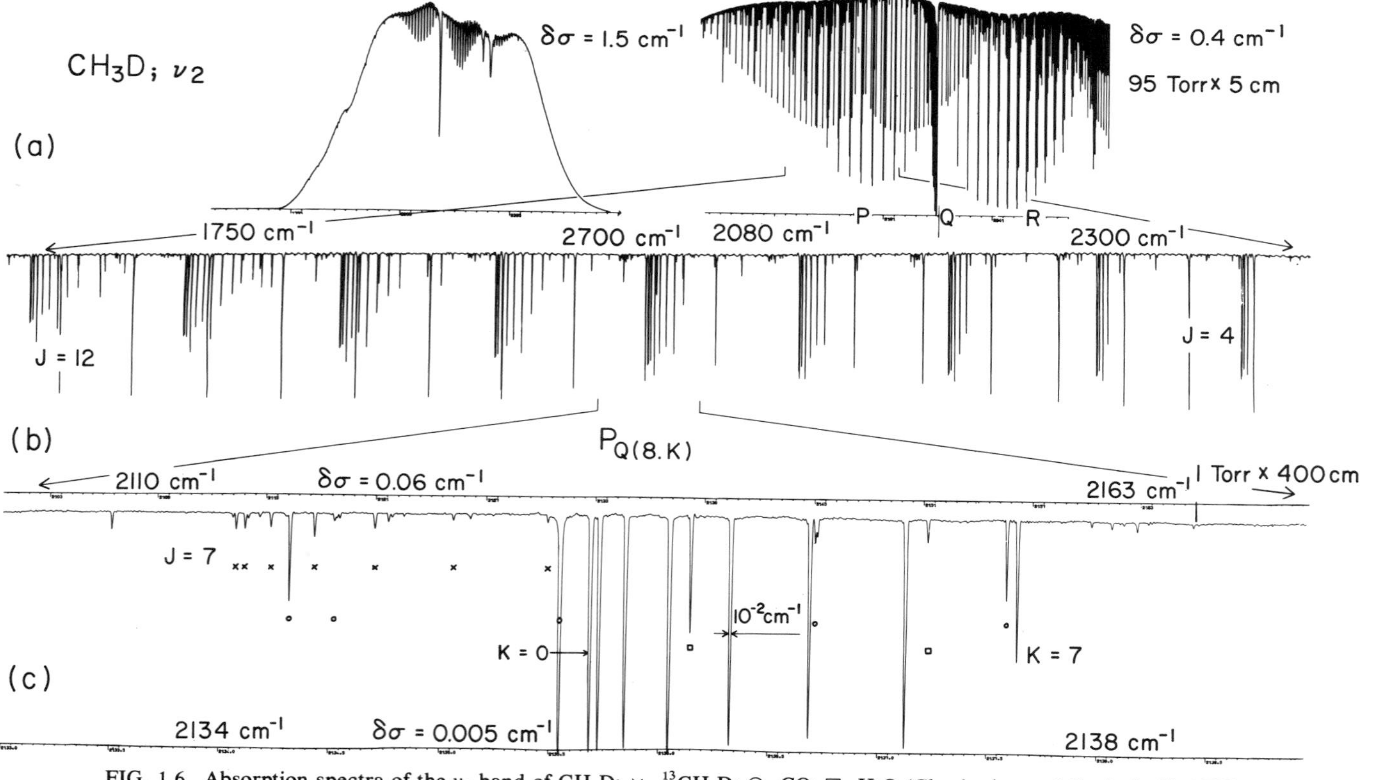

FIG. 1.6. Absorption spectra of the ν_2 band of CH_3D: ×, $^{13}CH_3D$; ○, CO; □, H_2O (Chackerian and Guelachvili, 1978).

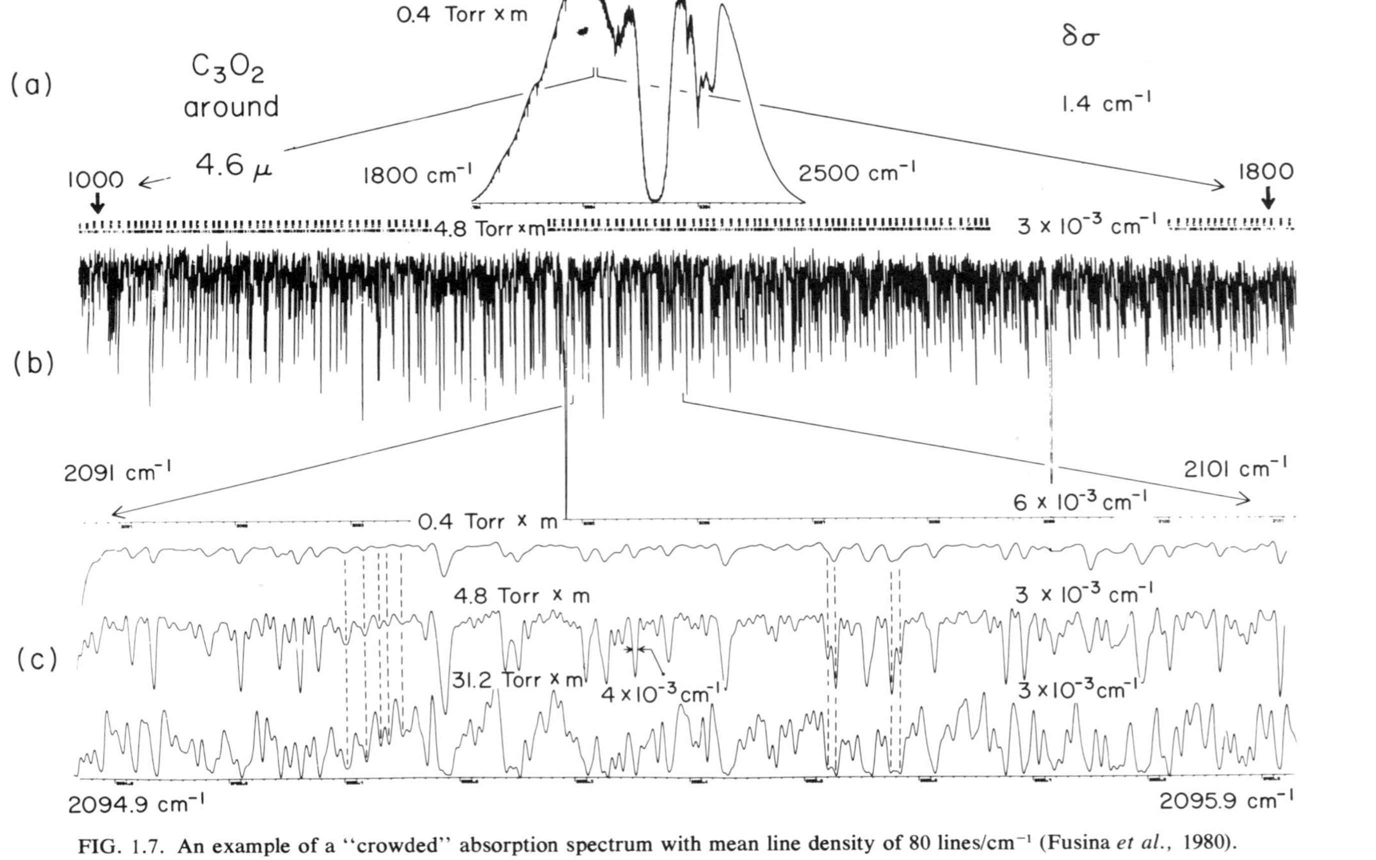

FIG. 1.7. An example of a "crowded" absorption spectrum with mean line density of 80 lines/cm^{-1} (Fusina *et al.*, 1980).

trace (b) gives the results of a line-finder program (J. Connes, 1971; Delouis, 1973) that has detected 800 lines in about a 10-cm^{-1} spectral interval. More sensitive detection would have given even more real lines. Actually 40,000 lines have been measured for an interval of 1000 cm^{-1}. On the lower wavenumber scale (c) (which is six times more expanded than trace (c) of Fig. 1.6), three different spectra with different absorbing conditions are reproduced, with respective resolutions equal to 6×10^{-3}, 3×10^{-3}, and 3×10^{-3} cm^{-1}. Dotted lines show that the small bumps of the low absorbing conditions spectrum are identified as being actual C_3O_2 transitions. The width of an almost isolated profile is measured to be 4×10^{-3} cm^{-1}. Even with higher resolution, the appearance of the spectrum would remain identical since it is Doppler limited.

Figure 1.8 represents only a very small fraction of an atomic spectrum (Morillon and Verges, 1975) which covers the range 3500–11,800 cm^{-1}. The transition is a Zeeman pattern of TeI. The source is an electrodeless tube, excited by an electric field at 2450 MHz with a power of about 100 W, containing isotopically pure tellurium. The interferogram with 850,000 samples is recorded during 10 hr, and the maximum path difference is 35.7 cm. Theoretical intensities of the π and σ Zeeman components are given in the figure. In this particular case, the magnetic field H is of the order of 10^4 Oe. The precision of the Landé factors is mainly limited in this spectrum by the accuracy of the measurement of H and not by the wavenumber determination.

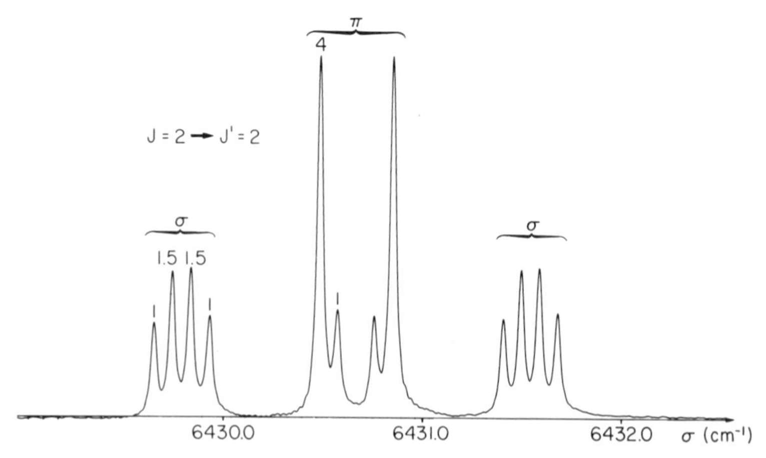

FIG. 1.8. Zeeman pattern in TeI. The apparatus function is unapodized. Convolution of the line shape and the $f(\sigma)$ profile removes the oscillation of the instrumental shape (Morillon and Verges, 1975).

Other examples could have been taken among astronomical spectra, (e.g., see Fig. 23 of P. Connes and Michel, 1975). In order to summarize the type of results given by the third generation interferometry, one may use the quality factor Q, which gives a general appreciation of any spectrum. It is defined as the ratio of the total energy corresponding to the entire spectrum, to the smallest measurable energy, the latter being the area of the line of width equal to the resolution whose intensity is of the same level as the noise (P. Connes, 1971; Guelachvili and Maillard, 1971). The best Q factors obtained by high-information Fourier spectrometry are generally of the order of 5×10^7.

1.5. Classification of the Systematic Errors in Fourier Transform Spectrometry

Let us write again Eqs. (1.1) and (1.2) of Section 1.2,B giving, respectively, the sine interferogram of a monochromatic line, the wavenumber of which is σ_0,

$$I(\Delta) = \sin 2\pi\sigma_0\Delta, \tag{1.1}$$

and the computed spectrum of this line through a sine Fourier transformation

$$B(\sigma) = \int_0^{\Delta_M} \sin 2\pi\sigma_0\Delta \sin 2\pi\sigma\Delta \, d\Delta. \tag{1.2}$$

The particular choice for these expressions (of the sine) has been explained in Section 1.4,A.1 as being due to the internal modulation mode of recording. It has, however, no special importance, and the general conclusions one can draw from Eqs. (1.1) and (1.2) are the same if one starts from cosine or exponential expressions.

From a general point of view the errors which may affect the interferogram given by Eq. (1.1) are possibly of three types.

First, one may have an *additional* term ε_1 corresponding to the addition of a systematic unwanted signal, varying or not, during the recording of the interferogram.

Second, an inteferogram phase distortion effect ε_2 may occur.

Third, the intensity of the interferogram may also be perturbed and will be expressed by ε_3. These last terms, ε_2 and ε_3, produce effects that depend on the interferogram itself. They consequently will be designated as *multiplicative errors*.

Summarizing, instead of getting a spectrum represented by Eq. (1.2), the actual spectrum $B(\sigma)$ of the monochromatic line will be given by

$$B'(\sigma) = \int_0^{\Delta_M} [(1 + \varepsilon_3) \sin 2\pi\sigma_0(\Delta + \varepsilon_2) + \varepsilon_1] \sin 2\pi\sigma\Delta \, d\Delta. \quad (1.7)$$

In the next paragraphs the effects due to ε_1, ε_2, ε_3 on $B(\sigma)$ are separately examined and briefly illustrated, and speculative general conclusions are drawn.

A. Additive Effects

We first consider the expression

$$B'(\sigma) = \int_0^{\Delta_M} (\sin 2\pi\sigma_0\Delta + \varepsilon_1) \sin 2\pi\sigma\Delta \, d\Delta. \quad (1.8)$$

With the unperturbed spectrum $B(\sigma)$, an additional feature $B_a(\sigma)$ expressed by

$$B_a(\sigma) = \int_0^{\Delta_M} \varepsilon_1 \sin 2\pi\sigma\Delta \, d\Delta \quad (1.9)$$

will appear in a form depending on ε_1.

Finally,

$$B'(\sigma) = B(\sigma) + B_a(\sigma).$$

1. $\varepsilon_1 = \varepsilon_{01} = const.$

This occurs, for instance, when an electronic amplifier in the sample measurement system has an incorrect but constant zero level. Integration of Eq. (1.9) is straightforward. It gives a result similar to Eq. (1.4), which is not surprising since ε_1 in Eq. (1.9) may be written

$$\varepsilon_1 \cos[2\pi(0)\,\Delta].$$

The additional spectrum $B_a(\sigma)$ will be located at the frequency zero. It is proportional to ε_1 and has the asymmetric shape shown in Fig. 1.2. This remains true for a broadband spectrum, which may be distorted if the zero frequency area is aliased due to the sampling theorem and covers a spectral zone of interest.

2. $\varepsilon_1 = \varepsilon_{01} \, sin(2\pi nt + \Phi_0)$

The perturbing phenomenon in this case is periodic (amplitude ε_{01}, frequency n). It may often be due to an additional modulated optical signal onto the detectors or to electrical parasitic 50 Hz pickup by the electronics. It is generally directly dependent on time and not on path difference.

Equation (1.9) becomes

$$B_a(\sigma) = \int_0^{\Delta M} \varepsilon_{01} \sin(2\pi nt + \Phi_0) \sin 2\pi\sigma\Delta \, d\Delta,$$

and since Δ is practically a linear function of the time, i.e.,

$$\Delta = vt,$$

then

$$B_a(\sigma) = \int_0^{\Delta_M} \varepsilon_{01} \sin(2\pi\alpha\Delta + \Phi_0) \sin 2\pi\sigma\Delta \, d\,\Delta, \qquad (1.10)$$

with $\alpha = n/v$ expressed in $(\text{length})^{-1}$ unit.

Expression (1.10) is similar to the relation discussed in Section 1.2,B.3. The additional perturbing spectrum $B_a(\sigma)$ appears then as a monochromatic line proportional to ε_{01} and located at α. Its shape depends on Φ_0 and generally does not have the correct instrumental profile. The above is merely to be used as an indication of this phenomenon. Also, the location in the spectrum must change when the function for the variation of Δ versus time is changed.

In the case of a broadband spectrum, this additional parasitic spectrum will obviously disturb the spectral region around α.

3. *General Case*

The term ε_1 may take any value depending either on the time or on the path difference. For instance, the fluctuations of the voltage level of an electronic amplifier of the measurement system will obviously be a function of time. As a function of Δ, one may have parasitic optical reflections on moving pieces in the interferometer, or parasitic actual interferograms due, for example, to thermal background. Additional noise belongs also to this category.

In the general case, the additional spectrum $B_a(\sigma)$ will cover a spectral range which depends on the origin of the phenomenon and generally will perturb the measured spectrum. It may be considered as a combination of different "monochromatic" effects discussed in the previous paragraph having different values of ε_{01} and α. As a consequence, $B_a(\sigma)$ in practice is always affected by a large phase error, and its appearance is an indication of this phenomenon. To be properly eliminated, this effect should be removed directly at its origin. An a posteriori correction is practically impossible. At least the untouched part of the perturbed analyzed spectrum, where no overlapping exists with the parasitic $B_a(\sigma)$, remains available, correctly measured, and may be used without any precaution, provided, of course, that no phase or intensity multiplicative errors discussed below have been introduced by additional interferograms. The spectrally clean area is easily determined when one deals with emission spectra. This is not the case for absorption analysis, where the background given by the white-light source is generally not flat because of the chromatic transparency of the whole apparatus. In that case it could still be difficult to determine with confidence the unperturbed interesting part of the analyzed spectrum.

B. Multiplicative Effects

Unfortunately multiplicative effects will perturb more severely the original spectrum than additive effects. However, they behave very specifically, and this should allow their detection in the final spectrum without too much difficulty. As said above, two types of multiplicative effects are to be considered; the phase and the intensity effects.

1. *Phase Distortions*

Discussion bears here on the following particular form of relation (1.7):

$$B'(\sigma) = \int_0^{\Delta_M} \sin 2\pi\sigma_0(\Delta + \varepsilon_2) \sin 2\pi\sigma\Delta \, d\,\Delta, \qquad (1.11)$$

It indicates that instead of getting a pure sine interferogram of the monochromatic source, one records a signal distorted by an error in the argument of the trigonometric function, explicitly on the path difference itself.

a. $\varepsilon_2 = \varepsilon_{02} = const.$ This is the well-known situation of an interferogram starting at $\Delta = \varepsilon_{02}$ instead of $\Delta = 0$. Equation (1.11) has already been developed in Section 1.2,B.3, Eq. (1.5). The effect of this constant phase error on the monochromatic spectrum at σ presents itself in two ways.

First, a decrease in amplitude of the correct instrumental profile is produced since it is multiplied by $\cos 2\pi\sigma_0\varepsilon_{02}$. Second, a *perfectly asymmetric* spectrum is added to this profile. Its importance depends also on the multiplicative factor $\sin 2\pi\sigma_0\varepsilon_{02}$.

Consequently, for a broadband spectrum the asymmetric distortion affecting the correct instrumental shape will vary according to σ with a periodicity $\Delta\sigma = 1/\varepsilon_{02}$. This effect is thus seen to be chromatic as well.

Care must be taken when building an interferometer to have no structural asymmetry, which, from the above point of view, is a defect that can prohibit obtaining good spectra.

b. $\varepsilon_2 = \varepsilon_{02} \sin(2\pi\beta\Delta + \Phi_0)$. In this case the error on the path difference is periodic. This could be produced, for instance, by a parasitic electrical 50-Hz signal affecting the servo control system, or by a slight error in mechanical motion when using a screw. These two examples are different. They are, respectively, time or path-difference dependent. In the first case the distorting effects will depend on the time used to record the interferogram, whereas in the second they will remain identical from one interferogram to the next.

Expression (1.11) becomes

$$B'(\sigma) = \int_0^{\Delta_M} \sin 2\pi\sigma_0[\Delta + \varepsilon_{02} \sin(2\pi\beta\Delta + \Phi_0)] \sin 2\pi\sigma\Delta \, d\Delta.$$

For the sake of convenience, suppose that $\Phi_0 = 0$ and also that ε_{02} is small. This does not affect the generality of the results.

$B'(\sigma)$ is expanded into the sum of the following two terms:

$$\begin{aligned} B'(\sigma) = & \int_0^{\Delta_M} \sin 2\pi\sigma_0\Delta \cos(2\pi\sigma_0 \varepsilon_{02} \sin 2\pi\beta\Delta) \sin 2\pi\sigma\Delta \, d\Delta \\ & + \int_0^{\Delta_M} \sin(2\pi\sigma_0 \varepsilon_{02} \sin 2\pi\beta\Delta) \cos 2\pi\sigma_0\Delta \sin 2\pi\sigma\Delta \, d\Delta. \end{aligned} \tag{1.12}$$

With the approximations

$$\cos(2\pi\sigma_0 \varepsilon_{02} \sin 2\pi\beta\Delta) \simeq 1$$

and

$$\sin(2\pi\sigma_0 \varepsilon_{02} \sin 2\pi\beta\Delta) \simeq 2\pi\sigma_0 \varepsilon_{02} \sin 2\pi\beta\Delta,$$

one obtains

$$\begin{aligned} B'(\sigma) = & \int_0^{\Delta_M} \sin 2\pi\sigma_0\Delta \sin 2\pi\sigma\Delta \, d\Delta \\ & + \pi\sigma_0 \varepsilon_{02} \int_0^{\Delta_M} \sin 2\pi(\sigma_0 - \beta)\Delta \sin 2\pi\sigma\Delta \, d\Delta \\ & - \pi\sigma_0 \varepsilon_{02} \int_0^{\Delta_M} \sin 2\pi(\sigma_0 - \beta)\Delta \sin 2\pi\sigma\Delta \, d\Delta, \end{aligned} \tag{1.13}$$

a sum of three expressions similar to Eq. (1.2) in Section 1.2,B.1. The three terms of Eq. (1.13) represent successively the normal apparatus function, i.e., the recovery of the monochromatic spectrum at σ_0, a similar response multiplied by the factor $\pi\sigma_0 \varepsilon_{02}$ and located at $\sigma_0 + \beta$, and again a similar response multiplied by the factor $-\pi\sigma_0 \varepsilon_{02}$ and located at $\sigma_0 - \beta$.

Two ghosts on either side of the correct spectrum thus represent the distortion, which shows up as *perfectly odd*. Depending on the period β, these two ghosts may or may not hinder the central feature. Representation of this periodic phase error is given on trace (a) of Fig. 1.9. Trace (b) represents the same phenomenon with $\Phi_0 = \pi/2$. In that case again the additional distortion is an odd function centered on σ_0.

This periodic phase effect is chromatic, but only for the relative intensity of the two ghosts, the ratio of ghost intensity to central peak intensity being given by $\pi\sigma_0 \varepsilon_{02}$. The greater the wavenumber, the higher is the ghost intensity. In the visible, the distortion of the spectrum is more important for the same phase effect on the interferogram than in the infrared. On the other hand, the distance β between the central peak and the side features is independent of σ.

c. General Case. The term ε_2 distorting the path difference may be expressed in the form of any time or path-difference dependent function. Practical examples may be electronic level variation or noise in the servo control loop, systematic mechanical friction during the movement of optical pieces, acoustic or ground vibration disturbing the control of Δ, and fluctuations of the wavenumber of the reference monochromatic beam varying with time or possibly with path difference. This last example may occur in the case where the reference beam comes from a servo-controlled

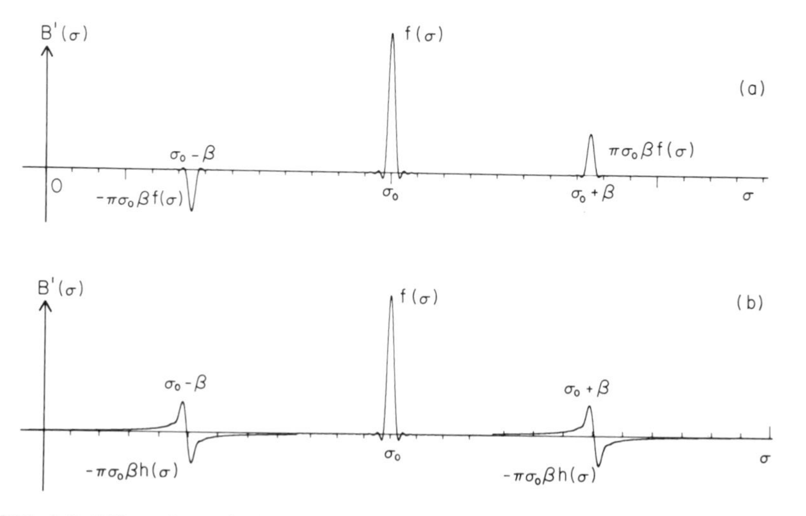

FIG. 1.9. Effect of a periodic error in path difference on the apodized instrumental line shape for (a) $\Phi_0 = 0$ and (b) $\Phi_0 = \pi/2$. In general the ghosts in the spectra are a combination of the two types shown in the figure.

laser cavity and is sensitive to backward reflections from the interferometer itself. Whatever the function ε_{02} may be, it may be expressed as a series of effects discussed in the previous paragraph. One can draw the general conclusion that if line shape perturbation exists due to phase distortion, its effect is expressed by an *odd function centered on* σ_0.

The approximate treatment used to go from Eq. (1.12) to Eq. (1.13) may be avoided by using Bessel's function J_k. Recalling that

$$\cos(t \sin \theta) = J_0(t) + 2 \sum_{k=1}^{\infty} J_{2k}(t) \cos(2k\theta)$$

and

$$\sin(t \sin \theta) = 2 \sum_{k=1}^{\infty} J_{2k-1}(t) \sin[(2k - 1)\theta],$$

with k an integer, one derives from Eq. (1.12) the following general expression [Eq. (1.14)] for the computed spectrum of a monochromatic line, the interferogram of which has been perturbed by a phase distortion $\varepsilon_2 = \varepsilon_{02} \sin 2\pi\beta\Delta$:

$$\begin{aligned} B'(\sigma) = {} & J_0(2\pi\sigma_0 \varepsilon_{02}) \int_0^{\Delta_M} \sin 2\pi\sigma_0\Delta \sin 2\pi\sigma\Delta \, d\Delta \\ & + \sum_{k=1}^{\infty} J_k(2\pi\sigma_0 \varepsilon_{02}) \int_0^{\Delta_M} [\sin 2\pi(\sigma_0 + k\beta)\Delta \\ & \qquad + (-1)^k \sin 2\pi(\sigma_0 - k\beta)\Delta] \sin 2\pi\sigma\Delta \, d\Delta. \qquad (1.14) \end{aligned}$$

Since $J_0(0) = 1$ and $J_k(0) = 0$ for k an integer, if $\varepsilon_{02} = 0$, Eq. (1.14) gives back the unperturbed spectrum. For $\varepsilon_{02} \neq 0$, Eq. (1.14) expresses first that the central unperturbed feature is in fact multiplied by a numerical factor $J_0(2\pi\sigma_0 \varepsilon_{02})$ smaller than one. This did not appear in Eq. (1.13). Also, on either side of the central peak, additional features located at $\pm\beta$ and at $\pm 2\beta$, $\pm 3\beta$, . . . , exist, with an intensity proportional to $J_1(2\pi\sigma_0 \varepsilon_{02})$, $J_2(2\pi\sigma_0 \varepsilon_{02})$, $J_3(2\pi\sigma_0 \varepsilon_{02})$, . . . Depending on the parity of k [because of the $(-1)^k$ factor], these features have antisymmetric (k odd) or symmetric (k even) shapes centered on σ_0.

Let us come back to more practical considerations. To get nonnegligible, even-type, phase perturbation at $\pm 2\beta$, one must have $2\pi\sigma\varepsilon_{02} \sim 1$. Then $J_0 \sim 0.8$, $J_1 \sim 0.5$, and $J_2 \sim 0.1$. However, the essential part of the phase perturbation is at $\pm\beta$ and has an odd behavior. Such a result supposes that

$$\varepsilon_{02} \simeq \lambda_0/2\pi,$$

which implies very poor control of the path difference and should not occur in practice.

2. *Intensity Distortions*

This multiplicative intensity effect corresponds to the particular following form of Eq. (1.7):

$$B'(\sigma) = \int_0^{\Delta_M} (1 + \varepsilon_3) \sin 2\pi\sigma_0\Delta \sin 2\pi\sigma\Delta \, d\Delta. \tag{1.15}$$

The term ε_3 introduces distortion in the intensity of the interferogram.

a. $\varepsilon_3 = \varepsilon_{03} = const.$ This corresponds, for instance, to an improper evaluation of the gain used to amplify the sampled signal given by the detector. Obviously the computed spectrum will have a form similar to the original spectrum. This effect is of no practical importance. In general it is completely ignored since one usually normalizes the intensity of the computed spectrum.

b. $\varepsilon_3 = \varepsilon_{03} \sin(2\pi\gamma\Delta + \Phi_0)$. The intensity of the interferogram is sinusoidally modulated with a periodicity γ depending directly (or through the time) on Δ. This could be illustrated by a small pure oscillation, during the recording, of a mirror used to send the output beam of the interferometer onto the detector.

Let us assume first that $\Phi_0 = 0$; Eq. (1.7) may then be written as

$$B(\sigma) = \int_0^{\Delta_M} (1 + \varepsilon_{03} \sin 2\pi\gamma\Delta) \sin 2\pi\sigma_0\Delta \sin 2\pi\sigma\Delta \, d\Delta.$$

This is easily expanded into

$$B'(\sigma) = \int_0^{\Delta_M} \sin 2\pi\sigma_0\Delta \sin 2\pi\sigma\Delta \, d\Delta + \frac{\varepsilon_{03}}{2} \int_0^{\Delta_M} (\cos 2\pi(\sigma_0 - \gamma)\Delta - \cos 2\pi(\sigma_0 + \gamma)\Delta) \sin 2\pi\sigma\Delta \, d\Delta \tag{1.16}$$

On both sides of the unperturbed profile $B(\sigma)$, located at $\sigma_0 - \gamma$ and $\sigma_0 + \gamma$, one finds then additional features represented by the second term of Eq. (1.16), similar to the relation giving $h(\sigma)$ already developed in Section 1.2,B.2. Accordingly, the sidebands have odd shapes, but as shown on trace (a) of Fig. 1.10, they are symmetric with respect to the center of the main peak.

When Φ_0 takes the value $\pi/2$ instead of zero, development of $B(\sigma)$ is still straightforward. The two features are then symmetric and still behave symmetrically about σ_0, as shown on trace (b) of Fig. 1.10. Depending on γ, the sidebands may overlap the central profile at σ_0, which appears as a symmetrical perturbation.

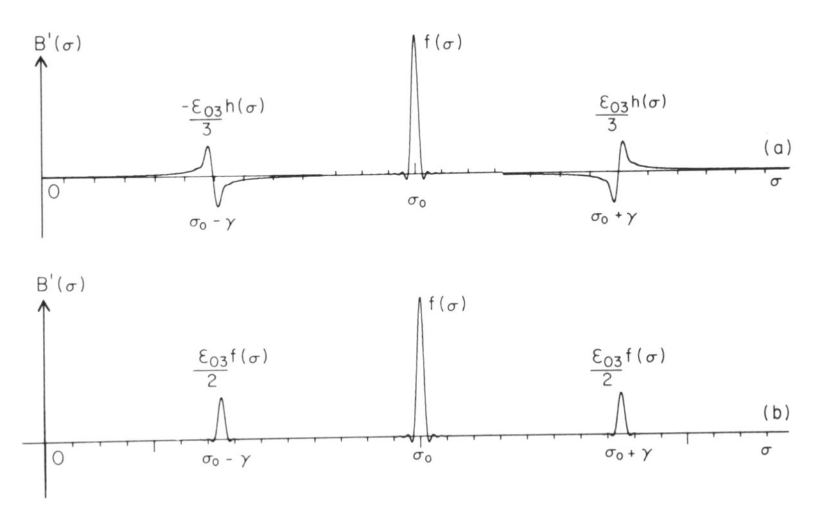

FIG. 1.10. Effect of a periodic error in the intensity of the interferogram on the apodized instrumental line shape, for (a) $\Phi_0 = 0$ and (b) $\Phi_0 = \pi/2$. In general, the ghosts in the spectra are combinations of the two types shown in the figure.

The intensity distortion given by Eq. (1.16) is not chromatic. Wherever the line may be, the distance, the amplitude, and the shape of the perturbation remain the same. This is not too surprising. Anticipating the next paragraph, the recording of an interferogram from $\Delta = 0$ to only $\Delta_{\max}$ may be considered as an intensity perturbation, and it is well known that the main effect on a monochromatic line, wherever this line may be, is its reconstruction with a constant half width called the resolution.

c. General Case. An actual example of an intensity distortion (this term is actually improper in that case) is the apodization procedure. It generally consists in multiplying the interferogram by a path difference-dependent function leading to a symmetrical modification of the line shape of the lines. Another well-known intensity distortion effect is a nonlinear detection of the samples already pointed out in Section 1.3,B.2 and developed in Section 1.6,C.2b. Also, if during the recording the interferogram signal disappears for a short period of time (due to source problem, for instance), whereas the recording procedure is not interrupted, one will also get intensity distortion.

Since an intensity-perturbing effect may be developed in a series, the terms of which are similar to the expression given in the previous paragraph where no approximation has been used, the previous conclusion remains valid. The distortions on the line shape are symmetric about the line center, and do not vary with the spectral location of the perturbed spectrum.

C. Conclusion

In Sections 1.5,A and 1.5,B the various systematic effects have been examined separately. An analysis of Eq. (1.7) including simultaneously the three types of errors may similarly be done. Additional cross terms between the two multiplicative errors ε_2 and ε_3 would appear, whereas the additive term ε_1 would remain independent of the others. For example, new features would be located at $\sigma_0 + (\gamma + \beta)$ and $\sigma_0 - (\gamma + \beta)$ and weighted by $\frac{1}{2}\varepsilon_{03}J_1(2\pi\sigma_0\varepsilon_{02})$. In fact, these second-order terms are less important than the others already reviewed, and when one is getting ready to build a Fourier interferometer, or one is suddenly faced with strange instrumental behavior, the previous considerations are adequate enough to begin tracking down the source of errors.

The correction of systematic effects must generally be done at their source once and for all. Their appearance in the spectra are often typical and give important indications of their origin. A posteriori corrections are often difficult and questionable when one wants accurate results. The conclusions arrived at for systematic distortions are naturally also valid for random effects such as noise.

Let us now summarize Section 1.5.

(1) The systematic distortions are divided into two groups: (i) the additive effects which do not modify the shape of the instrumental function and appear under the form of an additional, generally out of phase, perturbing spectrum, and (ii) the multiplicative effects producing severe changes in the instrumental profile. This second group encompasses two well-separated effects: (j) the phase errors which are chromatic and produce antisymmetric sidebands and (jj) the intensity errors, achromatic and symmetrically distorting the apparatus function.

(2) All the systematic effects may be time or path-difference dependent. Their appearance in the spectra will then depend on the experimental recording conditions. For the time-dependent case, if one varies the recording time without changing the stepping mode, the effects behave differently in the spectrum, while in the second case they remain the same. Intermediate situations seldom occur. Suppose that the recorder (or the tape-reading program) drops one sample out of n samples. In that special case, changing the stepping rate (i.e., the recording time) without changing the sampling interval, for instance, will produce identical errors in the computed spectrum. However, these errors will appear different after changing the sampling interval. They are then interval-dependent.

(3) One can say that there is actually no Fourier spectrum free of these systematic effects, and one must of course minimize these as much as possible. Measurement of the wavenumber and intensity of the lines is

always affected by these errors, especially the multiplicative errors. The *internal standard method* already discussed in Section 1.3,B eliminates this difficulty. Generally speaking, provided these effects are small, they are practically independent of σ_0. The standard and the measured profiles have the same distorted shape which allows for easy errorless calibration along the two axes by comparison.

1.6. Review of Some Systematic Effects in the Fourier Spectra

The aim of this section is to give a descriptive approach of the systematic errors. Instead of remaining with the general point of view of Section 1.5, the choice is made to go into more detail, starting usually from a specific type of spectrometer, i.e., the high-information type instrument already discussed in Section 1.3. Although the examination of these errors will be centered mainly on a particular type of instrument, this should not restrict its general interest since strong analogies exist among all Fourier spectrometers.

No pretense of exhaustively treating the subject is intended here. This does not mean that small effects will not be considered when they may have a rather high possibility of occurrence. Indeed, Fourier spectrometry has now reached the capability of high accuracy, and even small perturbations must be considered.

The present review will not be ordered according to the general categories of errors given in Section 1.5. A more experimental, concrete point of view is taken. One considers first the core of the instrument, i.e., the interferometer itself, and its relationship with its peripheral components. Then the errors of the path difference control system are examined. The last part deals with the measurement of the data samples, i.e., the specific possible effects due to the source and the recording system.

A. Instrument

Distinction must first be made here between the interferometer itself, composed of beam splitter–mixer plates and two reflectors, and the optical external components relaying the interferometer to the analyzed and reference sources and to the detectors.

1. *Interferometer*

a. Optical Quality of the Components. This is a well-known question (J. Connes, 1961) that concerns the optical tolerances of the reflecting surfaces. If one supposes the highly unfavorable case where the cross section of the interferometer is divided into two equal parts, one giving rise to a path difference Δ and the other $\Delta + \lambda_0/2$, then the interferogram of the line of wavenumber $\sigma_0 = 1/\lambda_0$ will always be equal to zero. This radical spectral attenuation may be mathematically expressed by the convolution of the interferogram function and the error function given by two equal Dirac peaks, at 0 and $\lambda_0/2$. More generally the interferogram is convolved with an error function $D(\Delta)$. This means that the spectrum is multiplied by the Fourier transform of $D(\Delta)$. The larger the error compared to the wavelength, the greater the distortion of the spectrum. It acts like a low-pass filter, where the longer wavelengths are less perturbed than the shorter wavelengths.

The best way to appreciate these imperfections is of course not to look at the spectrum but at the interference pattern given by the interferometer. If these imperfections are too large, the best solution is to better polish the imperfect components.

It is important that the interferogram start at $\Delta_0 = 0$, and great care must be taken to estimate this initial path difference. If it is done by an auxiliary white-light beam, which generally does not cover the same section of the aperture as the analyzed light, this could produce a constant phase error due to optical imperfections. The best way to start an interferogram at $\Delta_0 = 0$ is to use the analyzed beam itself for the initial determination.

b. Structural Stability. As a mechanical assembly, the interferometer may be sensitive to various mechanical instabilities. Effects due to temperature variations are well known. A heavy moving cat's-eye may bend the base on which the optical elements are attached. Also, resonances are often encountered. These deformations may cause optical misalignment. Retroreflectors, tilt-compensated systems, and room temperature stabilization can remove these difficulties. Also, when the deformations have similar effects on the reference and the analyzed beams, they are perfectly corrected by the servo control system. This strucutre stability problem is, in fact, very important in the early stage of the construction of an apparatus, but once it has been solved (e.g., for the optical quality imperfections in the preceding paragraph), it is solved "forever." I remember our first model of beam splitter/mixer plate mechanical support for the third generation apparatus. In spite of its appreciable weight, it behaved like rubber. Almost ten years later, its final internal stressless form has not shown any signs of deformation. For that reason we will not discuss this question any further. The remaining minor effects are typically relevant to drift problems discussed later in this chapter.

c. Asymmetry in the Optical Configuration. The two paths of the divided beams must be as identical as possible. A symmetrical structure for the apparatus is, of course, necessary: same number of mirrors in each arm and equal thickness of the beam splitter and mixer plates. Also, the coatings in each arm must be identical, and care must be taken to coat symmetrical optical pieces uniformly and simultaneously.

Let us consider an asymmetrical configuration. Instead of the geometrical path difference Δ being independent of σ, one gets Δ' given by

$$\Delta' = \Delta + n_1(\sigma)\Delta_1, \tag{1.17}$$

where the chromatic term $n_1(\sigma)\Delta_1$ corresponds to nonsymmetrical path in refractive material as well as various phase shifts due to different reflecting coatings on the mirrors or beam splitter. It is independent of Δ.

According to Section 1.5,B.1, Eq. (1.17) should produce phase distortion effects, which will then antisymmetrically perturb the instrumental line shape. The chromatic variations of this deformation will depend on $n_1(\sigma)\Delta_1$. Over a wide spectral range this systematic effect appears more evident. If it is too strong, the best solution consists again, as in the two preceding paragraphs, in a radical correction of the structure of the interferometer itself.

Let us consider a numerical calculation of the remaining uncorrected effect. If e_S and e_M represent, respectively, the thickness of the beam splitter and mixer plates made of Infrasil (see Fig. 1.11), then

$$\Delta_1 = e_S - e_M;$$

also $n_1(10{,}000\ \mathrm{cm}^{-1}) \simeq 1.4505$ and $n_1(2800\ \mathrm{cm}^{-1}) \simeq 1.4060$.

Equality of thickness of Δ_1 equal to $\lambda/4$ ($\lambda = 6328$ Å) is feasible. Then from 10,000 to 2800 cm^{-1} the chromatic variation of path difference will be

$$(6328/4) \times (1.4505 - 1.4060) \simeq 70\ \text{Å}.$$

If the path difference zero is well determined for $\sigma = 10{,}000\ \mathrm{cm}^{-1}$, then the constant term ε_{02} in Section 1.5,B.1a is equal to 70 Å for 2800 cm^{-1}.

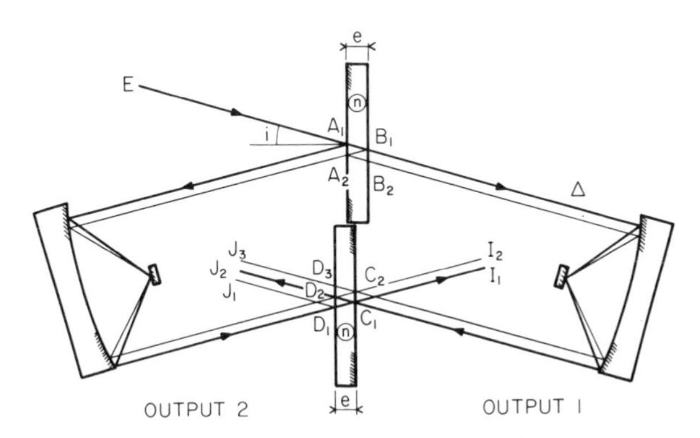

FIG. 1.11. Parasitic interferences due to additional reflections on the second face of the splitter and mixer plates, in a Connes-type interferometer.

According to Eq. (1.28) in Section 1.6,B.1a, the error in the measurement of a line of half width 5×10^{-3} cm^{-1} could be equal to about 30×10^{-3} cm^{-1}. This very important value corresponds to a maximum for this type of distortion. It is obviously inadmissible, and phase-error correction techniques (J. Connes, 1961; Vanasse and Sakai, 1967) should be applied.

d. Air Index. The effect of the air on the wavenumber measurement is well known, and a posteriori correction using the Edlen formula (Edlen, 1953) is applicable. In fact, for accurate spectra it is better to evacuate the interferometer. Indeed, the ideal a posteriori correction assumes accurate knowledge of the index of air and especially of having good control of atmospheric conditions during a measurement. Evacuating the interferometer chamber is an ideal solution to this latter difficulty, and has the additional advantage of eliminating noise due to turbulence and unwanted absorption by atmospheric molecules.

The determination of path difference by using the fringes of a monochromatic reference line, in the same medium as the radiation being analyzed, decreases the air-index effect, which needs only a differencing (or ratio) correction, contrary to other path-difference measuring schemes, e.g., moire fringes.

The air-index effect will obviously affect the knowledge of the path difference and may be considered a phase effect. It does not distort the lines, which remain symmetrical since the path difference is well determined, but it is of a chromatic nature.

Suppose that $n(\sigma_R)$ and $n(\sigma_0)$ are, respectively, the air index of the reference line and of the analyzed monochromatic line, the vacuum wavenumbers of which are σ_R and σ_0, respectively. Since the path difference is determined by the reference line, instead of measuring the geometrical value Δ, obtained when σ_R is used, one has the air optical value of the path difference Δ_{air} with

$$\Delta_{air} = \Delta/n(\sigma_R). \tag{1.18}$$

The interferogram $I(\Delta)$ of the monochromatic line recorded in air will then be $\sin 2\pi\sigma_{air}\Delta_{air}$.

Since

$$\sigma_{air} = n(\sigma_0)\sigma_0,$$
$$I(\Delta) = \sin 2\pi n(\sigma_0)\sigma_0\, \Delta/n(\sigma_R).$$

If the Fourier transform of $I(\Delta)$ is performed assuming that the interferogram was recorded under vacuum, the computed spectrum will be

$$B(\sigma) = \int_0^{\Delta_M} \sin 2\pi n(\sigma_0)\sigma_0 \frac{\Delta}{n(\sigma_R)} \sin 2\pi\sigma\Delta \, d\Delta.$$

The line σ_0 is then located at $[n(\sigma_0)/n(\sigma_R)]\sigma_0$.

An error $d\sigma_0$ in the measurement of the wavenumber σ_0 exists and is given by

$$\frac{d\sigma_0}{\sigma_0} = \frac{n(\sigma_0) - n(\sigma_R)}{n(\sigma_R)} = \frac{dn}{n} . \tag{1.19}$$

Generally, the greater the distance between σ_0 and the reference line, the greater is the error, which of course cancels for $\sigma_0 = \sigma_R$.

As a numerical application, suppose the measurement is done in air of a line located at 14,000 cm^{-1} using a reference at 2800 cm^{-1}, and the above transformation procedure is used.

From Edlen (1953),

$$n(2800) = 1 + 272{,}724 \times 10^{-9} \quad \text{and} \quad n(14{,}000) = 1 + 275{,}788 \times 10^{-9}.$$

Equation (1.19) gives

$$d\sigma \simeq 43 \times 10^{-3}\ \text{cm}^{-1},$$

which is far from negligible. The a posteriori corrections using the Edlen formula are then indispensable.

When the interferometer is in vacuum, these corrections become unnecessary. Indeed, even with a rough primary vacuum of 1 Torr, if one admits a linear dependency of the index versus the pressure, one obtains in the previous case an error equal to 6×10^{-5} cm^{-1}.

e. Parasitic Optically Induced Interferograms. This will be the last point treated under the interferometer section. It concerns the existence of additional reflections in the refractive material making up the beam splitter/mixer plates. Figure 1.11 gives a schematic representation of a Connes-type interferometer. Generally, one considers only the beams represented by thick lines, i.e., the interferences between $EA_1B_1C_1I_1$ and $EA_1D_1C_1I_1$ corresponding to the output 1, and the interference between $EA_1B_1C_1D_2J_2$ and $A_1D_1C_1D_2J_2$ corresponding to the output 2. In fact, the reflections on the second face of the plates may produce parasitic contributions to this interference system, and one must take into account a larger number of interfering beams. These additional beams are drawn in Fig. 1.11, where only the paths involving one glass reflection r_v are taken into consideration. It would be too lengthy to give here the exact development that leads to the final interference formulas. However, one can easily find the following general results.

Let δ be the additional path difference due to the thickness e of the plates, the material of which has a refractive index n. Then $\delta = 2ne \cos i/n$ is, due to n, chromatic (i is the incidence angle). Assuming that $r_v^2 \simeq 0$ (this does not restrict the final conclusions), the modulated part of the interferogram $I_1(\Delta)$ of the output 1 is equal to

$$I_1(\Delta) = AI(\Delta) + \alpha[I(\Delta - \delta) + I(\Delta + \delta)] \tag{1.20}$$

instead of $I(\Delta)$ in the usual case, and the modulated part $I_2(\Delta)$ of output 2 is

$$I_2(\Delta) = -BI(\Delta) - \beta_1 I(\Delta - \delta) - \beta_2 I(\Delta + \delta) - \beta_3 I(\Delta + 2\delta) \quad (1.21)$$

instead of $-I(\Delta)$. A and B are of the order of 1; α, β_1, and β_2, of the order of r_v, and β_3 is proportional to r_v^2. The third term of Eq. (1.21) will be neglected.

In other words, Eqs. (1.20) and (1.21) express the existence of additional parasitic interferograms shifted in path difference by $\pm\delta$.† The two additional interferograms of the output 1 have equal amplitude, whereas the corresponding features of the other output are generally unequal, with $\beta_1 \neq \beta_2$. Since the final interferogram $I'(\Delta)$ is the difference of $I_1(\Delta)$ and $I_2(\Delta)$, the computed spectrum will be given by the Fourier transform of

$$I'(\Delta) = (A + B)I(\Delta) + (\alpha + \beta_1)\mathrm{I}(\Delta - \delta) + (\alpha + \beta_2)I(\Delta + \delta) \quad (1.22)$$

and not of $I(\Delta)$.

The multiplicative $(A + B)$ factor has no importance (see Section 1.5,B.2a). Quite the contrary—the two other additive terms will distort the spectrum. They cannot be taken as simple additive interferograms because they are intimately related to $I(\Delta)$, behaving like echoes, the delays of which $\pm\delta$ are, in addition, strongly chromatic. Let us see in more detail the distortions that they produce.

Suppose again that the interferogram is that of a monochromatic line σ_0. Relation (1.22) becomes

$$I'(\Delta) = (A + B) \sin 2\pi\sigma_0\Delta + (\alpha + \beta_1) \sin 2\pi\sigma_0(\Delta - \delta) + (\alpha + \beta_2) \sin 2\pi\sigma_0(\Delta + \delta),$$

and the spectrum

$$B'(\sigma) = \int_0^{\Delta_M} I'(\Delta) \sin 2\pi\sigma\Delta \, d\Delta$$

is given by

$$B'(\sigma) = (A + B + (2\alpha + \beta_1 + \beta_2) \cos 2\pi\sigma_0\delta) \int_0^{\Delta_M} \sin 2\pi\sigma_0\Delta \sin 2\pi\sigma\Delta \, d\Delta + (\beta_2 - \beta_1) \sin 2\pi\sigma_0\delta \int_0^{\Delta_M} \cos 2\pi\sigma_0\Delta \sin 2\pi\sigma\Delta \, d\Delta, \quad (1.23)$$

expressing that the normal apparatus function $f(\sigma)$ (see Eq. (1.2) of Section 1.2,B.1) is multiplied by a factor modulated by $\cos 2\pi\sigma_0\delta$. The second term of Eq. (1.23) represents the addition to the normal instrumental

† Taking into account the beams of intensity proportional to r_v^2, r_v^3, . . . , one could find other additional interferograms at $\pm 2\delta$, $\pm 3\delta$, ...

profile of the antisymmetric function $h(\alpha)$ of Section 1.2,B.2 modulated by $\sin 2\pi\sigma_0\delta$. As a result, both the intensity and the symmetry of the line may be distorted, but in a "compensating" way. The greater the perturbation on the intensity, the smaller is the perturbation on the symmetry, and vice versa.

For a broadband emission spectrum, the distortion of the intensity will not be apparent, whereas the phase effect will be perhaps detectable due to its periodicity $1/\delta$. On the other hand, an absorption spectrum obviously will have its baseline modulated with the amplitude $2\alpha + \beta_1 + \beta_2$, as it would be with an additional channel spectrum (Section 1.6,C.1d). The situation is, however, different because of the antisymmetrical term. Figure 1.12 represents the distortion of the background of an ideal absorption spectrum, which should be located at the intensity 1. In fact, to the constant level $A + B$, a "rolling" function is added. It has a periodicity of $1/\delta$ (although it depends on n, over a small spectral range δ it may be considered constant). The sine line represents the actual position of the background. The distortion of the profile is maximum at $1/\delta(k \pm \frac{1}{4})$ (k integer).

Obviously parasitic interferograms will affect both the intensity and the wavenumber measurement. They deserve particular attention. If they do appear, they generally give a pronounced feature at $\Delta = \delta$ (easily detected) in the interferogram. Figure 1.13 shows an actual feature obtained with our instrument from a broadband absorption spectrum. The refrac-

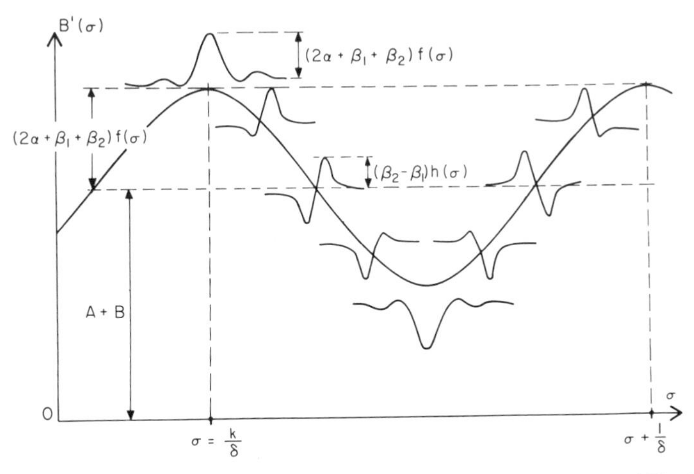

FIG. 1.12. The spectrum should be equal to one. Parasitic interferences of Fig. 1.11 shift the mean level to $A + B$, which is additionally modulated with a periodicity of $1/\delta$.

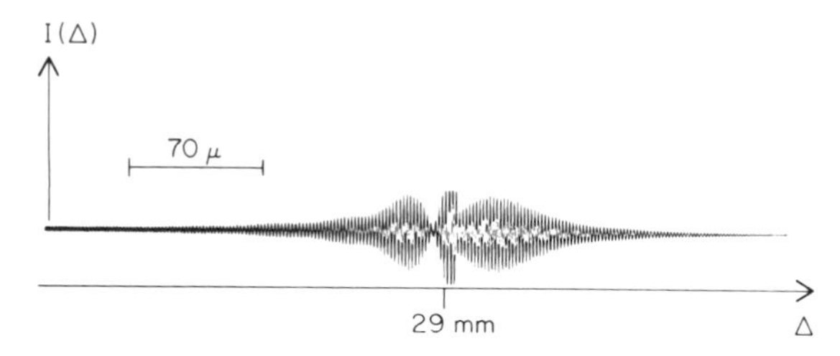

FIG. 1.13. Actual parasitic interferences due to beam system of Fig. 1.11.

tive material has an index $n \simeq 1.5$ and a thickness of 10 mm, with an angle of incidence $i = 22°$. According to

$$\delta = 2ne \cos(i/n),$$

the central peak of the feature should be at 29 mm, which is the actually measured location. The strong chromaticity due to the quartz destroys the symmetry of the parasitic interferogram at δ.

Several types of corrections may be applied. The first one consists in a radical attenuation of the parasitic interferences. This may be accomplished by depositing an antireflective coating on the bare face of the plates; but this is effective only over a narrow† spectral range. Wedging the beam splitter and mixer plates allows geometrical separation between the parasitic and main interferograms. This is an elegant solution. It has, however, to be done with caution since incorrect optical adjustment may produce chromaticity of the type already discussed in Section 1.6,A.1c. Fabrication of the refractive optical pieces also becomes more difficult than that of parallel plates, which in itself is already a delicate task.

Depending on the particular conditions in which the spectrum is recorded, it may occur that use of the whole admissible *étendue* of the interferometer is not obtained. If the angular diameter of the entrance beam is small compared to the ring pattern, the parasitic interferogram may possibly be enhanced. Indeed, since $\delta = 2ne \cos(i/n)$ depends on i, the terms $I(\Delta \pm \delta)$ of Eqs. (1.20) and (1.21) may vanish for a given range of i. In the example shown in Fig. 1.13, a small angular diameter for the entrance beam had been chosen. For normal field of view coverage, everything being similar, the parasitic interferogram disappeared completely.

A third point which allows us to be free at least from wavenumber systematic errors is suggested by Eq. (1.23) giving $B'(\sigma)$. If $(\beta_2 - \beta_1)$ equals zero, then no odd distortion occurs. This is obtained when one records only the interferogram $I_1(\Delta)$ expressed by Eq. (1.20). This removes the second term of $B'(\sigma)$. The modulation of the background,

† That is, compared to the range generally obtainable with the Fourier method.

however, remains and may be considered as channel spectra (Section 1.6,C.1d).

2. *Input and Output Relay Optics*

a. Finite Entrance Beam. The interferometer analyzes light which actually covers a finite *étendue*. The interferogram is then the sum of various interferograms, each from different point sources defined by an annulus of a given angular diameter centered on the optical axis. Suppose that the entrance hole is seen from the collimator under the solid angle Ω (one could also define the *étendue* by the output light collector and the size of the detector). Then two different effects occur on the theoretical spectrum. First, it is convolved with a rectangular function of width $\sigma\Omega/2\pi$. The second important effect is a wavenumber shift. A line profile centered at σ_0 is shifted to $\sigma_0 - (\sigma_0\Omega/4\pi)$. It is shifted toward a lower wavenumber by an amount proportional to σ_0. From a physical point of view, these two effects may be qualitatively explained as a blurring phenomenon due to the simultaneous observation of various interferograms, the path differences of which are decreased proportional to the angle of observation of each.†

As a consequence one must adapt the solid angle Ω to the desired resolution. The width of the rectangular function $\sigma_0\Omega/2\pi$ should not be greater than the resolution $1/2\Delta_M$. This gives the well-known relation

$$\sigma\Omega\Delta_M = \pi,$$

which essentially indicates that it is useless to pursue the recording of the interferogram as soon as $\Delta = \Delta_M$, for which the ring pattern covers one complete interference order.‡ Another consequence is the need to correct the wavenumber axis, and this may be the source of systematic errors, as shown by the following numerical example.

Let r and f be, respectively, the radius of the entrance hole and the focal length of the collimator (Fig. 1.14). Then

$$\Omega = \pi r^2/f^2.$$

Suppose that r and f are actually measured, with an error Δr and Δf. The numerical a posteriori correction of the wavenumber shift, which should be

$$d\sigma = \sigma_0\Omega/4\pi,$$

† Mathematical formulation is possible using the same procedure as in the next section. Also see Terrien (1958), J. Connes (1961), Vanasse and Sakai (1967).

‡ It is possible to overcome this limitation in *étendue* by introducing in the interference system additional compensating path differences; see Baker (1977).

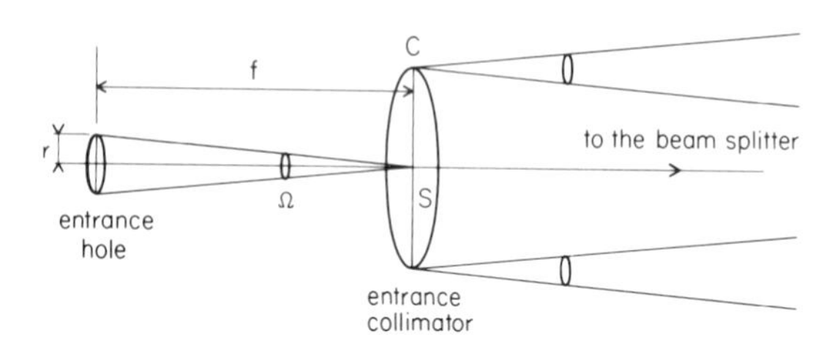

FIG. 1.14. The étendue of the entrance is ΩS, where Ω is the solid angle covering the entrance hole from the collimator whose area is S.

will be in error by $\Delta\sigma$ such that

$$\Delta\sigma/d\sigma = d\Omega/\Omega = 2(\Delta r/r + \Delta f/f).$$

Inserting the value for $d\sigma$, one gets

$$\Delta\sigma/\sigma_0 = (r/2f^2)\,[\Delta r + r\,(\Delta f/f)].$$

With $\sigma_0 = 10{,}000\ \text{cm}^{-1}$, $\Delta_M = 200$ cm, $f = 200$ cm, the optimum diameter of the entrance hole located at the focal plane of the collimator is 0.4 cm. The wavenumber correction is then $2.5 \times 10^{-3}\ \text{cm}^{-1}$. Assuming f to be perfectly known, Δr equal to ± 0.1 mm produces an error on the a posteriori correction equal to $\pm 0.1 \times 10^{-3}\ \text{cm}^{-1}$, which is not at all negligible for metrological determinations.

In fact, the best way to correct the finite aperture shift is to use the *internal standard method* (Section 1.3,B.1), which additionally corrects, simultaneously, the similar optical effects occurring in the reference beam.

b. Off-Axis Entrance Beam. Contrary to the previous paragraph, the entrance beam is not centered on the optical axis of the interferometer. In other words, the entrance hole is not concentric to the ring pattern. This has an effect on the shape as well as on the position of the line.

If one supposes first a zero *étendue* beam whose angular position versus the optical axis is i, then a monochromatic interferogram is written

$$\sin(2\pi\sigma_0\Delta \cos i).$$

Provided that i is small, the computed spectrum will be located at

$$\sigma_0(1 - \tfrac{1}{2}i^2).$$

The line will be shifted by

$$\Delta\sigma = \sigma_0\,\tfrac{1}{2}i^2$$

toward a lower wavenumber.

For a finite *étendue,* the off-axis misadjustment will be responsible for both a shift in the measured wavenumber and a distortion in the profile of the line. Figure 1.15 represents the focal plane of the collimator entrance mirror, whose focal length is f. The optical axis of the interferometer is given by O. T_1 and T_2 represent the same entrance hole in the on-axis and off-axis positions, respectively. For T_2 the exact formulation of the instrumental effect produced is not simple. One can, however, approach a solution by approximating the entrance surface using the circular sectors $P_1, P_2, \ldots, P_6$, all having O as the center of curvature. Suppose an elementary sector, e.g., P_1, is defined by r_1, r_2, and $2\theta_1$. With a monochromatic source σ_0, the corresponding interferogram will be given by

$$I_{P_1}(\Delta) = \int_{r_1}^{r_2} \int_{-\theta_1}^{+\theta_1} \sin 2\pi\sigma_0\Delta \left(1 - \frac{r^2}{2f^2}\right) r \, dr \, d\theta, \tag{1.24}$$

where r defines the position of the elementary emitting surface, and where the previously defined angular position $i = r/f$ is again assumed to be small.

Integration of Eq. (1.24) leads to

$$I_{P_1}(\Delta) = \theta_1(r_2^2 - r_1^2) \frac{\sin 2\pi\sigma_0\Delta(r_2^2 - r_1^2)/4f^2}{2\pi\sigma_0\Delta(r_2^2 - r_1^2)/4f^2} \sin 2\pi\sigma_0\Delta[1 - (r_1^2 - r_2^2)/4f^2]. \tag{1.25}$$

The interferogram $I_{P_1}(\Delta)$ has an intensity proportional to the area $r_2^2 - r_1^2$ of the sector, and has a sine behavior multiplied by a sinc term. As a result, the Fourier transform will be the convolution of the two individual Fourier transforms of

$$\frac{\sin 2\pi\sigma_0\Delta(r_2^2 - r_1^2)/4f^2}{2\pi\sigma_0\Delta(r_2^2 - r_1^2)/4f^2} \quad \text{and} \quad \sin 2\pi\sigma_0\Delta[1 - (r_2^2 + r_1^2)/4f^2].$$

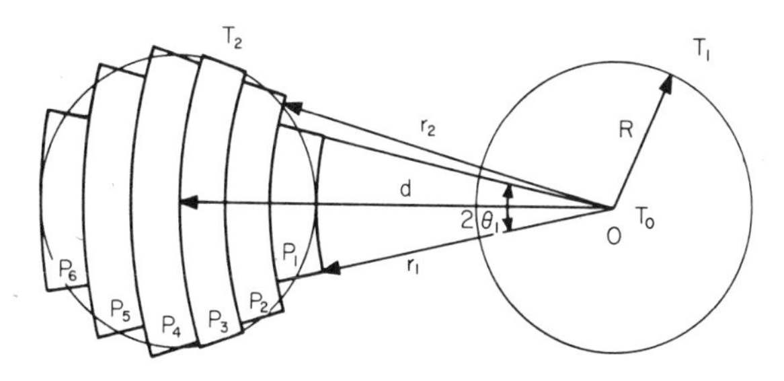

FIG. 1.15. Entrance holes: T_0 point source on the optical axis O of the interferometer; T_1, centered on O; and T_2, same hole in an off-axis position.

Consequently, after computation, the line will be located at σ smaller than σ_0, such that

$$\sigma_0 - \sigma = [(r_2^2 + r_1^2)/4f^2]\, \sigma_0.$$

It will also be convolved by a rectangular function, the width of which is

$$[2(r_2^2 - r_1^2)/4f^2]\sigma_0,$$

and the amplitude of which is proportional to the area $\theta_1(r_2^2 - r_1^2)$.†

The resulting effects on the spectrum due to T_2 may be considered as the sum of the effects due to the six sectors $P_1, \ldots, P_6$. Figure 1.16 represents the synthetic spectra when T_1 or T_2 are illuminating the interferometer, T_0 giving the ideal position of the line. The area of the T_1 spectrum is equal to the area of the T_2 spectrum.

The computed profile has no symmetry and is drawn toward the low wavenumber side. This behavior is chromatic. The same optical misadjustment will produce effects more pronounced for larger wavenumber. A posteriori correction could be difficult and meaningless. It is obviously better to center the entrance hole on the ring pattern, provided the coatings of the beam splitter allow its observation in the visible. The distortion due to off-axis illumination may often be taken as a slight phase error on the beginning of the interferogram, and this should not be forgotten. Also, correct adjustment of the reference beam is necessary. If it is not, the same types of effect will occur on the spectra (Section 1.6,B.1b,c).

Suppose for a numerical example that the off-axis shift is small enough so that the shape will remain practically symmetrical. What is the error in the measurement of the wavenumber, when the centering is done within $\frac{1}{4}$ the angular diameter of the first ring observed at $\Delta = 1$ m with $\sigma_0 \simeq 6328$ Å? In such a condition the angular diameter is about 2.3×10^{-3} rad. For the simple case of a point source, the relative error is

$$\Delta\sigma/\sigma_0 = \tfrac{1}{2}[(2.3/4) \times 10^{-3}]^2 = 1.65 \times 10^{-7}.$$

This corresponds to 1.6×10^{-3} cm^{-1} at $\sigma_0 = 10{,}000$ cm^{-1}, and it is not at all negligible. As in various previous examples, the *internal standard method* (Section 1.3,B.1) takes this type of error into account and avoids the need for a special correction.

c. Defocused Entrance Beam. In the above considerations, the entrance aperture was supposed to be in the focal plane of the collimator mirror. The purpose of this section is to examine the effects caused by poor collimation.

† In all the expressions given here one could have introduced Ω instead of $\pi r^2/f^2$.

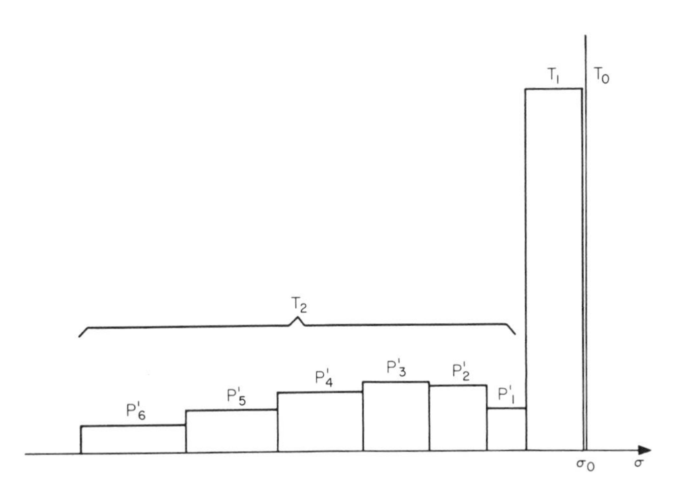

FIG. 1.16. The same monochromatic source at σ_0 is calculated at T_0, T_1, or T_2, respectively, corresponding to the different entrance holes of Fig. 1.15.

Figure 1.17 gives a schematic representation of the collimator C of diameter $\theta = 2R$. Its focus F is at Δf from the entrance hole T_0, which is on the optical axis of the interferometer represented by M_1 and M_2. Assuming that T_0 is a point source and that C is perfect, instead of a plane wave front one has approximately a spherical one entering the interferometer, with a radius of curvature depending on the sign and the amplitude of Δf. It is converging in the example of Fig. 1.17, where T_0 is imaged by C on T'. The fringes localized at infinity that are produced from the same ray will correspond, for a given geometrical path difference Δ in the interferometer, to different interference orders depending on the incidence angle $i(\rho)$, which itself depends on the distance ρ (Fig. 1.17). The inter-

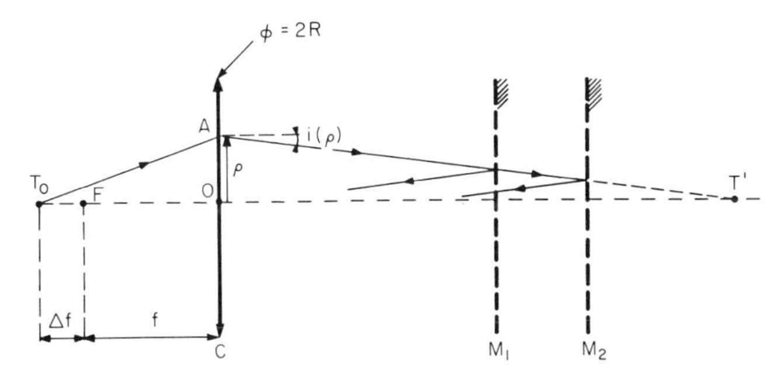

FIG. 1.17. The interferometer is represented by M_1 and M_2. It is illuminated by a point source T_0, which is not at the focus F of the collimator C.

ferogram of a monochromatic line σ_0 will then be written

$$\sin 2\pi\sigma_0\Delta[1 - \tfrac{1}{2}i^2(\rho)],$$

provided $i(\rho)$ is small enough to let $\cos i \simeq 1 - \frac{1}{2}i^2$. Since $i(\rho) \simeq \rho\ \Delta f/f^2$, the complete interferogram above corresponding to the entire collimator may be written as

$$I(\Delta) = 2\pi \int_0^R \sin 2\pi\sigma_0\Delta \left[1 - \frac{1}{2}\left(\frac{\Delta f}{f^2}\right)^2 \rho^2\right] \rho\, d\rho, \qquad (1.26)$$

which is similar to Eq. (1.24) and gives

$$I(\Delta) = \pi R^2(\sin \pi\sigma_0\Delta\alpha R^2/\pi\sigma_0\Delta\alpha R^2) \sin 2\pi\sigma_0\Delta(1 - \alpha R^2/2) \quad (1.27)$$

with $\alpha = \frac{1}{2}(\Delta f/f^2)^2$.

As a result, one has the two different effects already expressed in the preceding section, i.e., first a shift in the location of the line, which is equal to $\sigma_0\alpha R^2/2$, and a displacement of the profile toward a lower wavenumber. This shift is proportional to $(R\Delta f)^2$ and inversely proportional to f^4. Second, the initial spectrum is convolved by a rectangular function, the width of which is $\Delta\sigma = \alpha R^2\sigma_0$. In Eq. (1.27) the first term πR^2 represents the area of the collimator C and indicates the obvious fact that the larger C is, the brighter is the interferogram for a given source brightness.

With $f \simeq 200$ cm, $R = 4$ cm, and $\Delta f = 2$ cm, at $\sigma_0 \simeq 10{,}000\ \text{cm}^{-1}$, the wavenumber is decreased by

$$d\sigma = \sigma\alpha R^2/2 = 10^4 \times \tfrac{1}{2}\ |2/(200)^2|^2 \times \tfrac{1}{2}\ 4^2 = 10^{-4}\ \text{cm}^{-1}.$$

This is not very significant. However, if one keeps the same relative error on f, i.e., $\Delta f/f \sim 10^{-2}$, but decreases the focal length by a factor of 4 ($f = 50$ cm), the shift introduced under these relatively common conditions is then equal to $1.6 \times 10^{-3}\ \text{cm}^{-1}$, which is not negligible.

The width of the convolving rectangle in the above two numerical examples is respectively equal to $0.2 \times 10^{-3}\ \text{cm}^{-1}$ and $3.2 \times 10^{-3}\ \text{cm}^{-1}$. In general, this is of no great importance when compared to the actual width of the lines and of the apparatus functions, which in general are the limiting factors.

Obviously, the same effect can influence the beam from the reference line. A posteriori corrections seem difficult. Precise autocollimation is required in order to get Δf as small as possible. The remaining distortions are corrected by the *internal standard method* (Section 1.3,B.1), provided that α is constant for all the analyzed wavenumbers. This is obtained when reflecting instead of refracting imaging systems are used, thereby avoiding chromatism.

A similar exposition could be developed for the effects due to optical aberrations of the input and output optics. For instance, a spherical aberration of the entrance system giving a spherical wave front instead of a plane entrance wave front would produce the same type of effects (shifts and broadening of the lines) as defocusing. Care must be taken with the optical quality of these mirrors, which otherwise may distort the fringe pattern. Provided again that mirrors instead of refractive material are used, the *internal standard method* allows an efficient correction of these distortions.

B. Path-Difference Control System

1. *Path-Difference Determination*

a. $\Delta = 0$. When the path difference zero is experimentally determined by the detection of the zero near the central peak of the odd interferogram, one is always faced with a slight residual phase error due to unavoidable perturbations (e.g., noise, vibration, turbulence, wrong bias level, chromaticity of the interferometer, parasitic flux). Since the resulting distortions have already been discussed (Section 1.5,B.1a), we shall give only numerical estimates. Suppose that ε_{02} is small compared to the observed λ_0. In such a case the relative error on the positioning of the line $d\sigma/\sigma_0$ is linearly proportional to ε_{02} and to the half width $\Delta\sigma$ of the measured profile (Delouis, 1973). Roughly

$$d\sigma/\sigma = 3 \times 10^{-8}\, \varepsilon_{02}\, \Delta\sigma, \tag{1.28}$$

with ε_{02} in angstroms and $\Delta\sigma$ in centimeters^{-1}.

With a servo control of Δ within ± 5 Å (see Guelachvili, 1978a) at 10,000 cm^{-1}, and $\Delta\sigma = 10^{-2}\ \text{cm}^{-1}$, the residual error on the position of the measured line is equal to $\pm 15 \times 10^{-6}\ \text{cm}^{-1}$, which is almost negligible. This shift is reduced further if one uses the *internal standard method.* However, one must point out here that this method cannot strictly compensate the effects of phase error. A posteriori correction may also be done, as already said in Section 1.6,A.1c.

b. Instrumental Effects on the Reference Line Beam. The role of the reference line is to measure the path difference. We have previously supposed that it was perfectly monochromatic, and that its wavenumber was perfectly known. Actually, errors of several types may occur which will in some way perturb the knowledge of the path difference. Essentially, the systematic effects due to the reference line are antisymmetric and chromatic, since they will produce pure phase distortion (Section 1.5,B).

If, instead of the geometrical path difference Δ, the reference line actually indicates a path difference

$$\Delta' = f(\Delta),$$

the interferogram $I(\Delta)$ of a monochromatic line σ_0 will be given by

$$I(\Delta) = \sin 2\pi\sigma_0 f(\Delta),$$

and the computed spectrum by

$$B'(\sigma) = \int_0^{\Delta_M} \sin 2\pi\sigma_0 f(\Delta) \sin 2\pi\sigma\Delta \, d\Delta. \tag{1.29}$$

An example of that type of error has been discussed previously in Section 1.6,A.1d which concerns the air-index perturbation, where $f(\Delta)$ is equal to $\Delta/n(\sigma_R)$. All the effects already treated for the analyzed beam (Section 1.6,A.2) affect Δ for the reference beam as well and must be taken into account. For instance, the finite entrance beam (Section 1.6,A.2a) will transform the wavenumber σ_R into $\sigma_R(1 - \Omega_{Ref}/4\pi)$, with Ω_{Ref} being the solid angle corresponding to the reference beam. Since Δ is determined by its relationship to

$$\lambda_R = 1/\sigma_R,$$
$$\Delta' = \Delta/(1 - \Omega_{Ref}/4\pi),$$

and Eq. (1.29) becomes

$$B'(\sigma) = \int_0^{\Delta_M} \sin 2\pi\sigma_0 \frac{\Delta}{(1 - \Omega_{Ref}/4\pi)} \sin 2\pi\sigma\Delta \, d\Delta.$$

As a result, the computed wavenumber will be $\sigma_0/(1 - \Omega_{Ref}/4\pi)$ instead of σ_0. This tends to compensate the similar effect on the analyzed beam. For equal solid angles, these wavenumber shifts cancel each other. The *internal standard method* efficiently corrects these perturbations, as already discussed, whereas the a posteriori correction supposes that both solid angles are known.

The systematic effects due to the reference line similar to all the instrumental points discussed in Section 1.6,A could be given here as it has been done above for the finite entrance beam. We prefer instead to look at the more specific effect, due to the fact that the reference line is actually a spectrum with possible *asymmetry* and *width*.†

† Frequency instabilities of the reference line due to many causes, e.g., malfunctioning of the source itself, feedback from the interferometer to the servo controlled laser cavity, and sound or ground perturbing vibrations will not be developed here. Effects due to phase errors are to be treated as in Section 1.6,B.1.

c. Asymmetry of the Reference Line. For the sake of simplicity, let us suppose that the reference line is, in fact, the sum of two different monochromatic lines at σ_R and $\sigma_R + \delta\sigma_R$ with intensities of 1 and α ($\alpha < 1$), respectively, as shown in Fig. 1.18a. Under such conditions, what will be the spectrum of a monochromatic line σ_0?

The interferogram $I_R(\Delta)$ of the reference line is then given by

$$I_R(\Delta) = \sin 2\pi\sigma_R\Delta + \alpha \sin 2\pi(\sigma_R + \delta\sigma_R)\Delta. \tag{1.30}$$

The values of Δ where the zero crossings of $I_R(\Delta)$ occur will not be equidistant as they are for the single line σ_R. Consequently, one immediately realizes that the stepping is affected and nonequal sampling intervals will be produced. This phase error appears as antisymmetrical and chromatic effects on the recorded spectrum.

Suppose that $I_R(\Delta)$ in Eq. (1.30) is equal to zero for

$$\Delta_n = n\lambda_R + \varepsilon_2(n) \qquad (n \text{ integer}). \tag{1.31}$$

Then, from Eqs. (1.30) and (1.31),

$$\sin 2\pi\sigma_R[n\lambda_R + \varepsilon_2(n)] = -\alpha \sin 2\pi(\sigma_R + \delta\sigma_R)\,[n\lambda_R + \varepsilon_2(n)],$$

or, assuming $\delta\sigma_R\varepsilon_2(n) \simeq 0$,

$$\sin 2\pi\sigma_R\varepsilon_2(n) = -\alpha \sin[2\pi\sigma_R\varepsilon_2(n) + 2\pi\delta\sigma_R n\lambda_R]. \tag{1.32}$$

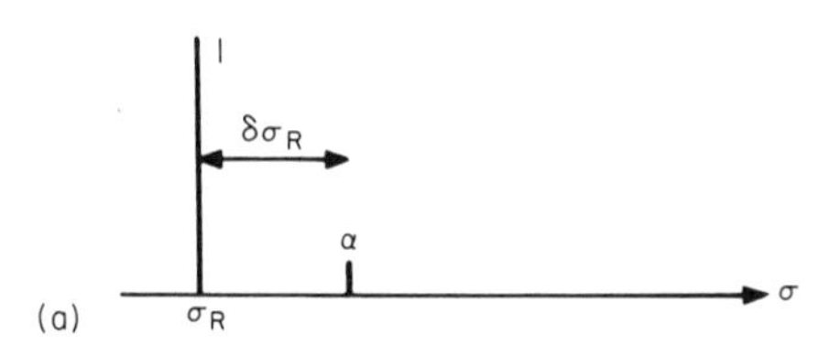

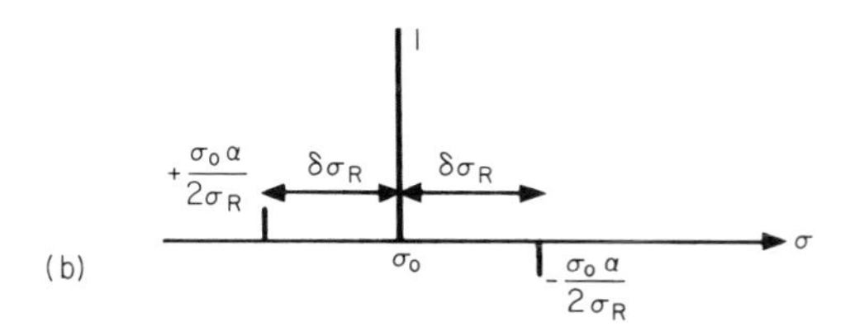

FIG. 1.18. (a) Asymmetry in the reference line, and (b) corresponding effect on the spectrum.

Expanding Eq. (1.32) leads to

$$\tan 2\pi\sigma_R\varepsilon_2(n) = -\alpha \sin 2\pi\delta\sigma_R n\lambda_R/(1 + \alpha \cos 2\pi\delta\sigma_R n\lambda_R).$$

For slight asymmetry we may use the following two approximations:

$$\tan 2\pi\sigma_R\varepsilon_2(n) \simeq 2\pi\sigma_0\varepsilon_2(n)$$

and

$$1 + \alpha \cos 2\pi\sigma_R n\lambda_R \simeq 1.$$

Finally,

$$\varepsilon_2(n) = -(\alpha/2\pi\sigma_R) \sin 2\pi\delta\sigma_R n\lambda_R. \tag{1.33}$$

Equation (1.31) represents the function $f(\Delta)$ already discussed in the previous section, $n\lambda_R$ being the correct Δ and $\varepsilon_2(n)$ the phase error.

From Eqs. (1.31) and (1.33), one has

$$f(\Delta) = \Delta[1 - (\alpha/2\pi\sigma_R) \sin 2\pi\delta\sigma_R\Delta].$$

According to Eq. (1.29), the interferogram of the monochromatic line σ_0 is given by

$$B(\sigma) = \int_0^{\Delta_M} \sin 2\pi\sigma_0 \left(\Delta - \frac{\alpha}{2\pi\sigma_R} \sin 2\pi\delta\sigma_R\Delta\right) \sin 2\pi\sigma\Delta \, d\Delta, \tag{1.34}$$

which is a relation similar to Eq. (1.12) developed in Section 1.5,B.1b.

Consequently, to the normal apparatus function two satellites will be added at $\pm\delta\sigma_R$ with relative amplitudes $\mp\sigma_0\alpha/2\sigma_R$. This is represented in Fig. 1.18b. The distortion appearing in the spectrum due to an asymmetrical reference line is then antisymmetric and chromatic. The larger σ_0, the greater are the intensities of the satellites, the amplitudes of which vary linearly with σ_0. Depending on the relative values of $\delta\sigma_R$ and of the resolution $1/\Delta_M$, the shape of the central peak will or will not be affected by the neighboring features.

d. Width of the Reference Line. Contrary to the asymmetry property, the width of the reference line should not introduce any phase error. The visibility of the fringes decreases when Δ increases, but the zero crossings are not displaced. However, slight errors in path difference determination may occur in a roundabout way. Indeed, the path difference is usually determined by detecting the zero crossings of the fringes, the constant component of which has been removed by an electronic device which triggers at a level ΔV. In Fig. 1.19, curves I and II represent the reference fringe at two different path differences. Curve I is near $\Delta = 0$, and the decrease of the intensity as seen in curve II is due to the width of the

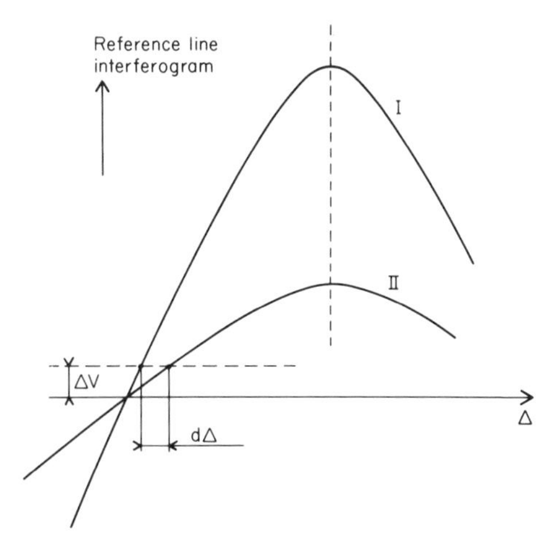

FIG. 1.19. Variation of the intensity of the reference line may induce phase error in the path-difference monitoring system.

reference line. As a consequence, a shift in the triggering occurs, and a phase error may occur if this point is not treated carefully when necessary.

If the intensity of the fringes is supposed to vary linearly with Δ, then

$$d\Delta(\Delta) = \alpha\Delta.$$

With α constant, depending on the experimental conditions, the recovered spectrum will be shifted by $\alpha\sigma$ (Section 1.6,B.1). This error is corrected by the *internal standard method*.

2. *Path-Difference Control*

a. Loss of the Phase. During the recording, if a malfunction of the servo system occurs at Δ, the control of the phase is lost, and this produces a break in the stepping sequence. The interferogram is then correct up to Δ_1. It continues from Δ_2 to Δ_M, generally, with a different phase ($\Delta_2 - \Delta_1 \neq k\lambda_0$). This is typically a "fatal error." It is practically always impossible to know the exact values of Δ_1 and Δ_2, although they can be roughly approximated. If, however, Δ_1 is not too small compared with the desired Δ_M, one may compute the spectrum corresponding to the correct first part of the interferogram, with a decreased resolution. Consequently, it is useful to know at least Δ_1. Additional knowledge of Δ_2 also allows a better understanding of the origin of the error.

From $\Delta = 0$ to Δ_1, the interferogram of a monochromatic line σ_0 is written as $\sin 2\pi\sigma_0\Delta$, whereas the second part is $\sin 2\pi\sigma_0(\Delta + l)$ from Δ_2

to Δ_M, with $l = \Delta_2 - \Delta_1$. The computed spectrum is given by

$$B'(\sigma) = \int_0^{\Delta_1} \sin 2\pi\sigma_0\Delta \sin 2\pi\sigma\Delta \, d\Delta + \int_{\Delta_1}^{\Delta_M} \sin 2\pi\sigma_0(\Delta + l) \sin 2\pi\sigma\Delta \, d\Delta. \tag{1.35}$$

If

$$l = k\lambda_0 \qquad (k \text{ integer}),$$

$B(\sigma)$ is, in fact, not perturbed, since Eq. (1.35) is then identical to Eq. (1.2). If

$$l \neq k\lambda_0,$$

the first term of Eq. (1.35) represents the normal apparatus function $f(\sigma)$ with a resolution equal to $1/2\Delta_1$. In the more general case, Eq. (1.35) may be written

$$B(\sigma) = \int_0^{\Delta_1} \sin 2\pi\sigma_0\Delta \sin 2\pi\sigma\Delta \, d\Delta + \int_0^{\Delta_M} \sin 2\pi\sigma_0(\Delta + l) \sin 2\pi\sigma\Delta \, d\Delta - \int_0^{\Delta_1} \sin 2\pi\sigma_0(\Delta + l) \sin 2\pi\sigma\Delta \, d\Delta.$$

Expanding the third term as in Section 1.2,B.3, one obtains

$$B(\sigma) = (1 - \cos 2\pi\sigma_0 l) \int_0^{\Delta_1} \sin 2\pi\sigma_0\Delta \sin 2\pi\sigma\Delta \, d\Delta - \sin 2\pi\sigma_0 l \int_0^{\Delta_1} \cos 2\pi\sigma_0\Delta \sin 2\pi\sigma\Delta \, d\Delta + \int_0^{\Delta_M} \sin 2\pi\sigma_0(\Delta + l) \sin 2\pi\sigma\Delta \, d\Delta. \tag{1.36}$$

The sum of the first two terms corresponds to an out of phase apparatus function with half width equal to $1/2\Delta_1$. The added third term is also an out of phase apparatus function, but with the maximum resolution. The experimental result will have a strange form. One may, however, extract the knowledge of Δ_1 through the width of the largest function. This may be easy when Δ_1 is sufficiently different from Δ_M, i.e., when the contributing function corresponding respectively to the first two terms and the third term of Eq. (1.36) have rather different resolutions. The parameter l, i.e., the length of the jump during the loss of phase, can possibly be determined through the assymmetry of the profiles. In fact, for a wide spectral range, l is better known by measuring on the spectrum, the distance $\Delta\sigma$ between successive similar profiles. Then, according to Eq. (1.36),

$$\Delta\sigma \, l = 1.$$

More information on l is also given by the wavenumber σ_0, whose profile has a normal shape. Then

$$\sigma_0 l = k \qquad (k \text{ integer}).$$

In fact, these above approximate determinations are usable only as indications of the occurrence of the phenomenon and of its parameters. It is practically impossible to use them for precise a posteriori correction of the distorted interferogram. Also, unfortunately this type of error can occur several times on the same interferogram, which consequently is of no use. In such conditions an appropriate phase error detector may be useful.

b. Imperfect Servo Control. Depending on the adjustment of the servo system to the experimental recording conditions, the control of the path difference may be more or less efficient. The above section treats the case when this efficiency completely vanishes. Actually, as already discussed in Section 1.4,A.1, an imperfect servo control leads to an error in the Δ determination and then to a phase error $\varepsilon_{02}(\Delta)$, with the resulting consequences described in Section 1.5, B.1. If, for example, the sampling interval is too long and the required stepping rate too high compared with the characteristics of the servo system, a shift in the path difference will occur. Also particular positions along Δ may be more sensitive to resonances, at which positions the control of Δ will be less precise. This should appear on the error signal itself, which deviates from zero when the sampling measurement is disturbed. In a relatively simple interferogram this may also be apparent, as shown in Fig. 1.20.

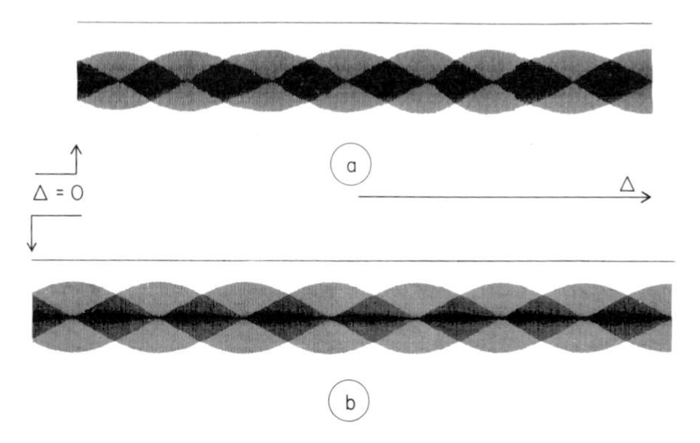

FIG. 1.20. Two different servo conditions for two similar actual interferograms: (a) the servo control system is working badly. The distances between consecutive zeros are not equal and introduce phase errors. (b) The servo works correctly. The zeros are equidistant.

Traces (a) and (b) of Fig. 1.20 represent the same interferogram recorded under different servo conditions. The source is monochromatic, and the sampling interval is large compared with the wavelength. The experimental conditions are identical to those described in Guelachvili (1977). In the figure the two interferograms begin at $\Delta = 0$ and are given only for a short displacement compared with the possible 6 m. Whereas trace (b), for which the servo control was well adjusted, is regular, interferogram (a) shows important phase fluctuations, as suggested by the loss of equidistant sampling, indicated by different distances between two consecutive zeros.

Another defect which must be mentioned here is the possible drift in the servo control. Drifts may come from a shift in the reference wavelength or electronics bias fluctuations due to temperature, for example. Also, a parasitic electrical signal in the servo loop, such as a 50-Hz signal, may be present. It should be detectable as antisymmetric ghosts in the spectrum (Section 1.5,B.1b). The location of these ghosts changes as the recording time is changed.

The possibility of the appearance of ghosts produced by a periodic phase error may also come from the nature of the servo system itself. The period of the error on the path difference is then equal to the wavelength of the reference line λ_R. (see, e.g., J. Connes *et al.*, 1970). As already indicated in Section 1.5,B.1c, these ghosts have a permanent behavior in the spectra whatever the recording conditions may be. They are always located at the same distance from the central line since the perturbation is path-difference dependent. These ghosts are more or less removable. If no care is taken, they can distort the level of the background when an absorption spectrum is analyzed and thus decrease the precision of the intensity measurement. They are less troublesome for the emission spectra since they appear more clearly there. However, although they present the same behavior as ghosts given by a grating, they are more easily controllable. More precisely, the distance between the central feature and the side ghosts is equal to σ_R, and generally the ghosts appear as "images" in the spectrum. The position of these images may be easily chosen by a judicious choice of sampling interval. The spectra of Fig. 1.22 may be used as an illustration of this possibility. The feature appearing at 2850 cm^{-1} corresponds to parasitic flux, from the reference line, in the spectrum. Naturally it spoils the region around 2850 cm^{-1}. At 4550 cm^{-1} one ghost of this line is present. It is, in fact, actually located at 5700 cm^{-1}, but its image is seen on the spectra since the free spectral range goes up to 5075 cm^{-1}. Changing this upper limit, i.e., the sampling interval, will change the location of this ghost.

The last point which should be mentioned in this section is the error due to the operator himself, who might choose inadequate sampling interval and internal modulation amplitude for the recording of a given spectral range.† Can we, in this case, also say that the servo control is imperfect?

C. Measurement of the Intensity of the Samples

1. *Source*

The analyzed source itself, its geometrical characteristics, its spectral behavior, and its time dependence, are possible origins of systematic effects which are examined here.

a. Inequality of Illumination. For various reasons, the optical source seen by the instrument may be of unequal brightness due to the source itself or to the relay optical system used for its image transfer to the entrance aperture. Suppose, for instance, that the entrance aperture is not uniformly illuminated by the light beam. The effects already discussed in Section 1.6,A.2a will then be expressed by different formulas. A simple example could be given by an entrance hole with uniform illumination everywhere except in a centered circular zone where no light is passing. This could occur in astronomical spectra where the entrance hole is conjugate to the primary mirror of the telescope. The a posteriori correction yielding the correct wavenumber would then be different than in Section 1.6,A.2a. If Ω_1 and Ω_2 are the solid angles corresponding to the dark center and the entire hole, respectively, the interferogram of a monochromatic line would be given by

$$I(\Delta) = \sin [\sigma_0\Delta \tfrac{1}{2}(\Omega_2 - \Omega_1)/\sigma_0\Delta \tfrac{1}{2}(\Omega_2 - \Omega_1)] \times \sin 2\pi\sigma_0\Delta[1 - (\Omega_2 + \Omega_1)/4\pi].$$

The resultant systematic shift $[(\Omega_2 + \Omega_1)/4\pi]\ \sigma_0$ is greater than the one normally expected with uniform illumination.

Since it corresponds to a simple configuration, the shift in the above example could be well-corrected a posteriori. This is generally not possible when the nonuniformity is an arbitrary function which additionally may be time dependent. The *internal standard method* (Section 1.3,B.1) is then most helpful.

b. Time Variation of the Source Parameters. As explained in Section 1.5,C.2, this intensity effect will symmetrically deform the apparatus

† It sometimes happens!

function. If, for instance, a 50-Hz modulation is added to the flux of interest, ghosts will be present on both sides of the central profiles (Section 1.3,C.2b). An interruption of the emission itself (e.g., clouds between the planet and the telescope) will influence the whole spectrum (J. Connes *et al.*, 1967). If the recording continues when the source is not measurable and if Δ_1 and Δ_2 correspond to the interval where the source assumed to be monochromatic disappears, the computed spectrum will be

$$B'(\sigma) = \int_0^{\Delta_1} \sin 2\pi\sigma_0\Delta \sin 2\pi\sigma\Delta \, d\Delta + \int_{\Delta_2}^{\Delta_M} \sin 2\pi\sigma_0\Delta \sin 2\pi\sigma\Delta \, d\Delta, \tag{1.37}$$

which is equivalent to

$$B(\sigma) = \int_0^{\Delta_M} \sin 2\pi\sigma_0\Delta \sin 2\pi\sigma\Delta \, d\Delta - \int_{\Delta_1}^{\Delta_2} \sin 2\pi\sigma_0\Delta \sin 2\pi\sigma\Delta \, d\Delta. \tag{1.38}$$

From the normal apparatus function given by the first term of Eq. (1.38), one subtracts the second term, which is equal to

$$\frac{(\Delta_2 - \Delta_1)}{2} \cos\left[2\pi(\sigma_0 - \sigma)\frac{\Delta_2 + \Delta_1}{2}\right] \frac{\sin 2\pi(\sigma_0 - \sigma)(\Delta_2 - \Delta_1)/2}{2\pi(\sigma_0 - \sigma)(\Delta_2 - \Delta_1)/2}; \tag{1.39}$$

i.e., a cosine function multiplied by a wide sinc function.

Let us first suppose that $\Delta_1 \simeq \Delta_2 = \Delta_i$. In other words, the source is not functioning during a short time (proportional to $\Delta_2 - \Delta_1$) compared to the total recording time proportional to Δ_M. Practically, the sinc of Eq. (1.39) remains equal to one whatever σ may be, and Eq. (1.39) may be written

$$\tfrac{1}{2}(\Delta_2 - \Delta_1) \cos 2\pi(\sigma_0 - \sigma)\,\Delta_i. \tag{1.40}$$

The distortion appears as a periodic function of σ centered [as expected because of the symmetry rule (Section 1.5,C.2)] on σ_0, with an amplitude $\frac{1}{2}(\Delta_2 - \Delta_1)$, while, according to Eq. (1.2), the amplitude of the first term of Eq. (1.38) is equal to $\frac{1}{2}\Delta_M$. The period of the perturbing function is $1/\Delta_i$.

If Δ_1 is different from Δ_2, the sinc part of Eq. (1.39) introduces modulation in the amplitude of the cosine function, which again has a period equal to $1/\Delta_i$, with $\Delta_1 + \Delta_2 = 2\Delta_i$.

For a broadband spectrum the distortion is given by the convolution of the spectrum, with the function represented by Eq. (1.39).

A posteriori correction of these unknown intensity fluctuations is difficult. A good solution consists in monitoring in real-time the gain of the amplifier measuring the samples, by the intensity of the source itself. The same technique is usable for controlling the integration time instead of controlling the amplifier gain.

Another source parameter which may be time dependent is the spectral composition; for example, in an absorption spectrum there could be a variation of the total amount of the analyzed gas during the recording of the interferogram. A practical example of another type of time-varying source spectral distribution is given by Fig. 1.21. The two traces represent two different spectra. The source is a discharge tube electrically excited by a direct current, as already described by Bailly *et al.* (1978). The two corresponding gas mixtures are, respectively, $^{12}C^{16}O_2$–N_2–He–$^{16}O_2$ and $^{13}C^{16}O$–N_2–He–$^{18}O_2$, and the observed transitions are due to various isotopic CO lines. In traces (a) and (b) the different isotopic species have particular specific instrumental shapes although the resolution is the same for the two spectra (5.4 mK). This is due to the formation procedure for the isotopes that are present at different times. For instance, the low-resolution details of the shape of the $^{12}C^{16}O$ lines in trace (b) appear strongly. It is certainly due to the nonemission (or fainter emission) of $^{12}C^{16}O$ in the beginning of the recording, where ^{12}C appears as an impurity. All the profiles are symmetrically distorted, since the perturbation is dependent on the intensity of the source (Section 1.5,C.2).

This type of systematic effect cannot be corrected a posteriori. The

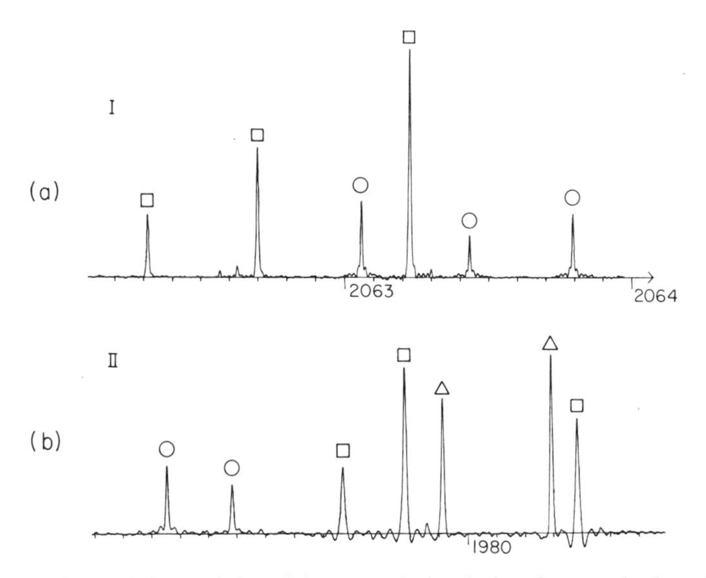

FIG. 1.21. Spectral time variation of the source during the interferogram leads to different symmetrically distorted profiles in the spectrum according to the isotopic species: □, $^{12}C^{16}O$; ○, $^{13}C^{16}O$; △, $^{13}C^{18}O$.

source itself must be better controlled. On the other hand, one could accentuate the phenomenon in order to obtain isotopic signatures in the spectrum (or for other reasons).

c. Additional Parasitic Source. The problem here is the same as for other spectroscopic techniques although it does not behave similarly. A rather common additional parasitic source is a periodic emission due, for instance, to a 50-Hz incandescent lamp, the flux of which directly falls on the detector. This case has already been treated in Section 1.5,A.2. In the spectrum a particular place will present a strong overwhelming perturbation, whose location depends on the frequency of the parasitic signal and the recording time of the interferogram.

Another type of additional parasitic source, well known by people working in the infrared, is the ambient temperature blackbody emission. Figure 1.22 gives an illustration of the occurrence of this thermal background. The five different traces correspond to five different spectra similarly recorded in the same spectral range with a globar source that is roughly a blackbody. The free spectral range is 0–5075 cm^{-1}. The filter has two peaks of transmission, at 2000 and 4000 cm^{-1}. The detector is a photovoltaic SbIn detector, whose lower sensitivity limit is 1650 cm^{-1}. The only parameter changing from one trace to the next is the temperature of the globar, which increases with the number identifying the spectra, as suggested by the relative strengths of the two peaks of the filter. The spectrum No. I has been inverted when compared to the others. In the two low-temperature spectra there appears clearly a feature that is faintly present in spectrum No. III. This feature corresponds to the emission of the surroundings at room temperature. It comes from the mirrors or any object seen by the detector. If no care is taken, this additional parasitic spectrum could induce severe distortions in the measurement of the intensity in a spectrum similar to No. III, and centered around 1800 cm^{-1}.

Figure 1.22 shows also additional parasitic spectrum: the reference line which is scattered in the direction close to that of the analyzed beam and appears at 2850 cm^{-1}. Since the spectra of Fig. 1.22 are normalized, this parasitic line appears with increased relative importance from trace V to I. Another feature is present at 4490 cm^{-1}. It is one ghost of the reference line, as already discussed in Section 1.6,B.2b. These emission features do not have the right instrumental profile since they are not following the correct optical path through the interferometer. Also, fortunately, they extend over a very much narrower range as the resolution increases.

d. Channel Spectra. This is a well-known phenomenon to all spectroscopists. Passing through parallel plates such as a window or a beam splitter (see Section 1.6,A.1e), the analyzed beam is separated into various

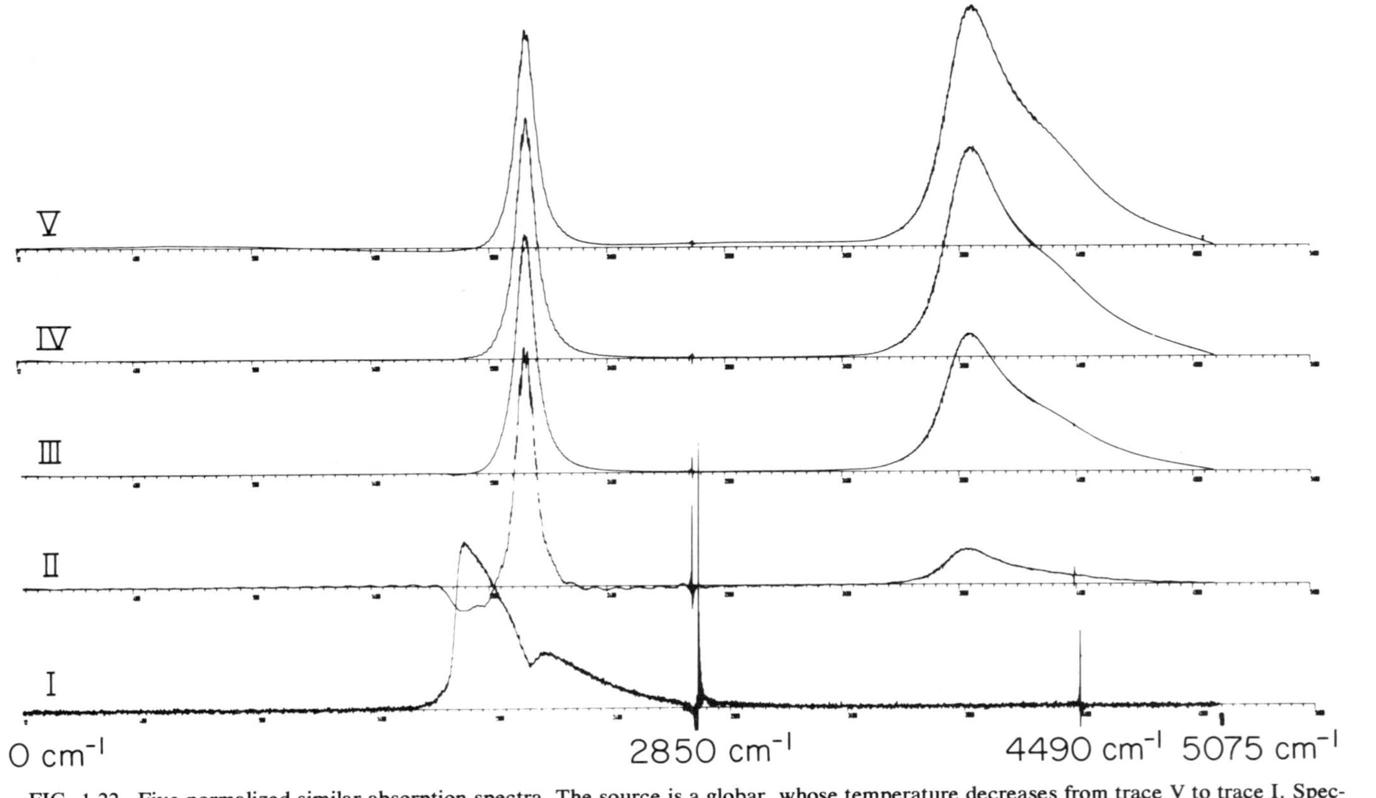

FIG. 1.22. Five normalized similar absorption spectra. The source is a globar, whose temperature decreases from trace V to trace I. Spectrum I has been inverted. Three types of distortions are present: thermal background, nonlinear detection, and parasitic signal.

coherent beams by parasitic reflections. This creates multiple-wave interference, which appears in the spectrum whatever the spectroscopic technique may be. The fringes are enhanced when the beam is parallel. The greater the wavelength, the more intense are the fringes. This already gives an indication of a way to remove these parasitic features from the spectra. As much as possible, when traversing the window, the beam must be uncollimated. This creates several interference orders in the interference pattern, and they tend to cancel each other. If the channel spectrum is experimentally unavoidable, it may still be corrected in the interferogram itself, where it exists as signatures at localized points. If $\Delta\sigma$ is the interfringe distance in the spectrum, the signature (enhanced interferogram signal) will be at $\Delta = 1/\Delta\sigma$. Its removal will essentially remove the fringes in the spectrum. Actually, several points must be cancelled in the interferogram since these channel spectra are generally not represented by a pure single frequency. The correction is not perfect but may, however, be considered fairly efficient. This is available for absorption spectra, where this interference pattern is usually clearly evident. In the case of emission spectra, it is more difficult to find both the channel spectrum and its corresponding signature in the interferogram. This is due to the lack of continuous background on which the fringes would be clearly evident. In other words (in a region around some Δ), the interferogram presents a more pronounced intensity modulation, which thereby might prevent the detection of the wanted data.

As a result, a shift in the position of the profile, depending on the slope of the multiplying function, can occur. Intensity distortions usually only occur with emission spectra since the role of the channel spectrum is then difficult to recognize. On the other hand, the intensity of absorption spectra can be determined since it is referred to the background, whatever this background may be, provided it varies slowly compared to the profile of interest.

2. *Recording System*

a. Gain Error and Parasitic Signals. On going from the detector to the recorder, errors in the measurement of the amplified signal are possible. For instance, the temperature of the detector may vary during the recording of the interferogram, producing a variable detectivity. Also, the different gains of the amplifier of the detector signal could be incorrectly estimated. Typically this type of error is of no great importance provided it is not too large. These errors multiply the interferogram by a slowly varying function of Δ, the Fourier transform of which will convolve the spectrum in a way similar to what the apodizing procedure does (J. Connes and P. Connes, 1966; Norton and Beer, 1976).

Parasitic electrical signals are more troublesome. One finds again here the already mentioned 50 Hz, which will behave like an additional source (Section 1.6,C.1c). Electrical pulses or spikes are also possible. They will produce in the spectrum a periodic modulation of amplitude proportional to the amplitude of the pulse and of period equal to $1/\Delta$, where Δ is the path difference at which the pulse occurred. The correction of this last perturbation is easy. The period in the spectrum can be accurately determined; it yields the corresponding value of Δ in the interferogram and allows the removal of the parasitic signal from the interferogram.

b. Detection Nonlinearity. Instead of $I(\Delta)$, the recording system is yielding $I'(\Delta)$, a nonlinear function of $I(\Delta)$. The characteristic of such a measurement is shown in Fig. 1.23. It produces distortions in the computed spectrum.

Let us use again a monochromatic interferogram

$$I(\Delta) = \sin 2\pi\sigma_0\Delta,$$

which, in fact, will be recorded as

$$I'(\Delta) = \sin 2\pi\sigma_0\Delta(\gamma_1 + \gamma_2 \sin 2\pi\sigma_0\Delta + \gamma_3 \sin^2 2\pi\sigma_0\Delta + \cdots). \quad (1.41)$$

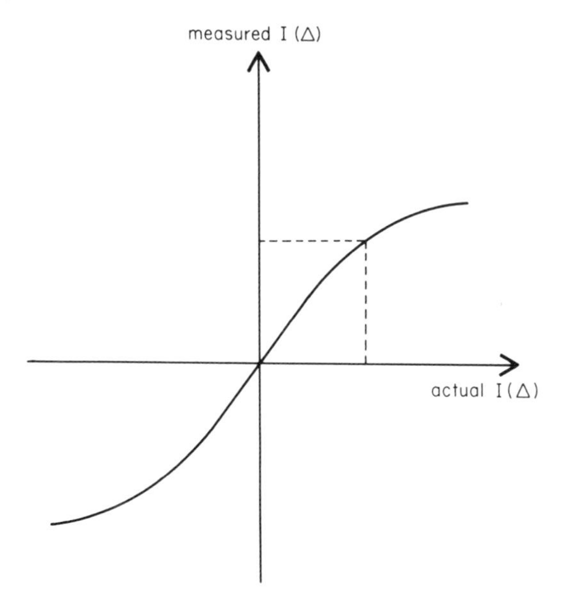

FIG. 1.23. Nonlinear recording system.

The calculated spectrum will be given by

$$\begin{aligned} B'(\sigma) &= \int_0^{\Delta_M} I'(\Delta) \sin 2\pi\sigma\Delta \, d\Delta \\ &= \gamma_1 \int_0^{\Delta_M} \sin 2\pi\sigma_0\Delta \sin 2\pi\sigma\Delta \, d\Delta \\ &\quad + \gamma_2 \int_0^{\Delta_M} \sin^2 2\pi\sigma_0\Delta \sin 2\pi\sigma\Delta \, d\Delta \\ &\quad + \gamma_3 \int_0^{\Delta_M} \sin^3 2\pi\sigma_0\Delta \sin 2\pi\sigma\Delta \, d\Delta \\ &\quad + \cdots, \end{aligned} \tag{1.42}$$

the first term of which represents the normal apparatus function $f(\sigma)$. The expansion of the second term of Eq. (1.42) gives

$$\frac{\gamma_2}{2}\int_0^{\Delta_M} \sin 2\pi\sigma\Delta \, d\Delta - \frac{\gamma_2}{2}\int_0^{\Delta_M} \cos 2\pi(2\sigma_0)\Delta \sin 2\pi\sigma\Delta \, d\Delta. \tag{1.43}$$

The third term of Eq. (1.42), according to the relation

$$\sin^3\alpha = \tfrac{1}{4}(3 \sin \alpha - \sin 3\alpha),$$

is equal to

$$\frac{\gamma_3}{4}\int_0^{\Delta_M} 3 \sin 2\pi\sigma_0\Delta \sin 2\pi\sigma\Delta \, d\Delta - \frac{\gamma_3}{4}\int_0^{\Delta_M} \sin 2\pi(3\sigma_0)\Delta \sin 2\pi\sigma\Delta \, d\Delta. \tag{1.44}$$

The first-order nonlinearity represented by Eq. (1.43) will produce two sidebands at $\sigma_0 = 0$ [supposing that the first integral contains $\cos 2\pi(0)\Delta$], and at $\sigma = 2\sigma_0$. These two satellites are symmetrically distorting the normal spectrum $f(\sigma)$ according to Section 1.5,B.2, as shown in Fig. 1.24. The second-order nonlinearity in the detection of the samples gives two other satellites at σ_0 and $3\sigma_0$ with symmetrical shape $f(\sigma)$, as represented in Fig. 1.24. Whereas the first-order perturbation is practically not affecting the line itself but creates satellites, the second-order perturbation, which does not displace the center of the profile, distorts the intensity by multiplying the normal apparatus function by $1 + \frac{3}{4}\lambda_3$.

More generally, the nonlinearity on $I(\Delta)$ may be expressed as

$$I'(\Delta) = I(\Delta)[\gamma_1 + \gamma_2 I(\Delta) + \gamma_3 I^2(\Delta) + \cdots],$$

leading to the spectrum:

$$\begin{aligned} B'(\sigma) &= \text{FT}\,[\gamma_1 I(\Delta)] + \gamma_2\,\text{FT}\,[I^2(\Delta)] + \gamma_3\,\text{FT}\,[I^3(\Delta)] + \cdots \\ &= \gamma_1 B(\sigma) + \gamma_2 B(\sigma) * B(\sigma) + \gamma_3 B(\sigma) * B(\sigma), \end{aligned} \tag{1.45}$$

where the asterisk represents convolution, and FT means Fourier transform.

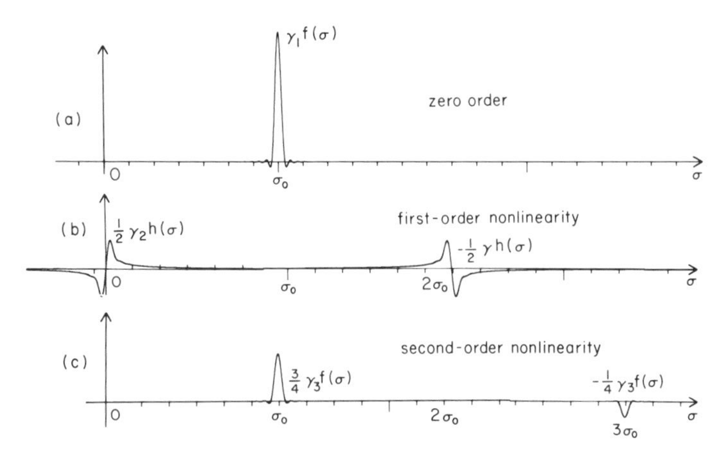

FIG. 1.24. The first- and second-order nonlinearities in the detection of the interferogram of the monochromatic line at σ_0 (a) are producing the features given in (b) and (c), respectively.

Applying Eq. (1.45) to a single-line spectrum $B(\sigma)$ will yield the previous result. This is shown in Fig. 1.25, where $B(\sigma)$, $B(\sigma) * B(\sigma)$, and $B(\sigma) * B(\sigma) * B(\sigma)$ are represented for $B(\sigma)$ being a monochromatic line at σ_0, assuming that Δ_M is infinite.

Figure 1.22 illustrates the nonlinear detection for a broad spectrum. The main features of this figure have been described above in Section 1.6,C.2c. The signal on the detector increases with the number identifying the spectrum. If nonlinear detection exists, it should then be more evident on the upper trace. Effectively one sees that the level of the spectrum, which should be zero outside of the two zones of transparency, is slightly varying. For instance, on the low wavenumber side it takes positive and negative values. This effect remains slightly visible on trace IV, whereas it completely disappears for smaller energy spectra.

The a posteriori correction of nonlinear effects is not easy to accomplish on the spectrum. The interferogram itself could be corrected, provided that the characteristics of the detector are well known. Another possibility is to remain in the linear region of the detector sensitivity curve by reducing either the spectral range or the source intensity, when possible. Introduction of a neutral density filter into the beam during the first part of the interferogram where the modulation is large around the path-difference zero is also a solution. Another possibility could be to estimate in real time the exact characteristics affecting the measured sample and simultaneously apply a correcting gain on the sample. A good check on the possible existence of remaining nonlinearities (Section 1.3,B.2) is to verify that the expected zero levels in the spectrum (e.g., the top of saturated

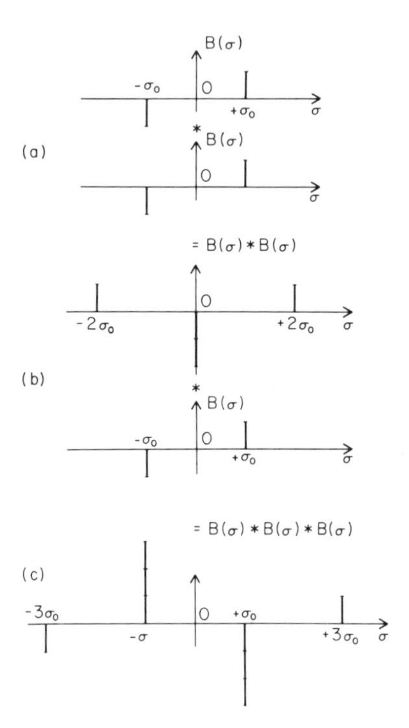

FIG. 1.25. (a) Monochromatic line spectrum, (b) effect on the spectrum of the first-order nonlinearity in the interferogram, (c) second-order effect.

absorption lines, or the opaque region of the filter as in Fig. 1.22) are actually zero.

c. Incorrect Recording or Reading Procedure. I have recently wasted a lot of time looking for an error having all the characteristics of a loss of phase, already treated here in Section 1.6,B.2a. Finally, I discovered that two cards had been interchanged in a program, due to an "overactive" IBM card reader. As a result, several points were missing in the interferogram, giving a "beautiful" systematic effect. This short paragraph has two motives. The first is to point out that recording or reading procedures must not be forgotten when strange spectra are obtained. The second, no less important motive is to settle an old score with this tactless card reader.

Acknowledgments

I warmly thank George A. Vanasse, who kindly corrected the first draft of this paper. I am also grateful to J. Collet, H. Calvignac, B. Chauveau, and M. Rey for their technical assistance in the realization of the figures.

REFERENCES

BAILLY, D., FARRENQ, R., and ROSSETTI, C. (1978). Vibrational luminescence of CO_2 excited by DC discharge. Rotational and vibrational constants, *J. Mol. Spectrosc.* **70,** 124.

BAKER, D. (1977). Field-widened interferometers for Fourier spectroscopy. *In* "Spectrometric Techniques" (G. Vanasse, ed.), Vol. I, p. 71. Academic Press, New York.

CHACKERIAN, C. JR., and GUELACHVILI, G. (1978). Absolute intensities of the ν_2 and the $2\nu_6$ bands of CH_3D, Paper TE2 at the *Symp. Mol. Spectrosc., 33rd, Columbus, Ohio.*

CONNES, J. (1961). Recherches sur la spectroscopie par transformation de Fourier, *Rev. Opt. Theor. Instrum.* **40,** *45,* 116, 171, 231. Translated into English by C. A. Flanagan, Navweps Rep. 8099, NOTS TP 3157.

CONNES, J. (1971). Computing Problems in Fourier Spectroscopy, A.F.C.R.L. Report No. 114-71-0019, p. 83.

CONNES, J., and CONNES, P. (1966). Near infrared planetary spectra by Fourier spectroscopy. I, Instruments and results, *J. Opt. Soc. Am.* **56,** *7,* 896.

CONNES, J., CONNES, P., and MAILLARD, J. P. (1967). Spectroscopie astronomique par transformation de Fourier, *J. Phys. Colloque C2* **28,** 120.

CONNES, J., DELOUIS, H., CONNES, P., GUELACHVILI, G., MAILLARD, J. P., and MICHEL, G. (1970). Spectroscopie de Fourier avec transformation d'un million de points, *Nouv. Rep. Opt. Appl.* **1,** No. 1, 3.

CONNES, P. (1958a). Spectromètre interférentiel a sélection par l'amplitude de modulation, *J. Phys. Radium* **19,** 215.

CONNES, P. (1958b). L'étalon de Fabry-Pérot sphérique, *J. Phys. Radium* **19,** 262.

CONNES, P. (1970). Astronomical Fourier spectroscopy, *Astron. Astrophys.* **8,** 209.

CONNES, P. (1971). High Resolution and High Information Fourier Spectroscopy, A.F.C.R.L. Report No. 114-71-0019, p. 121.

CONNES, P., and MICHEL, G. (1971). Real-Time Computer for Fourier Spectroscopy. A.F.C.R.L. Report No. 114-71-0019, p. 313.

CONNES, P., and MICHEL, G. (1975). Astronomical Fourier Spectrometer, *Appl. Opt.* **14,** 2067.

DELOUIS, H. (1971). Fourier Transformation of a 10^6 Samples Interferogram, A.F.C.R.L. Report No. 114-71-0019, p. 145.

DELOUIS, H. (1973). Mise au point d'une chaine de programmes permettant le calcul et l'exploitation automatique de spectres obtenus par transformation de Fourier, Thèse d-Etat No. 1088, Orsay.

DURANA, J. F., and MANTZ, A. W. (1979). Laboratory studies of reacting and transient systems. *In* "Fourier Transform Infrared Spectroscopy" (J. R. Ferrero and L. J. Basile, eds.), Vol. 2, p. 1, Academic Press, New York.

EDLEN, B. (1953). The Dispersion of Standard Air. *J. Opt. Soc. Am.* **43,** *5,* 33.

EVENSON, K. M., and PETERSEN, F. R. (1976). Laser frequency measurement, the speed of light and the meter. *In* "Laser Spectroscopy of Atoms and Molecules" (H. Walther, ed.), Vol. 2, p. 349. Springer-Verlag, Berlin and New York.

FELLGETT, P. (1958). A propos de la théorie du spectromètre interférentiel multiplex, *J. Phys. Radium* **19,** 187.

FELLGETT, P. (1967). Conclusions on multiplex methods, *J. Phys. Colloque C2* **28,** 165.

FINK, U., and LARSON, H. P. (1979). Astronomy: Planetary atmospheres. *In* "Fourier Transform Infrared Spectroscopy" (J. R. Ferrero and L. J. Basile, eds.), Vol. 2, p. 243. Academic Press, New York.

FUSINA, L., MILLS, I., and GUELACHVILI, G. (1980). Carbon suboxide: The infrared spectrum from 1800 to 2600 cm^{-1}, *J. Mol. Spectrosc.* **79,** 101.

GUELACHVILI, G. (1972). Spectroscopie de Fourier avec transformation de 10^6 points. II. Mise sous vide et automatisation. *Nouv. Rev. Opt. Appl.* **3,** 6, 317.

GUELACHVILI, G. (1973). Spectromètrie de Fourier sous vide, à 10^6 points. Application à la mesure absolue de nombres d'ondes et à une nouvelle détermination de C. Thèse d'Etat, No. 1095, Orsay.

GUELACHVILI, G. (1977). Near infrared wideband spectroscopy with 27 MHz resolution, *Appl. Opt.* **16,** 2097.

GUELACHVILI, G. (1978a). High-accuracy doppler-limited 10^6 samples Fourier transform spectroscopy, *Appl. Opt.* **17,** 1322.

GUELACHVILI, G. (1978b). Excited states of CO and isotopic species through their $\Delta v = 1$ sequence spectra. Paper WE4 at the *Symp. Mol. Spectrosc., 33rd, Columbus, Ohio.*

GUELACHVILI, G., and MAILLARD, J. P. (1971). Fourier Spectroscopy from 10^6 Samples. A.F.C.R.L. Report No. 114-71-0019, p. 151.

JACQUINOT, P. (1958). Caractères communs aux nouvelles méthodes de spectroscopie interférentielle; Facteur de mérite, *J. Phys. Radium* **19,** 223.

LOWENSTEIN, E. V. (1966). The history and current status of Fourier transform spectroscopy, *Appl. Opt.* **5,** 845.

MERTZ, L. (1958). Spectromètre stellaire multicanal, *J. Phys. Radium* **19,** 233.

MORILLON, C., and VERGES, J. (1975). Observation et classification du spectre d'arc du tellure (TeI) entre 3678 et 11761 cm^{-1}, *Phys. Scripta* **12,** 129.

MURPHY, R. E., COOK, F. H., and SAKAI, H. (1975). Time-resolved Fourier spectroscopy, *J. Opt. Soc. Am.,* **65,** 5,600.

NORTON, R. H., and BEER, R. (1976). New apodizing functions for Fourier spectrometry, *J. Opt. Soc. Am.* **66,** 3,259.

PINARD, J. (1969). Realisation d'un spectromètre par transformation de Fourier à très haut pouvoir de résolution, *Ann. Phys.* **4,** 147.

SAKAI, H. (1977). High resolving power Fourier spectroscopy. *In* "Spectrometric Techniques" (G. A. Vanasse, ed.), Vol. I, p. 1. Academic Press, New York.

STRONG, J. D., and VANASSE, G. A. (1958). Modulation interferentielle et calculateur analogique pour un spectromètre interférentiel. *J. Phys. Radium* **19,** 192.

TERRIEN, J. (1958). Observations photoélectriques a l'interféromètre de Michelson, *J. Phys. Radium* **19,** 390.

VANASSE, G. A. (1973). Fourier Spectroscopy: A Critical Review, A.F.C.R.L. Report No. 74-0092.

VANASSE, G. A., and SAKAI, H. (1967). Fourier spectroscopy. *In* "Progress in Optics" (E. Wolf, ed.), Vol. VI, p. 259. North-Holland, Amsterdam.

Chapter **2**

First-Order Optical Design for Fourier Spectrometers

JAMES B. BRECKINRIDGE
RUDOLF A. SCHINDLER

JET PROPULSION LABORATORY
CALIFORNIA INSTITUTE OF TECHNOLOGY
PASADENA, CALIFORNIA

ISBN 0-12-710402-X

2.1. Introduction

Advanced instrument development is divided into three major sections: design approach, detail design, and fabrication. An incorrect detail design leads to an instrument impossible to fabricate. Similarly an awkward design approach, even with good detail design, may lead to fabrication difficulties and an instrument so difficult to use that it is worthless. Design approach, detail design, and fabrication all need nearly equal weight of effort to create an instrument of significant value to the scientific community.

Design approach for a Fourier transform spectrometer with emphasis on optics is the subject of this chapter. The advantages and disadvantages of Fourier transform spectrometers are well understood, and these points are not directly addressed. The purpose here is to provide the reader with the skill to examine a scientific problem for which he wants data and to match the optical performance of the instrument to his requirements. Problems of signal-to-noise ratio (SNR) and detectors are not directly addressed. Particular attention is given to high-resolution (>0.05 cm^{-1}) Fourier transform spectrometers for use in the visible and near infrared.

A. Background

An explosion of applications for Fourier transform spectrometers has taken place recently. Some are on instrument development; others concern scientific interpretation of recorded data. The number of papers published in the open literature using the key word *Fourier transform spectroscopy* (FTS) is given in Table 2.I. Clearly, the contents of this chapter cannot be a summary of the first-order designs of all Fourier spectrometers.

B. Contents

The body of this text is divided into six sections. Section 2.2, Foundation for Design, provides a brief background to outline the scientific usefulness of Fourier spectrometers and presents a calculation giving the

TABLE 2.I.

TABULATION OF PAPERS PUBLISHED WITH KEY WORD *Fourier Transform Spectroscopy*, FROM 1969 THROUGH 1979 WITHIN SCIENCE ABSTRACTS (INSPEC) DATA FILE

Year	Approximate number of papers
1969	10
1970	35
1971	93
1972	98
1973	131
1974	134
1975	168
1976	194
1977	191
1978	300
1979	310

optical path difference (OPD) required to resolve a Doppler-broadened spectral feature. Section 2.3 introduces fringe contrast as a measure of interferometer modulation efficiency and discusses beam-splitter, wavefront errors, and polarization contributions to the fringe contrast. Section 2.4 emphasizes the importance of proper mechanical design in a presentation of elasticity considerations, kinematic mounts, atmospheric turbulence, and scan carriage mechanisms. Section 2.5 provides a review of optical designs and throughput, introduces the $y\bar{y}$ diagram as a tool, and summarizes the channel spectra as unwanted modulations in the spectra. Section 2.6 contains new information on tilt-compensated optical configurations for Fourier transform spectrometers, describes an analysis technique, and gives a detailed discussion of the synthesis of a particular configuration. Section 2.7 is brief and relates an interferometer approach to the alignment of cat's-eyes.

C. APPLICATIONS REVIEW

The instrument which saw one of its original successful applications in the field of infrared planetary astronony (Review by Connes, 1970) has now been extended for use from the ultraviolet (Brault, 1976) through the far infrared (Griffiths, 1975), the submillimeter (Infrared Physics, 1979), and into the microwave (Richards, 1977). High resolution (0.001 cm^{-1}) IR laboratory spectroscopy is now routine (Guelachvili, 1978, Genzel and Sakai, 1977). The technology of the Fourier transform spectrometer is a general research topic only in the extreme regions of its use, UV and microwave. However, technological developments continue to improve the quality of the data throughout useful spectral regions.

Fonck *et al.* (1978) report on the design of an all-reflecting Fourier transform spectrometer using a diffraction grating beam splitter. This unique design approach may enable Fourier transform spectroscopy to be used in those spectral regions for which the more classical beam splitters do not exist, e.g., for the ultraviolet.

The techniques and analysis described here are presented within the framework of moderate-to-high-resolution interferometers for scientific measurement over 0.3–16 μ. As the observed wavelength shortens, the stability of the optical and mechanical components becomes increasingly important for good signal-to-noise ratios.

2.2. Foundation for Design

Instruments for acquiring data for scientific research are designed based on a measurement strategy and the environment of the experiment. Some measurement strategies require a general survey instrument; others, an instrument for a specific task. An instrument designed for a specific spectral feature is a laser heterodyne spectrometer. The profile of a particular spectral feature is measured with very high resolution and over a very small free spectral range. Applications to remote sensing of the atmosphere using the laser heterodyne spectrometer are discussed by Hinkley (1976). The Fourier transform spectrometer is a survey instrument, used to measure many spectral features simultaneously over thousands of wavenumbers at resolutions to 0.001 cm^{-1}. In this section we examine the relationship between a set of required scientific measurements and the initial design of a Fourier transform spectrometer. Fundamental parameters for the design are the optical path difference and the throughput or *étendu*. This section shows an example of how a priori knowledge, or a prediction of the nature of the source, enables the definition of the optical path difference.

A. Science Background

The interferometer measures the radiant emittance or absorptance of matter. Liquids, solids, and gases all have absorption and emission spectra which can be measured with Fourier transform spectrometers.

Here we select as an example the problem of measuring the spectrum in the 2–16 μ wavelength region of a gas in the earth's stratosphere and an experimental configuration used successfully by Farmer (1974) and Farmer *et al.* (1980). The experiment is to observe the solar spectrum at large solar zenith angles to record a superposition of the solar and stratospheric absorption spectrum from a platform either in the stratosphere or in orbit.

The physics of line formation is well understood [see Jefferies (1968) or Penner (1959], and this knowledge is used to estimate the spectrum to be measured. If there were no a priori knowledge of the nature of the source producing the radiation whose spectra is to be measured, then one might assume that spectra of the highest resolution and the highest signal-to-noise ratio (which consequently have the highest data content) are also the optimum spectra to be used for scientific analysis. This is not necessarily the case.

In many situations a priori knowledge exists, and the expensive resources of signal-to-noise ratio and resolution can be optimized to fit a particular scientific measurement. However, there will always remain those situations of exploratory science where no one predicts the outcome of a scientific measurement and, perhaps, the most exciting discoveries are made.

B. OPD Required to Resolve a Doppler-Broadened Spectral Feature

The optical path difference required to resolve spectral features in emission or absorption for a feature broadened only by Doppler effects and not by collisional, electromagnetic, or some nearest neighbor phenomena is calculated to illustrate a baseline design approach to identify the optical path difference or retardation.

The shape of a Doppler-broadened spectral line, in general, is given by

$$I = b^{-1}(\ln 2/\pi)^{1/2}\,(\exp\{-[(\nu - \nu_0)/b]^2 \ln 2\}), \tag{2.1}$$

where

$$b = (mc^2/2\pi\, kT)^{-1/2}(\ln 2/\pi)^{1/2}\nu_0\,, \tag{2.2}$$

and the Doppler half width (full width at half maximum) in units of centimeters^{-1} is given by Eqs. (3–30) in Penner (1959) to be

$$\Delta\nu = \nu_0[2\,kT(\ln 2)/m\,]^{1/2}, \tag{2.3}$$

where k is the Boltzmann constant, c is the speed of light, and the spectrum line is from a gas whose molecular weight is m and which is at thermodynamic equilibrium at temperature T.

A spectral feature in absorption and a sampling comb are shown in Fig. 2.1. The separation of the samples in spectrum space is ΔS. If this spectrum were recorded by a Fourier transform spectrometer, the maximum optical path difference, or retardation, X_0 would be given by

$$X_0 = 1/\Delta S. \tag{2.4}$$

We define a spectral line to be resolved if it has two samples per full width at half maximum, or

$$2\,\Delta S = \Delta\nu. \tag{2.5}$$

Combining Eqs. (2.3), (2.4), and (2.5), we find

$$X_0 = c/\sigma_0[m/2 \ln 2kT]^{1/2}, \tag{2.6}$$

where σ_0 is the frequency in wavenumber units cm^{-1}. If mass M is expressed in atomic mass units (amu), and X_0 in meters, then

$$X_0 = 2.8 \times 10^4(1/\sigma_0)(M/T)^{1/2}. \tag{2.7}$$

Equation (2.7) gives the optical path difference required to resolve a Doppler-broadened absorption or emission feature from a gas. An example is the application of Eq. (2.7) to features in the earth's upper atmosphere, where the average temperature is 240°K. Figure 2.2 gives a plot of the optical path difference required to resolve a Doppler broadened spectral feature at 240°K shown on a graph of molecular weight as a function of wavenumber (centimeters^{-1}) for path differences 0.04–10 M. Also shown are the locations of absorption features for N_2O, CO_2, and CH_4. This figure shows that lines that form the CO_2 band near 3000 cm^{-1} are resolved with a 4-meter path difference, whereas those in the adjacent N_2O band are not resolved.

If an interferogram of an absorption spectrum of the CO_2 lines near 3000 cm^{-1} were to be recorded with an ideal interferometer, the fringe modulation would be zero at 4-m optical path difference. The interferogram for the adjacent N_2O lines would not have dropped to zero at 4 m, but rather there would still be modulation in the signal. Qualitatively, if a spectral

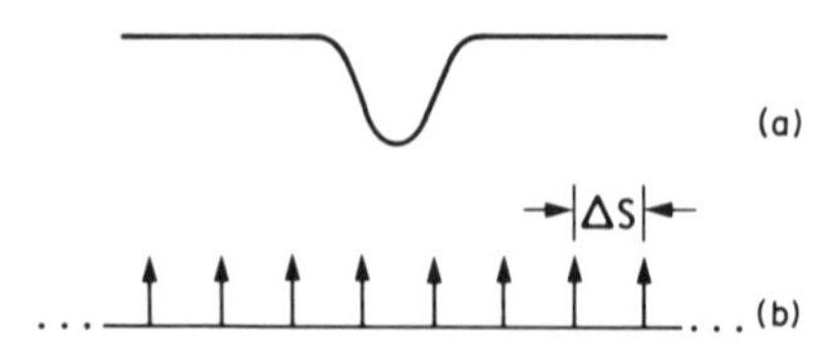

FIG. 2.1. (a) Absorption line spectrum $f(\nu)$ multiplied by (b) a periodic sampling function $\Sigma_{n=-\infty}^{\infty} \delta(\nu - n\,\Delta s)$. Sampled spectrum is $\Sigma_{n=-\infty}^{\infty} \delta(\nu - n\,\Delta S)f(\nu)$.

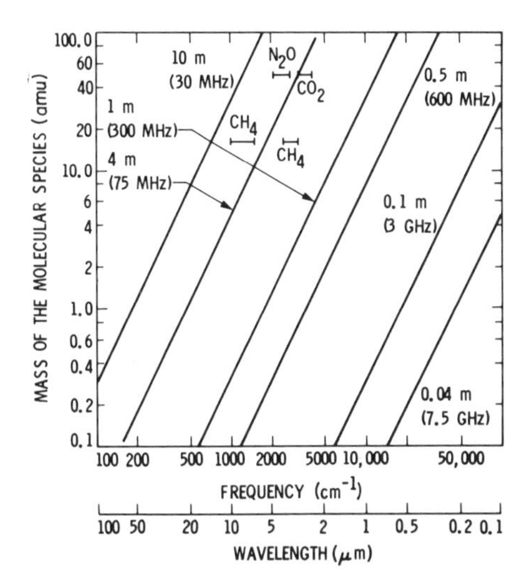

FIG. 2.2. Optical path difference required to resolve a Doppler-broadened spectral feature at 240°K on a graph of molecular weight as a function of wavenumber in centimeters^{-1} for 0.04–10 M retardations.

feature is greatly underresolved, a considerable amount of power exists in the modulation of the interferogram at the truncation of the scan. This truncation stops the interferogram data abruptly, and the Fourier transform of this clipped interferogram shows ringing on the spectral features. The amount of ringing and its effect on resolution are usually modified during data reduction by apodization. Apodization of spectra has been discussed by many authors; see, for example, Norton and Beer (1976), Harris (1978), Jacquinot and Roizen-Dossier (1964), and Jacquinot (1955). Note that if the optical path difference is such that all of the spectral features are resolved, then the fringe visibility drops to zero at the end of the scan; the spectrum will contain no ringing, and there is no need to apodize. In this case the spectral content "self-apodizes" the spectrum.

In many situations, the absorption line will be unresolved. However, fairly accurate estimates of the equivalent width or area of the line can be made (see Jansson *et al.*, 1970).

2.3. Fringe Contrast

The efficiency of a Fourier transform spectrometer depends on both the instrument's transmittance (throughput) as well as the modulation efficiency or the fringe contrast. (Throughput is discussed in detail in Section

2.5.) The degradation of fringe contrast resulting from the interferometer optical configuration is discussed in detail in Section 2.6. There, fringe contrast and those parameters that affect it, e.g., wave-front errors, coating efficiency, reflectance, transmittance in each arm, and the effects of polarized light, are discussed in detail for a narrow pencil beam down the axis of the instrument. The effects of wave-front tilts and wave-front errors are examined by opening up that pencil beam to fill the apertures and discussing cat's-eye retroreflectors, fabrication and alignment errors, and tilts.

A. Beam splitter, Transmittance

Figure 2.3 shows the standard configuration for a Fourier transform spectrometer. Radiation (after division) is recombined at the beam splitter, where light of intensity I_1 is from the one arm and light of intensity I_2 is from the other. The coherent addition of these two signals as a function of x, the retromirror position, is given by $I(x)$ for wavenumber σ and phase ϕ, where

$$I(x) = I_1 + I_2 + 2(I_1 I_2)^{1/2} \cos(2\pi\sigma x + \phi). \tag{2.8}$$

Michelson (1927) defined fringe contrast V to be

$$V = (I_{\max} - I_{\min})/(I_{\max} + I_{\min}), \tag{2.9}$$

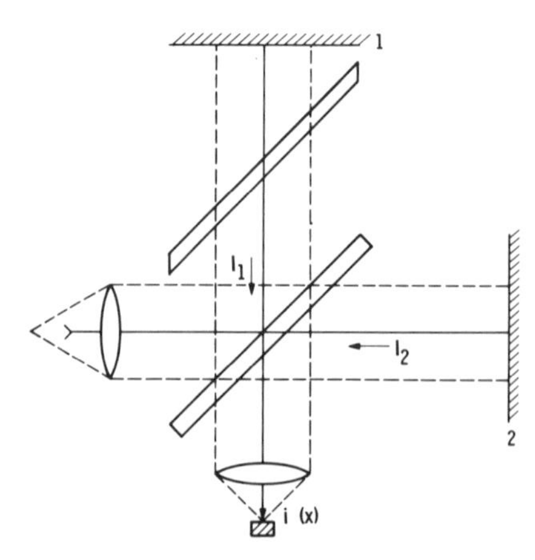

FIG. 2.3. Layout of a Fourier transform spectrometer operating in collimated light and with flat mirrors.

where I_{max} and I_{min} are the maximum and minimum intensities, respectively, in the fringe pattern. From Eq. (2.8), $I_{max} = I_1 + I_2 + 2(I_1 I_2)^{1/2}$; $I_{min} = I_1 + I_2 - 2(I_1 I_2)^{1/2}$; and, with Eq. (2.9),

$$V = 2(I_1 I_2)^{1/2}/(I_1 + I_2). \tag{2.10}$$

In electrical engineering, a sinusoidal signal is characterized by the modulation index M. Figure 2.4 shows a sinusoidal signal similar to the interferogram of a monochromatic source. If the equation $I(x) = A + (B/2) \cos(2\pi\sigma x + \phi)$ is used to characterize the function shown in Fig. 2.4, then $M \equiv B/2A$, and we see that $V = M$.

Equation (2.8) can be rewritten to include the fact that the intensities I_1 and I_2 vary as a function of wavelength, and that in the real case, the interferogram is sampled at discrete intervals x_j. In general, the intensity $i(x_j)$ in the interferogram measured at the jth point is

$$i(x_j) = A(\sigma) + \sum_{i=1}^{N} \frac{B(\sigma_i)}{2} \cos[2\pi x_j \sigma_i + \phi(\sigma_i)], \tag{2.11}$$

where A is a constant, $B(\sigma_i)$ is the power spectral density in the neighborhood of the frequency σ_i of the source radiation, and $\phi(\sigma_i)$ is a frequency-dependent phase term. The spectral information is contained in the term $B(\sigma_i)$, which is modulated by the cosine term; the term $A(\sigma)$ contributes noise, but no useful signal. It is therefore important that the term $A(\sigma)$ not dominate the recorded intensity.

If the interferometer stays in good alignment during the entire scan, the fringe contrast near zero optical path difference is a useful measure of efficiency. If the optical path distance in each interferometer arm is the same for all the observed wavelengths, the phase term ϕ is not a function of wavenumber, and near zero path difference this contrast is

$$V = \sum_{i=1}^{N} \frac{B(\sigma_i) \cos[\phi(\sigma_i)]}{2A(\sigma)}. \tag{2.12}$$

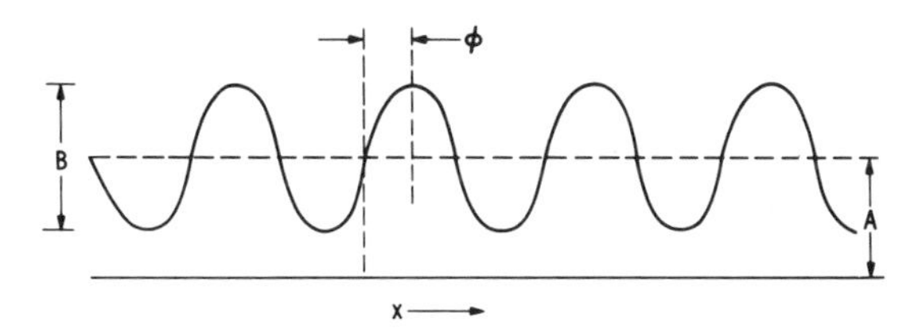

FIG. 2.4. Interferometer output intensity as a function of position for monochromatic light input represented by a cosinusodial signal of peak-to-peak amplitude B, offset A, and phase ϕ.

Building the interferometer such that the $\cos[\phi(\sigma_i)]$ term is approximately unity is discussed in Section 2.5,D. Set $\cos[\phi(\sigma_i)] = 1$ to let Eq. (2.12) become $V = B/2A$. Hence under certain conditions the equal-path fringe contrast is a direct measure of the efficiency of the interferometer. Designing the system to maximize the equal-path fringe contrast assures that most efficient use is being made of the incoming light.

The relationship between equal-path fringe contrast and the transmissivity and reflectivity of optical components within the interferometer is now calculated. Returning to Fig. 2.3, assume that the incident wave front has complex amplitude A. That portion of the beam that is transmitted has amplitude $A\tau_1$, where τ_1 is the transmissivity of the beam splitter. The transmitted light reflects from the mirror, whose reflectivity is r_3, and when it reflects from the beam splitter to join the other beam and exit the system, it leaves with amplitude given by $Ar_1r_3\tau_1$. Returning to the original beam, we find that Ar_1 strikes mirror 2, and leaving the system is $Ar_1r_2\tau_1$.

The intensity of the combined beams is

$$I = (Ar_1r_2\tau_1)^2 + (Ar_1r_3\tau_1)^2 + 2A^2{r^2}_1{\tau^2}_1 r_2r_3 \cos(2\pi\sigma x + \phi). \quad (2.13)$$

We assume that the beam-splitter surface is a lossless dielectric and not an absorption coating such as a metal. Discussions of the properties of partially absorbing thin films are given by Smith (1978), MacLeod (1978), and Raine and Downs (1978). A partially absorbing thin film has application for violet and ultraviolet beam splitters.

Comparing Eq. (2.13) with Eq. (2.8), we identify $I_1 = (Ar_1r_2\tau_1)^2$ and $I_2 = (Ar_1r_3\tau_1)^2$ and from Eq. (2.10) find

$$V = 2r_2r_3/(r_3 + r_2)^2. \quad (2.14)$$

Hence for the configuration shown in Figure 2.3 with a dielectric beam splitter, the fringe contrast is independent of the transmittance and reflectance of the beam splitter and depends on the transmittance of each interferometer arm.

Another interferometer configuration, used by Connes (1970) in several of his interferometers, and by Pinard (1969), Sakai (1977), Brault (1974), and others is shown in Fig. 2.5. This is characterized by using a cat's-eye retroreflector to laterally shift the beam, and using a beam recombiner separate from the beam splitter. Both outputs are accessible for the positioning of detectors, and two inputs are accessible.

The visibility for the fringe pattern seen at each arm is not the same function of T and R. Let S_L be the signal output on the left, and S_R be the

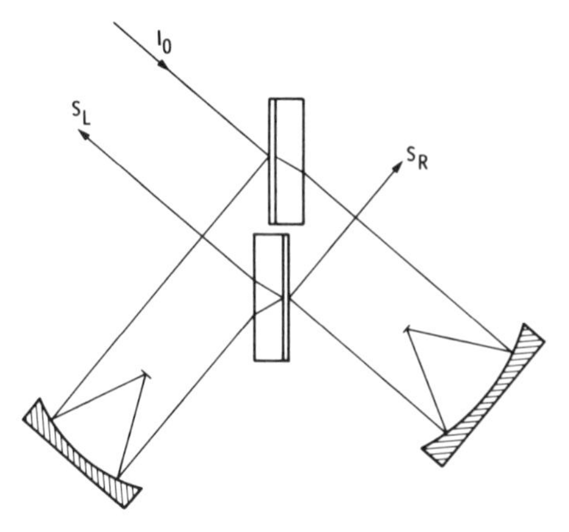

FIG. 2.5. An optical configuration used in several successful high-resolution Fourier transform spectrometers. It is shown in the single input/dual output mode with intensity I_0 entering, and signals S_L on the left and S_R on the right.

signal output on the right; then

$$S_L = I_0T^2 + I_0R^2 + 2I_0TR\cos(2\pi\sigma x + \phi) \tag{2.15}$$

and

$$S_R = I_0RT + I_0RT - 2I_0TR\cos(2\pi\sigma x + \phi), \tag{2.16}$$

where T and R are the transmittance and reflectance of the beam splitter, and the beam splitter and recombiner are assumed to be nonabsorbing.

The visibility in the left arm, V_L, and that for the right arm, V_R, are given by

$$V_L = 2TR/(T^2 + R^2) \qquad \text{and} \qquad V_R = 1. \tag{2.17}$$

Therefore, for the single-pass, dual output Fourier transform spectrometer configuration shown in Fig. 2.5, the fringe visibility, or contrast of the equal-path fringe at one of the outputs, is independent of the beam-splitter coating reflectivity and transmittance; whereas the other output has the dependence given in Eq. (2.17).

The configuration shown in Fig. 2.5 is not as precisely balanced as indicated by Eqs. (2.15) and (2.16). To keep the optical path difference through the dispersive material that forms the substrates of the beam splitter/beam recombiner nearly equal so that the interferogram is symmetric (i.e., $\phi \approx 0$) or linear with frequency σ, in Eq. (2.12) the beam-splitter coatings are oriented as shown in Fig. 2.5 and the substrates are of nearly equal thickness. The signal output on the left is calculated with the energy entering the cat's-eye in the lower right by transmitting in the order: air/beam splitter/substrate/air, both when the beam enters the cat's-eye and when it leaves the cat's-eye. The signal output on the left is calculated with the energy entering that cat's-eye in the lower left by reflecting from the beam splitter in the reflection sequence: air/beam splitter/air, and when it leaves the cat's-eye, it passes in a different reflection sequence: air/substrate/beam splitter/air. Close examination of the terms in the calculation of the signal on the right S_R shows that the reflections and transmittances are the same. The subtle differences in the reflection coefficients in the S_L calculations may result in asymmetries which cannot be corrected by changes in substrate thickness.

The efficiency of the optical configuration shown in Fig. 2.5 for output on the left-hand side is now calculated. Let the signal be represented by $I(x) = A + (B/2)\cos(2\pi\sigma x + \phi)$ (see Fig. 2.4); then by inspection of Eq. (2.15) we see that $A = I_0(T^2 + R^2)$ and $B = 4I_0RT$. For the output signal on the left side of the configuration shown in Fig. 2.5, the system efficiency e is given by

$$e = 4RT. \tag{2.18}$$

The efficiency of the interferometer depends on the $4RT$ product of the beam-splitter for the case of no absorption or polarization effects and a dielectric beam splitter.

It is important in the first-order optical design to realize those aspects of the design where specifications can be relaxed. Here we show that the system efficiency or modulation efficiency is relatively insensitive to imbalances in the radiation returning to the beam splitter from the interferometer arms. Figure 2.6 shows the interferometer efficiency from Eq.

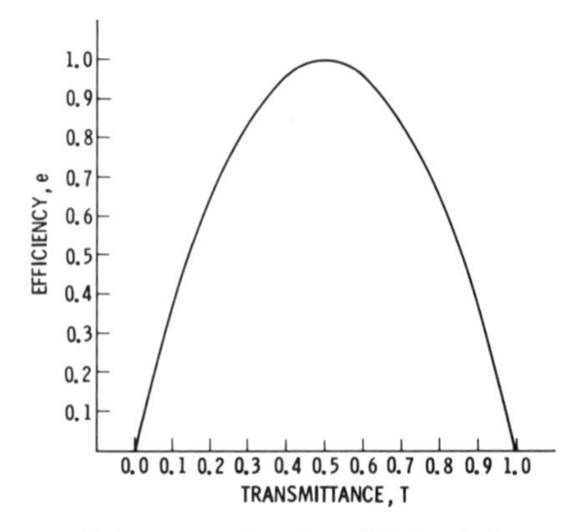

FIG. 2.6. Interferometer efficiency as a function of dielectric beam-splitter transmittance.

(2.18) as a function of the transmittance of the beam splitter. An important point to note is that even with a transmittance between 30 and 70%, implying a reflectivity between 70 and 30%, the efficiency is greater than 80%.

Equations (2.16) and (2.15) show that for the dual output/input optical configuration shown in Fig. 2.5, one of the outputs is the phase complement of the other. Such is the case if the beam-splitter/beam-recombiner thin film coatings are pure dielectrics and not absorbing films. In some spectral regions, dielectric thin films lack an efficient $4RT$ product over an entire desired bandwidth, e.g., 350–2000 nm. In this case very thin films of aluminum or silver are found to be useful (Brault, 1974). A metal coating of such thickness that the white-light transmittance is between 70 and 80% was found to behave enough like a dielectric for the $4RT$ product to be approximated by the curve shown in Fig. 2.6. With thicker coatings the asymmetries in the outputs are obvious, with colored bull's-eyes apparent at the equal-path fringe position of the interferometer whose configuration is shown in Fig. 2.5.

In many design approaches for interferometer optical configurations, the compensator plate covers the beam splitter to make the beam-splitter coating sandwiched between two substrates. An advantage of this approach is that the sometimes soft and delicate beam-splitter thin film coating is protected against abrasions and decay due to environmental factors. A disadvantage is that the thin film designer is constrained to optimize the beam-splitter design with materials whose index is greater than 1.5. The availability of an air/thin film interface gives the designer more degrees of freedom to optimize the design.

This efficiency is calculated assuming that there are no wave-front errors. In practice wave-front quality is determined by the materials selected and the skills of the optician. Often the desired wave-front error cannot be achieved because of a requirement such as optical transmission in the wavelength region of interest. For example, potassium bromide, with its low index of refraction and good transmission properties in the infrared, makes a highly desirable candidate for an infrared beam splitter. However, it is very difficult to figure the soft surface. The next section discusses wave-front error contributions.

B. Wave-Front Error Contribution

In the preceding section the fringe was discussed assuming the interferometer to be illuminated with an infinitely narrow pencil of light. The reflectances of mirror and beam-splitter surfaces and the transmittance of the beam splitter were considered contributors to the modulation efficiency of the interferometer. A narrow pencil of light is unrealistic. The radiation must get through the system (see Section 2.5), and consideration must therefore be given to beams that fill the apertures. The Fourier transform spectrometer configured as shown in Fig. 2.7 takes a single area in object space and maps it into two wave fronts at the beam splitter. If the recombining wave fronts are coplanar, then the fringe visibility or modulation efficiency is given by the system transmissivities and reflectivities, as discussed in Section 2.2. However, alignment errors and mirror figuring errors are major contributors to a degradation in fringe contrast.

Williams (1966) and Katti and Singh (1966) discuss wave-front errors caused by mirror misalignments in Fourier spectroscopy, primarily in terms of the tilt of plane surfaces. In Section 2.6, optical configurations that, to first order, are tilt compensated, are discussed. However, even with tilt errors removed from the wave front, residual figuring or polishing errors cause a degradation in fringe contrast. For example, some applications require compact instruments with high-speed (low-f-number) parabolas for cat's-eye primaries, which are very difficult to make without wave-front errors.

Figure 2.8 shows two wave fronts. The wave front indicated by a dashed line comes from one arm of the interferometer, whereas that indicated by the solid line comes from the other. These were recombined at a beam splitter and are shown now traveling toward a detector to the right. That light which falls outside the overlap region gives a single uniform signal with path difference and therefore does not contribute to the modulation. This misalignment yields a decrease in the equal-path fringe contrast.

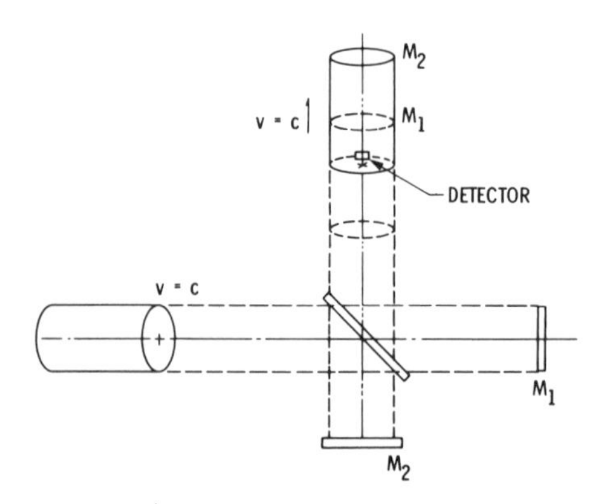

FIG. 2.7. A volume of complex amplitude signal in object space is transformed by the ideal interferometer into two volumes at image space. One volume is shown reflected from mirror M_1, the other from mirror M_2. Each is identical, but one is retarded by the OPD compared to the other. Both are immersed in a typical detector that is sensitive to the modulus square of the product of the complex signals.

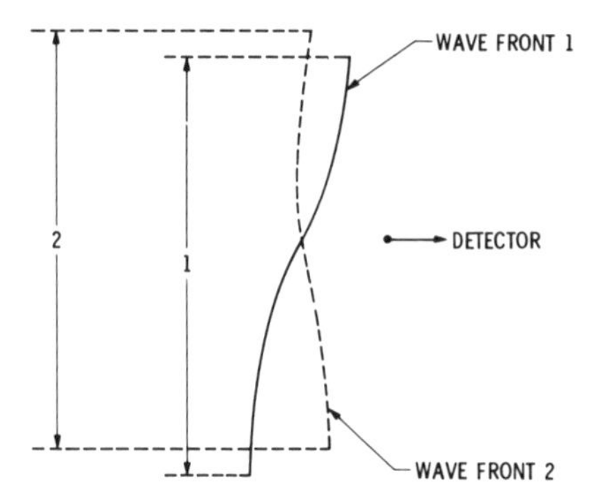

FIG. 2.8. Wave fronts from each interferometer arm recombine. Noncoplanarity shows realistic case of wave-front aberrations and misalignments.

The visibility [cf. Eq. (2.9)] is expressed as

$$V = \frac{1}{A} \int \frac{1 + \cos[kX(A)]}{2} \, dA,$$

where A is the area of the superposed wave fronts, $k = 2\pi/\lambda$, and $X(A)$ is the local OPD error in units of wavelengths of light (Norton, 1978). The distance $X(A)$ is not the OPD or retardation of the interferometer, but rather the OPD error between the two wave fronts.

In most interferometers the OPD error $X(A)$ is small, i.e., $kX \ll 1$ and $\cos(kX) = 1 - (kX)^2/2$. The fringe visibility is then

$$V = 1 - \frac{1}{4A} \int [kX(A)]^2 \, dA$$

or

$$V = 1 - \frac{k^2}{4N} \sum_{i=1}^{N} [X_i(A)]^2 \tag{2.19}$$

for N uniformly spaced rays, each ray with a known OPD error $X_i(A)$ at position A on the surface.

Williams (1966) noted that tilt errors in the Fourier transform spectrometer introduce an amplitude distortion which is a function of wavenumber on the reconstructed spectrum. Goorvitch (1975) showed that a Michelson interferometer with misaligned mirrors requires a quadratic term for phase correction. The problem of nonlinear phase errors introduced by misalignment of a Michelson interferometer is discussed in detail by Schröder and Geich (1978). In summary, the signal-to-noise ratio is optimized, data reduction is simplified, and the experiment enhanced if, during the first order design, efforts are expended to minimize the presence of tilt errors and other wave-front aberrations. In Section 2.4 mechanical design methods to minimize tilts are discussed. Section 2.6 provides a discussion of tilt-compensated configurations.

C. Polarization and Fringe Visibility

The vector nature of light plays a major role in interference. It was recognized late in the last century that two waves orthogonally polarized (Shurcliff, 1966) will not create interference fringes even though they are of the same wavelength. Strong (1958) summarizes the interference of polarized light.

In general, the transmittance and fringe contrast of a Fourier transform spectrometer depends on the state of polarization of the incoming radiation. The transmission with wavelength as a function of polarization is more uniform for the FTS than that measured for a diffraction grating spectrograph by Breckinridge (1971). In many applications, the FTS is not used for precision spectroradiometry, and the instrument's polarizing property is a concern only for throughput modulation efficiency, or fringe contrast, calculations.

The polarization of radiation has been used to advantage for Fourier spectroscopy in some regions of the spectrum. Martin and Puplett (1969) devised an interferometer for use in the far infrared which uses a polarizing, wire-grid beam splitter. In the far infrared, efficient beam-splitter materials are very difficult to find. The Martin and Puplett configuration has been very successful and is used in a number of laboratories (Richards, 1977; Parker, *et al.*, 1978; Mathers, 1977). Here our interests are not in the far infrared, but rather the near infrared and visible, where polarization, if not considered in the first-order optical design, may cause a degradation of the equal-path fringe contrast.

If the light entering the interferometer remains unpolarized as it passes through it, then on recombining at the beam splitter, the maximum equal-path fringe contrast can be obtained. The equal-path fringe contrast is lowered if the radiation in one arm is partially polarized with one preferential direction and the light in the other arm is partially polarized orthogonally to that in the first arm. The fringe contrast is also reduced if the radiation is not absorbed by an equal amount in each arm.

A theoretical analysis of the polarization effects in Fourier spectroscopy using a coherency matrix representation was given by Fymat (1972). The discussion here takes a simpler viewpoint and presents the effects in terms of modulation efficiency or equal-path fringe contrast.

An analysis of the effects of polarization follows. Figure 2.3 shows the standard flat mirror Fourier transform spectrometer with a compensator plate in one arm. The maximum fringe contrast possible with this arrangement is found using Born and Wolf (1975), who show that for a single dielectric/air interface, the intensity transmittance for light polarized parallel to the plane of incidence $\tau_{\parallel}$ is given by

$$\tau_{\parallel} = \frac{\sin 2\theta_i \sin 2\theta_t}{\sin^2(\theta_i + \theta_t) \cos^2(\theta_i - \theta_t)}, \tag{2.20}$$

and the intensity transmittance for light polarized perpendicular to the plane of incidence $\tau_{\perp}$ is given by

$$\tau_{\perp} = \frac{\sin 2\theta_i \sin 2\theta_t}{\sin^2(\theta_i + \theta_t)}, \tag{2.21}$$

where θ_i is the angle of incidence of the wave front on the dielectric interface and θ_t is the angle of the wave front inside the dielectric, after refraction. The dielectric surface has a different intensity transmittance for light polarized parallel than for light polarized perpendicular (Shurcliff, 1966) to the plane of incidence.

The ratio of the transmittance in perpendicular light $\tau_{\perp}$ to that in parallel light $\tau_{\parallel}$ is given by dividing Eq. (2.21) by Eq. (2.20), to get

$$\tau_{\perp}/\tau_{\parallel} = \cos^2(\theta_i - \theta_t). \tag{2.22}$$

On leaving the medium, the wave front passes another dielectric/air interface, and the total transmittance of a parallel plate at angle θ_i to the incoming beam ratio is

$$\tau_{\perp}/\tau_{\parallel} = \cos^4(\theta_i - \theta_t). \tag{2.23}$$

The loss in fringe contrast is found by assuming that the radiation entering the interferometer is composed of equal intensities of orthogonally linearly polarized light. Circular polarization is neglected, and all reflection surfaces are assumed to respond as dielectrics. If the optical configuration requires metal mirrors at nonnormal incidence, then the presence of circularly polarized light might be important, particularly if the angle of incidence is as large as 45° (see Born and Wolf, 1975, Chapter 13.2).

Consider an observer positioned within the beam-splitter substrate looking out into each arm of the interferometer. In the classical configuration shown in Fig. 2.3, the observer sees one arm with the free-standing compensator (path 1) and the other arm containing the beam-splitter substrate (path 2). From Eq. (2.23), that light which follows path 1 returns to the beam splitter with transmittance ratio

$$\tau_{R1} = (\tau_{\perp}/\tau_{\parallel})^2 = \cos^8(\theta_i - \theta_t). \tag{2.24}$$

TABLE 2.II.

TABULATION OF TRANSMITTANCE IN INTERFEROMETER ARMS 1 AND 2 (FIG. 2.3) FOR LIGHT POLARIZED PERPENDICULAR AND PARALLEL AND THE MAXIMUM FRINGE CONTRAST

Arm	$T_{\parallel}$	$T_{\perp}$
1	0.98	0.67
2	0.99	0.82
Maximum fringe contrast	99%	82%

That light which follows path 2 returns to the beam splitter with transmittance ratio

$$\tau_{R2} = \cos^4(\theta_i - \theta_t). \tag{2.25}$$

The interferometer shown in Fig. 2.3 has the angle of incidence θ_i on both the compensator and the beam splitter equal to 45°. For a crown glass substrate whose index of refraction is 1.52, the ray normal to the wave front is transmitted at angle θ_t (calculated from Snell's Law) equal to 27.7°. From Eq. (2.24), the polarization transmittance ratio from interferometer retromirror i τ_{R1} is 0.6911, and that for retro mirror 2 is found from Eq. (2.25) to be $\tau_{R2} = 0.8314$. Since these two numbers are not equal, a decrease in fringe contrast in addition to that which is calculated from the normal incidence intensity transmittance will occur.

If half of the light incident is polarized parallel and the other half polarized perpendicular, a calculation of the polarization transmittance of the two arms shows the maximum fringe contrast to be approximately 90% in place of 100%. The calculated transmittance in parallel light and perpendicular light for arm 1 and arm 2 of the interferometer is shown in Table 2.II.

To minimize the effects of the polarization of light by the optics, reflections from metal and dielectric mirrors should be as close to normal incidence as possible. Therefore, the optical design approach selected for the interferometer should allow as close to normal incidence reflection as possible.

2.4. Mechanical Design

An understanding of the constraints placed on the mechanical system by a requirement to record interferograms for high signal-to-noise ratio spectra is discussed in this section. Accurate interferogram sampling, kinematic mounts, uses of elasticity, and scan carriage mechanisms are emphasized.

At a scale of the order of the wavelength of visible light, the entire world is in continuous motion. Fringe sampling stabilities of the order of one millifringe or about 6.3×10^{-9} m are required to achieve a good signal-to-noise ratio in the red. A block which appears rigid to the eye becomes a quivering block of jello when viewed with a system whose sensitivity is of the order of the wavelength of light.

Two quite distinct design approaches are used to achieve this stability. One is to configure the mechanical system so that it is as rigid as possible. This is done by giving great care to the layout, using very careful annealing procedures for the metal parts so they do not creep, and paying close attention to mountings and temperature compensation designs. The other approach is to allow the components to move and tilt (on a very small scale) and to compensate for this motion using an electronic servo system. In both situations, better-quality spectra are obtained if considerable attention is given by the designer to the use of kinematic or semikinematic mounts and the use of elasticity in the first-order design of the interferometer.

A. Background

A Fourier transform spectrometer optical system is divided into two sections for analysis. Those sections least sensitive to small-scale displacements are the foreoptics and postoptics. The section most sensitive to small displacements is the interferometer, or the portion between the beam splitter and the beam recombiner. A portion slightly less sensitive is the part between the beam splitter and compensator. Here we discuss the constraints placed on the interferometer system.

Figure 2.9, which shows a FTS, is used to describe sampling errors. The need is to hold the spacing between the beam splitter BS and mirror A, and the beam splitter BS and mirror B, in such a way that mirror B can be positioned at regular intervals of optical path difference and the intensity at detector D measured at each interval. Two well-known techniques for moving the mirror are step and integrate (Connes, 1963; Schindler, 1970a), and moving the mirror at constant velocity with sampling at constant time intervals (Brault, 1973; Sakai, 1977). In both situations the uniformity of the samples is limited by the system mechanical stability.

The techniques of linear systems described by Gaskill (1978) and

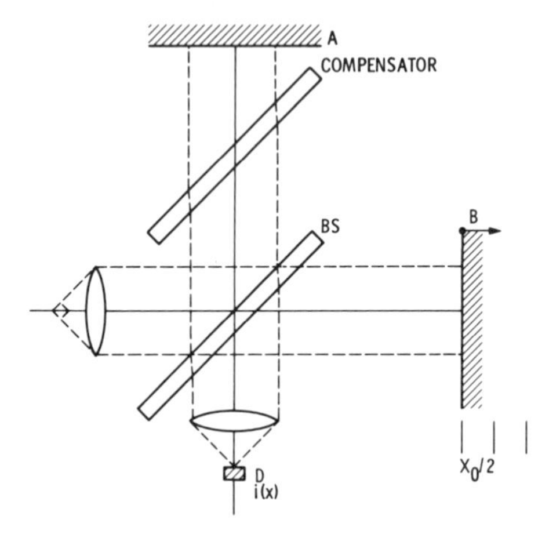

FIG. 2.9. Layout of typical flat-mirror FTS showing mirror motion of $X_0/2$ to give sample interval in scan of X_0.

Bracewell (1965) will be used to analyze the Fourier transform spectrometer sampling.

Let the unsampled interferogram be represented at the detector D in Fig. 2.9 as a function of the position x of mirror B, by $i(x)$. The sampled interferogram is then

$$i(x_n) = i(x) \sum_{n=-l}^{n=+l} \delta(x - nx_0), \tag{2.26}$$

where x_0 is the spacing between samples, and the sampling interval runs from $-l$ to $+l$.

In Eq. (2.26) the continuous function $i(x)$ is multiplied by an array of δ functions which extends from $-l$ to $+l$ where the scan is over length $2l$. For the ideal interferometer, the real part of the complex Fourier transform of $i(x_n)$ is the spectrum of the light coming from the source S. The spacing of the samples x_0 must obey the sampling theorem (Bracewell, 1965); otherwise, interpretation of the data is complicated.

Mechanical vibrations cause sampling errors (i.e., x_0 is not the same for each sample), and they contribute noise to the recorded interferogram. In general, the effect is twofold: the integration time between steps is changed, and the sampling interval is not uniform. An approximate representation of the recorded interferogram signal in the presence of these noise sources is, from Eq. (2.26),

$$i(x_n) = i(x)\underset{\sim}{a} \Sigma\, \delta(x - nx_0 + \underset{\sim}{\varepsilon}), \tag{2.27}$$

where we have used a random variable $\underset{\sim}{a}$ whose expectation is 1.0 to model the effects of variations in integration time, and $\underset{\sim}{\varepsilon}$, a random variable whose expectation is 0 to model the effects of variations in the sampling error.

The point of this section is not to continue with a detailed analysis of sampling errors (Chapter 1) but rather to emphasize the importance of good mechanical design to minimize sampling error contributions by mechanical mechanisms and resonances.

Experience by the authors and others (Brault, 1973; Norton, 1978) indicates that in absorption spectroscopy, to achieve a signal-to-noise ratio of 1000 at a given wavelength λ, the rms sampling error must be equal to or less than a milliwavelength. If we assume a Gaussian-distributed random variable to represent sampling errors, then the standard deviation of the random variable $\underset{\sim}{\varepsilon}$ must, at 1-μ wavelength, be 10 Å. It is difficult, but not impossible, to stabilize the optical path-difference intervals to this accuracy.

For a continuous scan interferometer, the effects of $\underset{\sim}{\varepsilon}$ and $\underset{\sim}{a}$ depend on the time scale involved for the error excursions. Damped mechanical bodies, when excited by an impulse, as for example when striking an assembly with a hammer, will show small excursions for high frequency and large excursions for lower frequency. Letting f = frequency, these excursions approximate a $1/f$ dependence. If the mechanism is resonant, the impulse energy causes it to oscillate at a well-determined frequency. This can be disastrous to the quality of the spectrum since a period in a signal superposed on the interferogram appears as Rowland or Lyman ghosts, which distort the spectrum and cause errors in line strengths, equivalent widths, and continuum position.

The optical bench of a Fourier transform spectrometer needs to be isolated from the vibrations of the outside world. Components on the bench as well as the bench itself must be well damped. It is necessary to keep the interferometer optics from vibrating relative to each other, to keep outside accelerations from causing differential motions among the interferometer components, and to remove unwanted fluctuations in the optical path. Sources of mechanical noise must not be mounted on the interferometer support structure.

B. Atmospheric Turbulence

Many Fourier transform spectrometers are operated in an open environment. Small-scale air temperature variations cause index of refraction variations which result in unwanted optical path modulations, degrading the signal-to-noise ratio of the spectra. These natural index of refraction variations are easily visualized in Schlierin experiments and in stellar speckle interferometry (Breckinridge *et al.*, 1979a).

At optical wavelengths, meteorological parameters are related (Straiton, 1964) to the refractive index changes through the relationship

$$\Delta n \times 10^6 = -77.6\, P_{DA}\, \Delta T/T_0^2,$$

where P_{DA} is the partial pressure of dry air in millibars, and T is the temperature in kelvins. Using constants from Grossmann–Doerth (1969), we find that at 293°K a 1°C temperature change yields an index change of $\Delta n \approx 10^{-6}$.

A slab of material of thickness h and refractive index n has an apparent thickness l related by $h = nl$. The change in apparent thickness for a 0.01°C temperature change for a slab of air of thickness of 1m is found from $dl = (h\, dn)/n^2$ to be $dl = 5$ millifringes at a wavelength of 2 μ in the near infrared.

A slight draft blowing across the instrument, the heat rising from the baseplate, or the turbulence created by moving a retroreflector through the air result in temperature variations in the volume of air through which the light beam passes. These temperature fluctuations cause index of refraction variations, which, in turn, result in sampling errors in the recorded data. Strohben (1971) gives a review article on the turbulence in the atmosphere. The volume of air with which we are concerned is that contained in each arm of the interferometer, between the beam splitter and the beam recombiner.

The fluctuations or variations in optical path difference produced by atmospheric turbulence are not periodic. Indeed, the power spectrum of these variations, i.e., the power spectrum of the random variable $\underset{\sim}{\varepsilon}$ used in Eq. (2.27), is close to zero for frequencies above 500 Hz. The cutoff depends, of course, on the amount of energy being pumped into the air. If the interferometer is covered by a well-insulated box (e.g., a styrofoam ice chest turned upside down), and care is taken in the first-order design to insure that there are minimum heat sources or sinks under the box, then the power spectrum of the atmospheric turbulence-induced optical path variations is close to zero for frequencies above 100 Hz and the amplitude of the excursion is low throughout the remaining frequencies, and errors of 3 millifringes rms (near $\lambda = 0.5\ \mu$) are obtained.

C. Elasticity Considerations in Instrument Design

Experience has shown that the best performing interferometers have their baseplates or support structures isolated from outside sources of vibration. Interferometer components and vibration sources, such as may be associated with the retroreflector scan motion, must be damped. Tilts of the order of hundredths of a wavelength of light or tilts totaling 5×10^{-9} m modify the fringe contrast. These undesirable tilts are introduced by unintentionally deforming rigid bodies.

Elasticity theory in instrument design is used both to isolate the support structure and to achieve small tilts. The subject is well treated by a number of authors in a variety of places. Jones (1961, 1962) described elasticity in optical instrument design and uses these methods (Jones, 1976, and Jones and Richards, 1954) for measurements.

An early example of the use of elasticity theory is given in Fig. 2.10, which shows a photograph of Rowland and his ruling engine for diffraction gratings. The instrument components are mounted on a stiff platform, and the platform is supported by three elastic or pliant members in a kinematic manner. This simple implementation reduces the effects of external distortions. Today the platforms of some laboratory instruments are mounted on specially designed rubber bladders filled with air (Newport Research Corporation, 1979) or more simply on rubber inner tubes to "float" the slab.

FIG. 2.10. Photograph of Henry Rowland at Johns Hopkins with his spindly-legged ruling engine for optical diffraction gratings. [Photograph from a collection by Dr. J. Strong, used with permission.]

This insulates the interferometer components from effects of external distortions. However, care must be taken to include damping; otherwise, the platform may oscillate and introduce unwanted accelerations.

D. Kinematic Mounts

First-order optical system design includes considerations of the mounting of optical elements and the environment of their use. Many authors discuss kinematic mounts for laboratory instruments. The books by Wilson (1952), Whitehead (1954), and Braddick (1963) are among the best. James and Sternberg (1969) give a special section on mounting optical elements.

Optical instruments used for remote sensing applications in general undergo wide temperature excursions either during their data taking sequence or during their travel to and from a remote site. Materials suitable for the excellent optical figures required for interferometer optical elements such as quartz or cer-vit do not have the same coefficient of thermal expansion as their mountings. In addition, the interferometer is composed of many different materials, and in general the system is not in thermodynamic equilibrium. If components (mirrors and mirror mounts) are properly mounted kinematically, thermal gradients that cause the warpage of structures will not cause mirror and beam-splitter surfaces to warp. Such deformations reduce the fringe contrast.

A completely free rigid body has six degrees of freedom: rotation about three orthogonal axes and displacement in three orthogonal directions. In general, if motion is to be controlled, a constraint is required for each degree of freedom.

Figure 2.11 shows a ball in a V-groove. The leg to which the ball is attached cannot move in the left-to-right direction and is constrained to move only into and out of the paper. It is well known that a three-legged table is more stable than a four-legged one. With the four-point contact on the floor, as the floor is twisted, the deformations are transmitted directly to the top of the table or else a table leg lifts up and the table becomes an unstable platform. With a three-point contact on the floor, a twist of the floor results in a tilt of the top of the table and no unwanted deformations are transmitted. A kinematic mount combines the motion constraining properties of a ball in a groove with the three-legged stability.

Figures 2.12 and 2.13 show kinematic mountings. In Fig. 2.12 each of the three legs has a ball mounted on the end that rests in a V-groove, each ball making two contacts in its groove. The grooves all intersect at the center. In Fig. 2.13 each of the three legs with a ball on it contacts a

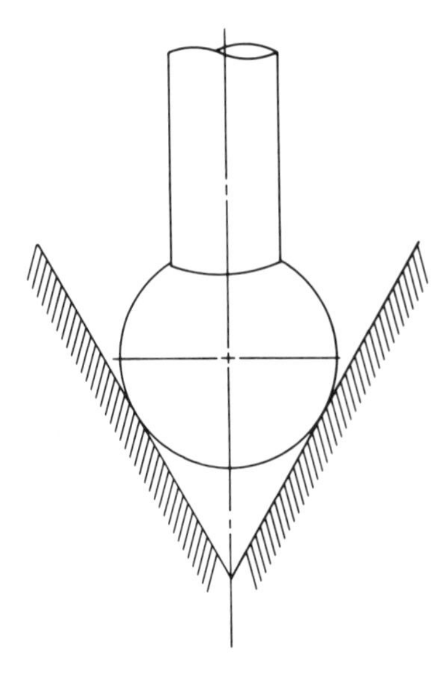

FIG. 2.11. Ball in a Vee groove. With gravity weighing it down, the ball is constrained to move along a line into and out of the page.

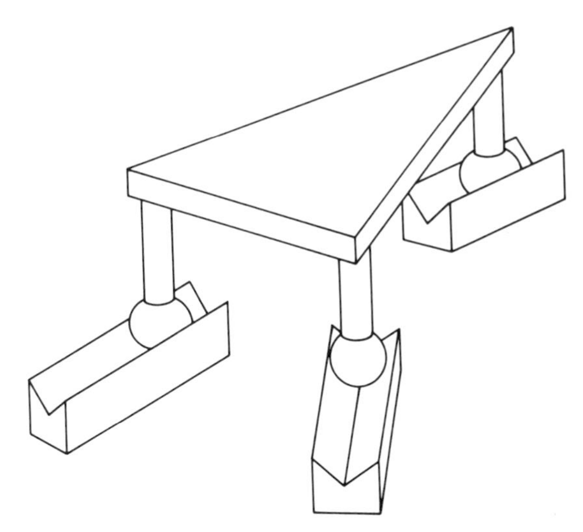

FIG. 2.12. A three-legged platform in one form of a kinematic mount. Balls at the end of the legs rest in grooves, which are aligned pointing to the center of the assembly. This kinematic amount is overconstrained.

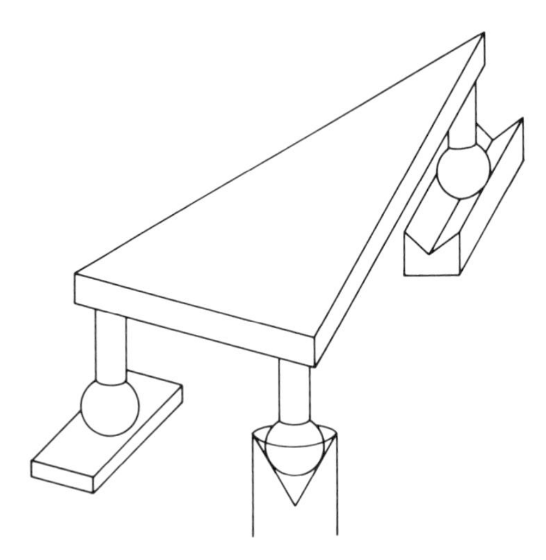

FIG. 2.13. A three-legged platform showing a kinematic mount. Balls at the end of the legs rest on a flat polished plane, a cone and a Vee groove, which is aligned pointing at the cone. This kinematic mount is not overconstrained.

different shaped surface. One rests in a cone or tapered hole and hence constrains the platform to rotate in a circle. Another rests in a V-groove oriented in such a way that a line drawn through the center of the ball and parallel to the groove passes through the center of the ball resting in the cone or tapered hole and hence constrains the platform to move only in rotation. The third ball makes contact at a single point on a smooth polished plate and hence supports the platform to a plane but does not constrain its motion on that plane. Distortions in the bottom plate are not transmitted to the top plate, which may support distortion-free optics. These mounts, when removed and then replaced, return very close to their original position. Since the balls make contact only at points or in a line (circle), thermal conductivity between the baseplate and the three-legged platform is very poor, and the platform is either convectively or radiatively cooled.

The types of mountings shown in Figs. 2.12 and 2.13 are unsuitable for use in an interferometer that must survive the rigors of launch or use in an airplane. One way to hold the three-legged platform down on the table and in contact with its supports without overconstraining the system is to stretch (tension) a spring from a point on the platform midway between the support legs to a similar point on the baseplate.

Mirror supports and beam-splitter mounts should be examined during the selection of the design approach for the instrument. The designer needs to remember that no matter how stiff his design, compliance to allow optical mounts to warp from thermal gradients and vibration and not distort optics must be taken into consideration. Tilts may be more desirable than mirror surface warpages and transmission of those vibrations whose frequencies are outside the range of the servo system.

E. Scan Carriage Mechanization

Many factors enter into a scan carriage mechanization, not the least of which is cost or resources expended for performance attained. A given mechanization may have scan velocity or positional stability far in excess of those required for the signal-to-noise ratio or for the sampled wavelength.

Factors other than cost include the wavelength to be sampled, maximum scan distance, allowable amplitude of spurious spectra, mass of the scanning element, and, most importantly, the environment in which the scanning mechanism must operate. Other important decisions are tradeoffs between constant velocity and step scanning, and the tilt allowable for the moving element.

Perhaps the most important design consideration in selecting a mechanism is the degree to which the moving mirror or other scanning element must be maintained parallel to the wave front of the stationary interferometer leg. In a plane-mirror Michelson configuration, the tilt tolerance is typically one or two arc-seconds, whereas with cat's-eye, the tilt tolerance on the carriage is typically a few arc-minutes. Arc-minute tolerances are readily obtained with rolling or sliding mechanisms having more or less standard machining tolerances, whereas arc-second tolerances require extremely flat carriageways which must remain flat under operational extremes of temperature and pressure. If bearings are used, the tilting caused by ball-diameter variations can be minimized by use of class 3 (± 0.000003 in.) ball bearings as well as the use of extremely parallel ways. One successful example of such a slide mechanism is used by Idealab, Inc. (Pritchard, 1969) for both Michelson and cat's-eye interferometers.

In a single-pass retroreflector (cat's-eye) system similar to that shown in Fig. 2.5, a factor related to the parallelism requirement must be considered. Although a retroreflector system can withstand a substantial degree of tilting, it can only withstand a limited amount of beam shearing, i.e., transverse offsets of the moving retroreflector centerline relative to that of the fixed retroreflector. The allowable offset tolerance is of the order of the square root of the product of the wavelength and the maximum path difference. Thus, for a path difference of 1 cm and wavelength 1 μm, the tolerance is 0.01 cm. Such tolerances are not difficult to achieve but require careful design. Section 2.6 provides a discussion of shear insensitive double pass systems.

Total required scan distance is an important design consideration because relatively tilt-free flexurally supported plane-mirror scanning systems have been implemented for operation in harsh environments. An example is the series of spacecraft interferometers by Hanel *et al.* (1971, 1975, 1977) used in earth and Mars orbiting missions. This system (Fig. 2.14) uses two sets of flexures, the inside of the first set being fixed and the outside set being fastened to three outriggers. The outside part of the second flexure set is also connected to the outriggers and to the inside by a beryllium shaft concentric with and integral to the moving-coil actuator and tachometer feedback coil. Total scan displacement is 0.5 cm. Another example is the aircraft-borne FTS (Walker and Rex, 1979) fabricated by Air Force Geophysics Laboratory (AFGL), which uses Bendix flexure pivots in a parallogram arrangement.

When scan distances are greater than a few centimeters, a rolling or sliding mechanism of some sort must be used because flexure systems become too large and unwieldy. An alternative to high-tilt-accuracy scanning is electronic tilt compensation of the interferometer's fixed mirror. Such systems have been successfully implemented in an AFGL by Walker and Rex (1979) interferometer and by Buijs (1979). The Walker and Rex system derives the error signal for tilt control by dividing the laser reference beam

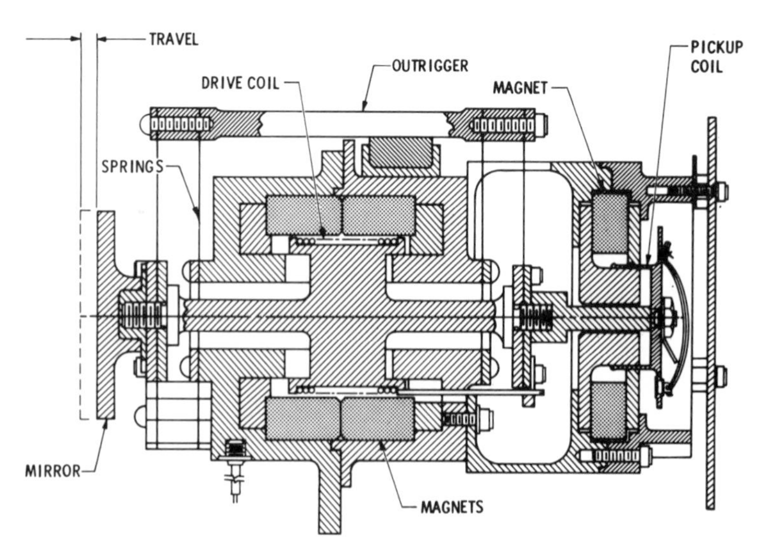

FIG. 2.14. Plane mirror actuator.

by a tetrahedral dividing mirror into three beams, each going to a different photodetector. When tilting is present, the detector receiving a greater signal due to the offset provides an error-correcting signal to a piezoelectric transducer. Buijs's system also uses three photodetector arrays. Initially, a search pattern scans the rotational axes until all three detectors are looking at the same fringe. Then control is turned over to a fringe phase comparison system, which provides two error signals at orthogonal points on the optical path. These drive limited angle dc motors couple to the interferometer's stationary mirror. Schindler (1975) describes details of a phase comparison system.

At least two cat's-eye scan carriage mechanizations use three wheels with ball bearings (Connes, 1966; Beer *et al.*, 1971). Although not usable in zero gravity or high vibration environments, they represent a simple implementation for a laboratory environment, and very good spectral data are obtained with them. The absence of friction in the scanning direction makes up for possible tilts introduced by wheel eccentricities, which are compensated in cat's-eye systems.

Two stepping FTS systems have been constructed using linear ball bearings for use in remotely controlled (balloon and aircraft) environments where the cat's-eyes needed to be transversely constrained to prevent damage (Schindler, 1970a,b). Although initial alignment is more difficult than that with the three-wheel implementations, friction is very low when the bearings are properly aligned and provision is made for lateral compliance between the two bearing shafts on opposite sides of the cat's-eye. Bendix rotational flexures are used to provide this lateral compliance and to make the cat's-eye resistant to rotations.

One of the most successful FTS systems in terms of signal-to-noise and freedom from spurious spectral artifacts is the Kitt Peak National Observatory's 1-m OPD cat's-eye interferometer (Brault, 1974) located at the McMath Solar Telescope. Both cat's-eyes are scanned under servo control actuated by linear induction motors and use a pressurized oil-bearing system to provide a very low degree of friction (Pearson, 1972). The pressurized oil comes through holes in the fixed meanite ways and totally supports the massive (100 kg) cat's-eyes. Scan distance for each cat's-eye in this single-pass system (see Fig. 2.5) is 25 cm. A low-vapor pressure oil (DC 704) is used so that the interferometer may be used in vacuum, unlike an air-bearing system. Another advantage is that oil is much less compliant (being a liquid), and the tendency to rotational oscillations due to the interaction of the different bearing points is less.

A variety of scan actuators are used in FTS systems; each has its advantages and disadvantages. The linear motion moving coil actuator, which combines the function of both motor and linkage to the scan carriage, is widely used. Others include dc torque motors with tachometer feedback, ac motors, and stepper motors. A lead screw or metal band linkage to the carriage with possible gear reduction are design approaches used to connect the motor to the scan carriage. In cat's-eye systems, piezoelectric transducers are often used to move the cat's-eye secondary mirror to compensate for unwanted small-amplitude, high-frequency scan fluctuations. They are used in the step and step–scan systems. The *Inchworm Translator* produced by Burleigh Instruments (1974), a device which uses three piezoelectric actuators to step a carriage along a precision shaft, offers interesting possibilities although it has not yet been used in an operational system FTS.

A moving-coil or audio-speaker actuator is used in many FTS systems. The force is given by the relationship $F = 2\pi R_m N_I B_0 I$, where R_m is the mean coil radius, N_I the number of turns, B_0 the magnetic field across the air gap, and I the coil current. A companion, mechanically coupled moving-coil tachometer provides rate feedback for servo-system damping and stabilization. The tachometer output voltage is $V = 2\pi R_m N_2 B_0 \dot{x}$, where N_2 is the number of turns (much greater than N_I) and $\dot{x}$ is the scan carriage velocity. Moving-coil actuators are used for scan distances ranging from fractions of a millimeter to over 25 cm. Advantages of this design approach are as follows:

(1) No need to convert rotational to translational motion, which circumvents concern for potential system resonances that occur in band-drives, crank-type linkages, and lead-screw drives.

(2) Fairly low power consumption except when used in a high-vibration environment or where the interferometer cannot be leveled.

A recent design trend is to use rare-earth samarium–cobalt magnets instead of the more popular Alnico 5. Because of the much higher coercivity of the rare-earth material, larger air gaps can be used, allowing more turns on the force coil and improving the power efficiency. The optimum high-permeability magnetic material for completing the magnetic circuit is one with a high-saturation flux density, e.g., Permendur 2V or Hyperco (24 kg). The actuator shown in Fig. 2.14 utilizes these materials. Precautions in the design of moving-coil actuators are needed to prevent transformer coupling between the force coil and tachometer coil since coupling causes the local force-tachometer servo loop to become unstable at open loop gain levels commensurate with adequate damping.

An alternative to long-stroke moving-coil drives is a lead-screw drive with an auxiliary short-stroke moving-coil actuator. The lead screw is used alone or in multiactuator systems wherein the position of a short-stroke moving-coil actuator is sensed by a potentiometer or linear-variable differential transformer. In such a system, the short-stroke actuator receives the laser fringe reference error signal and performs small high-frequency error corrections, and the lead screw performs the large-amplitude stroke and error corrections. The advantages of the multiactuator scheme are as follows:

(1) It is lighter for scan distances greater than about 5 cm.

(2) For systems in a nonlevel environment where the interferometer needs to operate in many directions, the inherent self-locking characteristic of the lead screw in remote sensing applications eliminates the need for a caging mechanism.

(3) In nonlevel operation, the heat dissipation from the moving coil may distort the scan carriage or defocus a cat's-eye, and the power consumption will be less at large angles.

Direct current torque motors, ac motors, and stepping motors have been used to actuate the lead screw. A dc torque motor is used in an FTS system provided by Air Force Geophysics Laboratory (Sakai, 1977). The JPL Mark II FTS system uses a synchronous ac motor driven by a variable frequency, digitally synthesized sine-wave generator in order to avoid commutation problems. A lead-screw-driven interferometer (Fig. 2.15) assembled to demonstrate the practicality of a 2-sec, 100-cm path-

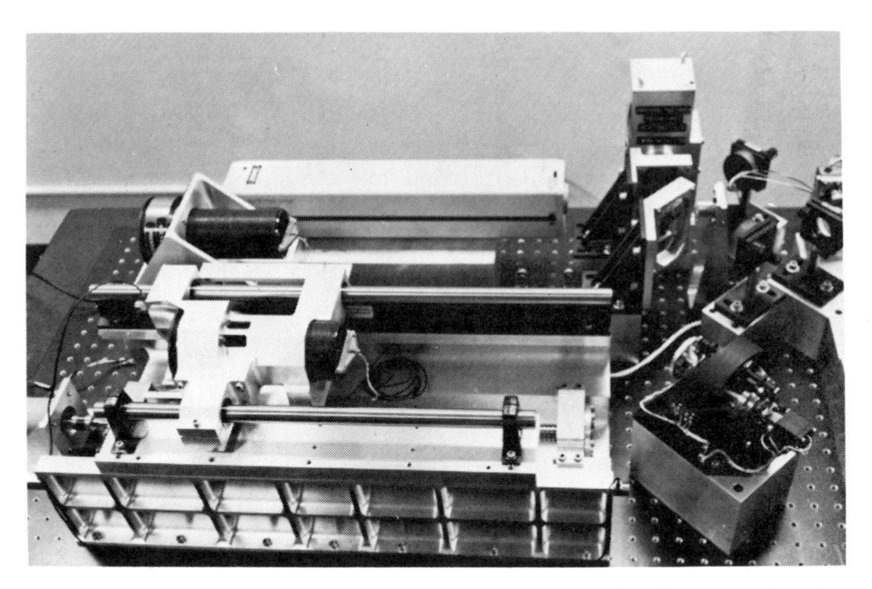

FIG. 2.15. Breadboard of a screw-driven moving cat's-eye interferometer. The plane mirror actuator is shown in the lower right.

difference scan for the ATMOS application, used the plane mirror actuator in Fig. 2.14 as the high-frequency actuator instead of the other cat's-eye, in order to avoid exciting resonances in the cat's-eye structure (this configuration is only possible in a double-passed cat's-eye system). Harries (1977) used a stepper-motor-driven lead screw for a long wavelength FTS system without additional actuators. In this case the accuracy of the lead screw and step positions of the motor were adequate to meet the sampling accuracy requirements at the longer wavelengths.

To avoid periodic variations in the optical path difference given a lead screw, the following precautions are needed:

(1) The lead screw shaft diameter must be adequate to avoid transverse shaft resonances (a problem only at higher rotational rates).

(2) Care must be taken to avoid overconstraining the scan carriage. Either the lead screw nut must have transverse movement capability, or the nut itself must be used as one of the carriage bearing points.

(3) The centerline of the motor must be concentric with that of the lead screw.

Band-driven actuators were used by Buijs (1979) for a cat's-eye and electronically tilt-compensated Michelson systems, respectively. They also have the advantage over the moving-coil actuator of being lighter and not dissipating heat into the carriage or cat's-eye, as well as lower-power consumption when the instrument is not leveled. A band drive is not subject to periodic lead error, and it is easier to prevent overconstraint of the carriage.

Elements of the band-driven actuator include a dc torque motor and tachometer, drive band, and spring restoring force to keep the band taut. Both of the above systems use a "wind up" spring, as in a tape measure, for the restoring force.

Piezoelectric transducers are used as actuators in several step–scan cat's-eye systems (Connes, 1966; Schindler, 1975; Beer *et al.*, 1971). The transducer provides small path-difference corrections to move the secondary mirror at the focus of the cat's-eye parabolic primary. The maximum excursion is limited to about ± 5 μm because of defocusing and voltage limitations. This excursion, combined with a moving-coil actuator having reasonably good frequency response, permits stepping rates greater than 2000 steps/sec. Two types of transducer configurations are employed. The first is to bond together a stack of piezoelectric disks, wiring the conductive faces of alternate layers in parallel to increase the voltage sensitivity (the poling direction for each consecutive disk must be reversed). Displacement of each disk per unit voltage change equals the d_{33} coefficient (5.93×10^{-4} μm/V for PZT-5H). Typically, 20 disks are used. The other configuration is a tubular transducer having the conductive coatings on the inner and outer walls and poled in the radial direction. Here the displacement per unit voltage is $d_{31} L/t$, where t is the wall thickness, L the transducer length, and d_{31} the transverse displacement coefficient (-2.74×10^{-4} μm/V for PZT-5H). Typically, $L/t \simeq 10$. The high-resonant frequencies of these transducers in the sizes required for the relatively small secondary mirror make them essentially the only practical way to obtain stepping rates and internal modulation at frequencies over 100 Hz with adequate positional stability. A piezoelectric transducer can also be used in a self-resonant (oscillator) mode to provide high-frequency internal modulation (typically 500 Hz) to generate the servo system position error signal. The resulting reference fringe signal is demodulated and, along with the low-frequency signal, which is in spatial quadrature with the demodulated signal, operates a bidirectional fringe counter.

The high-frequency performance of piezoelectric transducers may not be required in constant velocity scan systems, where a moving-coil actuator driving a plane mirror provides adequate frequency response. In a cat's-eye system, the moving-coil actuator may excite resonances internal to the cat's-eye. In this case, the piezoelectric transducer allows the system position loop gain to be greatly increased, thereby reducing velocity fluctuations.

Design precautions for piezoelectric transducers include the following:

(1) The basic resonant frequency of the tube or stack should be at least ten times greater than the required servo bandwidth because of the mechanical Qs of these transducers.

(2) The resonant frequency and Q of the secondary mirror–transducer support must be taken into account because of the reaction force from the transducer mirror. The support itself must have high stiffness in the axial direction. This includes focusing the adjustments on the support, which must be designed to be extremely rigid and have low Q.

Sternberg and James (1964) show a Michelson interferometer spectrometer which does not require extreme precision in the manufacture of its moving components. First-order optical system design requires an understanding of the system's mechanical limitations for interferometer scanning and stability of alignment.

2.5. Optimize Throughput

A. Background

First-order optical design includes system analysis to insure that as much of the energy from the source as needed to give the required signal-to-noise ratio fall on the detector. In Section 2.4, we saw how wave-front tilts, beam-splitter materials, and optical configurations contribute to degrading of the modulation efficiency of an interferometer.

In this section the ray trace equations are used to show that the area–solid angle product must be preserved throughout the interferometer. Also a technique used to find the optimum position of pupils within a high-resolution Fourier transform spectrometer and a wavelength-dependent transmittance (channel spectra) are described.

B. Throughput

Unlike a grating or prism spectrometer, which require slits to observe continuum sources, the Fourier transform spectrometer is configured to receive radiation from area sources. A measure of this capability is called the *luminosity, étendu,* or *throughput advantage* (Jacquinot and Dufour, 1948, and Jacquinot, 1954).

Figure 2.16 shows an object of height $\bar{y}$, a pupil of height y, and the marginal and chief rays of the system. The marginal ray is the ray traced from the center of the object through the rim or edge of the pupil. The chief ray is the ray traced from the extremity or edge of the object and passing through the center of the pupil. The object and pupil planes are separated by d. The chief ray makes angle $\bar{u}$ with the center of the pupil, and the marginal ray makes angle u with the center of the object. At a distance t through a medium of index n' an arbitrary plane is positioned.

For the marginal ray, after refraction or reflection at a surface of power ϕ into a medium indicated by a prime, we have, from simple geometric considerations,

$$n'y' = nu - y\phi. \tag{2.28}$$

The power ϕ is given by

$$\phi = (n - n')/R,$$

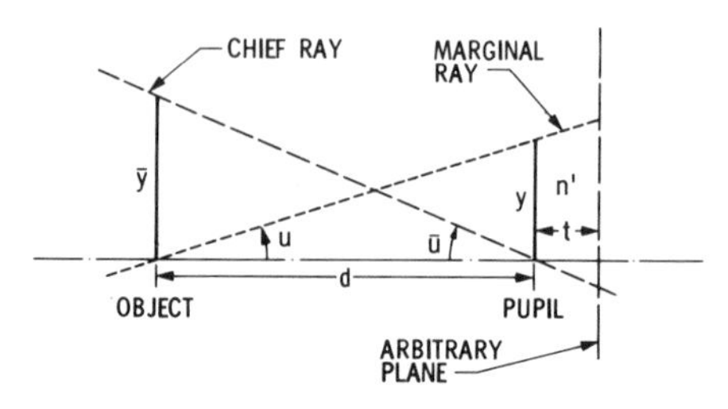

FIG. 2.16. Object and pupil relationships defining chief ray height and angle and marginal ray height and angle.

where R is the radius of curvature of the surface. For reflection in air $n = -n'$ and $\phi = -2/R$. The chief ray after refraction or reflection at a surface of power ϕ into a medium indicated by a prime is

$$n'\bar{u}' = n\bar{u} - \bar{y}\phi. \tag{2.29}$$

Similarly, a pair of equations, one for the marginal ray height, the other for the chief ray height, are written for an arbitrary plane separated by distance t in medium of index n'. For the marginal ray,

$$y' = y + (t/N')(N'u'), \tag{2.30}$$

and for the chief ray,

$$\bar{y}' = \bar{y} + (t/N')(N'\bar{u}'). \tag{2.31}$$

We combine Eqs. (2.28)–(2.31) at a plane across any surface within the optical system to give

$$(n'\bar{u}' - n\bar{u})/(-\bar{y}) = \phi = (n'u' - nu)/(-y) \tag{2.32}$$

or

$$n(\bar{u}y - u\bar{y}) = n'(\bar{u}'y - u'\bar{y}) = \mathrm{H}. \tag{2.33}$$

Equation (2.33) shows that there is a property of an optical system that is true at any plane within the system. It is called the Lagrange invariant, and the symbol H is often used as a shorthand representation. Equation (2.33) is rewritten with the object plane expressed on the left-hand side ($y = 0$) and the pupil plane on the right-hand side ($\bar{y} = 0$) to give

$$n\bar{y}u = ny\bar{u}, \tag{2.34}$$

where we have set $n = n'$ since both the object and the pupil are immersed in a medium of the same refractive index.

The primary interest of radiometry is the measurement of radiant power falling on a detector from a source of a certain radiant emittance (W/m^2). To identify an area on the object shown in Fig. 2.16, the chief ray height at the object $\bar{y}$ is rotated around, out of the plane in a circle. This area, A_0, is $A_0 = \pi\bar{y}^2$. Similarly, the marginal ray angle u is a solid angle, given by Ω_0 as $\Omega_0 = \pi u^2$, where we assume angle u is small.

The solid angle Ω is generally expressed in units of steradians. A steradian is the solid angle subtended by $1/(4\pi)$ of the surface area of a sphere. The size of a solid angle in steradians is given by calculating the area of the portion of the surface of a sphere included within the solid angle and dividing that area by the square of the radius of the sphere.

The rays drawn in Fig. 2.16 show that for small u, $u = y/d$, and the area of the pupil plane A is $A = \pi y^2$. The solid angle of the pupil as seen from the object is then $\Omega_0 = A/d = \pi u^2$. If we assume the medium to be air, and let the index of refraction n be 1, then multiplying both sides of Eq. (2.34) by π and squaring them, we find $(\pi\bar{y}u)^2 = (\pi\bar{y}u)^2$. This is rewritten to give

$$A_0\Omega_0 = A_p\Omega_p, \tag{2.35}$$

which is a statement of the well-known conservation of area–solid angle product through optical systems.

The image-forming properties of the Fourier spectrometer are somewhat unique. The light is divided into two arms at the beam splitter and then recombined after traveling the optical path lengths, sometimes different by over a meter. Depending on the application, either the image plane or the pupil plane is relayed onto the detector. An array detector (Huppi, 1979) is used to increase the optical system effective throughput without degrading spectral resolution. If the image plane is relayed onto the detector, an opportunity exists for studying the spatial distribution of spectral information.

C. Ray Trace of Fourier Spectrometers

A ray trace of the marginal and chief rays through any optical system is required to assure that a loss in signal at the output plane be not attributable to vignetting. The long optical path lengths required for high-resolution Fourier transform spectrometers require careful analysis, and the insertion of field-lens type power surfaces are necessary to control the marginal ray to avoid vignetting and overfilling the beam splitter.

An understanding of the locations of image planes and pupils within the optical system of a Fourier transform spectrometer is obtained using the first-order optical design tool, the $y\bar{y}$ diagram (Delano, 1963; Lopez-Lopez, 1973). Figure 2.17 shows a perspective view of a marginal ray height y, chief ray height $\bar{y}$, and skew ray for that region within an optical system between the object plane and the pupil plane. The $y\bar{y}$ diagram is a plot of the marginal ray height as a function of the chief ray height. Reference to Figs. 2.16 and 2.17 shows that when $\bar{y} = 0$, the value of y is the ray height at the pupil plane, and when $y = 0$, the value of $\bar{y}$ is the ray height at one of the image planes, or the object plane.

The $y\bar{y}$ diagram approach to first-order design of optical systems is detailed by Delano (1963), Lopez-Lopez (1973), and Shack (1975). It is not the intent here to discuss quantitative details of the utility of such an approach to first-order design, but rather to use the approach in a semiqualitative manner to identify locations of pupils and image planes within the optical system. For example, three optical systems of general interest are given in Fig. 2.18 to show the usefulness of these diagrams for

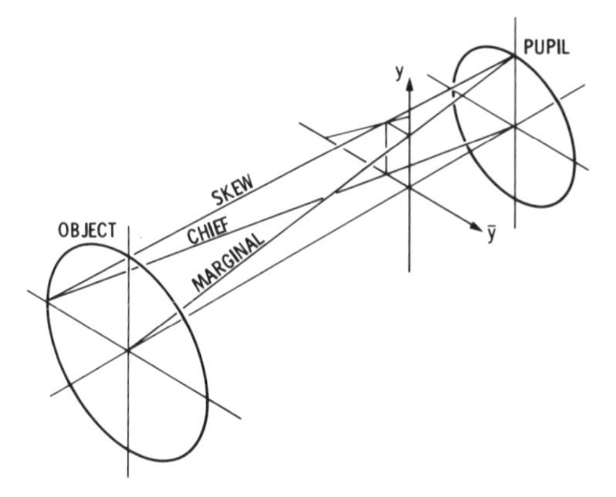

FIG. 2.17. Perspective view of object and pupil relationships to define the chief ray height $\bar{y}$ and angle $\bar{u}$ and the marginal ray height y and u at an arbitrary plane, shown here as $y\bar{y}$ located between the object and pupil.

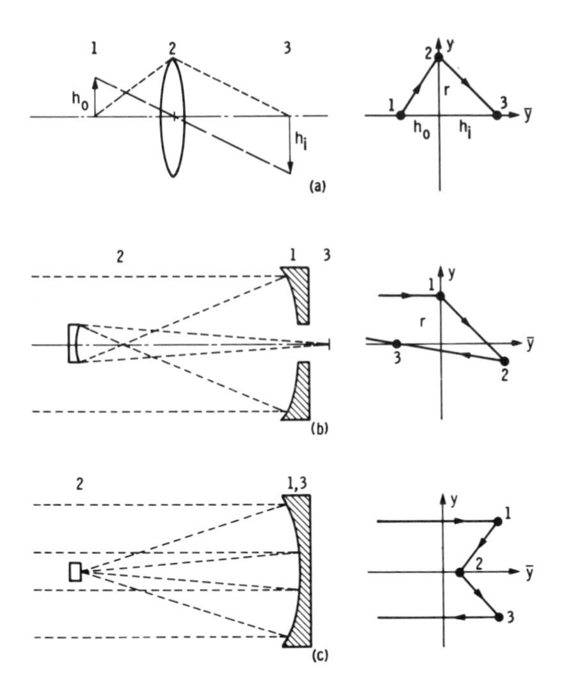

FIG. 2.18. (a) An image relay, (b) a Gregorian telescope, and (c) a focal cat's-eye shown in layout and in the $y\bar{y}$ diagram.

qualitative discussion. Figure 2.18a shows an image relay which relays an object height h_o ($\bar{y} = h_o, y = 0$) through a lens of radius r ($\bar{y} = 0, y = r$) to an image of height h_i ($\bar{y} = h_i$, $y = 0$). Figure 2.18b shows a Gregorian telescope imaging system whose pupil is of radius r. The ray from a distant source strikes the primary mirror (here used as the pupil at 1). The focus in front of the secondary is indicated on the left-hand side of the $y\bar{y}$ plot by the line passing from 1 to 2, crossing the $y = 0$ line. The power element at 2 bends the rays back to form an image at point 3 on the $y = 0$ line. Figure 2.18c shows the afocal cat's-eye retroreflector. Note that there is an image plane ($y = 0$) on the cat's-eye secondary. The $y\bar{y}$ diagram enables a designer performing first-order design to locate all pupils and image planes and calculate the pupil areas for a required brightness at the final image plane.

Foreoptics for a Fourier spectrometer serve two functions. One is to control both the marginal and chief rays into the interferometer to minimize the vignetting, which may be a problem particularly for a cat's-eye, long-path length configuration. The other is to provide an image plane for a stop to define that portion of object space which is entering the interferometer.

Figure 2.19 shows the optical path design approach used for the 1-m OPD interferometer at Kitt Peak Natioal Observatory. The numbered elements refer to Fig. 2.20. This relatively complicated path is shown, in relative units of y and $\bar{y}$, in Fig. 2.20. A skew ray comes in from the left in Fig. 2.20 (the $y\bar{y}$ diagram), strikes the pupil of the foreoptics at point 1, and crosses the $y = 0$ line, where an image plane stop is used to define that position of object space admitted to the interferometer. The radiation travels onto a collimator at 2, which collimates the marginal ray. The beam splitter is shown at point 3, where an image of the pupil is located. Points 4, 5 and 6 describe the afocal cat's-eye. The power of the cat's-eye

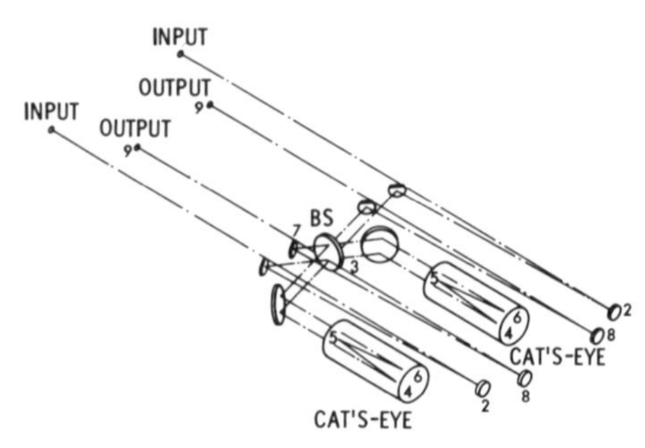

FIG. 2.19. Optical path layout of the 1-m OPD interferometer at the McMath solar telescope of the Kitt Peak National Observatory, showing the folded configuration required to fit it into a vacuum tank and the dual input/output capability.

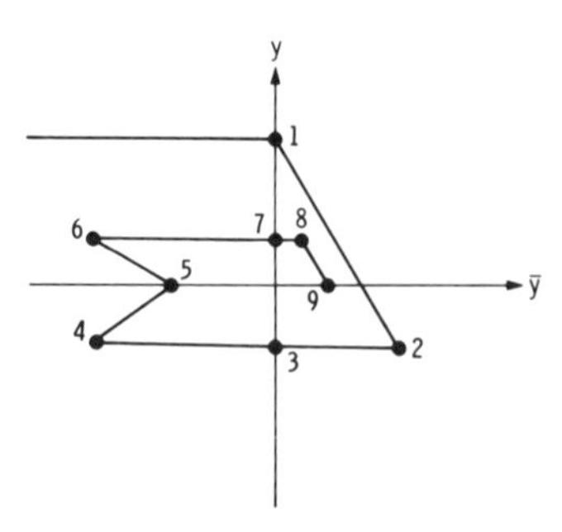

FIG. 2.20. The optical path given in Fig. 2.19 shown in terms of relative units of y and $\bar{y}$.

secondary is such that the image of the pupil at plane 3 (the beam splitter) is relayed onto the "beam recombiner" at plane 7. An optical element at point 8 relays an image onto the detector at point 9. In some instruments a field lens is positioned at point 9, to relay the pupil onto the detector. The latter is done if there is undesirable spatial structure in the object.

The foreoptics are designed to relay an image of the telescope pupil onto the beam splitter. With the cat's-eyes at their equal path, or neutral position, the cat's-eye optics relay an image of the beam splitter onto the beam recombiner. This arrangement leads to a beam splitter of minimum size for maximum throughput.

Wave-front distortion in cat's-eye retroreflectors is related to the curvature (concave or convex) selected for the cat's-eye retroreflector (Ezhevskaya and Sinitsa, 1978 and Beer and Marjaniemi, 1966).

D. CHANNEL SPECTRA

The channel spectrum is an interference phenomenon produced by multiple reflections of a wave front of light within the plane parallel sides of a transparent medium such as crystal or glass. Strong (1958) discusses the orgin of channel spectra, and P. R. Griffiths (1975) discusses their effects on scientific data recorded with FTS.

The multiple reflections within the interferometer cause the interferogram to be a superposition of several interferograms. Generally, two are obvious. The larger is the desired one, and a smaller one, offset from the larger by the optical path thickness, is the unwanted interferogram. The measured interferogram is not a simple linear superposition of these two offset inteferograms, but rather optical partial coherence effects cause the two interferograms to be only partially correlated.

Spectra produced from these interferograms show modulation of the continuum or the instrument transmittance as a function of wavelength. These modulations are particularly annoying for the case in which the line profile has approximately the same spacing as the channels. This can be the case for good signal-to-noise ratio high-resolution spectroscopy, where channel spectra completely mask the wanted scientific information.

An instrument design approach to remove the effects of channel spectra is to design the optical paths so that plane-parallel elements are replaced with wedged elements. The entrance window, beam splitter, compensators, and all transmission optics need to be wedged. To avoid problems with transverse chromatic aberration, the wedge angle should be as small as possible. The angle between the primary transmitted ray and the multiply reflected ray, in the same general direction, is $S = 2n\phi$, where ϕ is the wedge angle of the substrate of index n. Both compensator and beam splitter need to be wedged and an alignment (rotation or translation) capability provided to tune the total optical phase distortion caused by optical dispersion to minimize the phase corrections discussed in Eq. (2.12). Section 2.6,D discusses a particular implementation for removing channel spectra and providing phase-dispersion tuning for an interferometer.

2.6. Interferometer Configurations

A class of configurations whose fringe contrast or modulation efficiency is insensitive to interferometer component tilt is discussed in detail. An in-depth analysis of one interferometer configuration is given to show a methodology for first-order design of a double-pass tilt-compensated, phase-tunable Fourier transform spectrometer for spectroscopy. The experiment environment and an understanding of the sensitivities of the interferometer suggest an implementation design philosophy and an approach to packaging the instrument.

A. Tilts

A major contributor to a loss in the equal-path fringe contrast, or modulation efficiency, is a wave-front tilt introduced by component misalignments which occur in one arm of the interferometer and are not canceled by a component tilt in the other arm (see Section 2.3). Optical designs that achieve tilt compensation for interferometer components, as well as designs for achieving tilt compensation for the entire system and a technique for analyzing complicated interferometer configurations for tilt sensitivity, are presented here.

First-order tilt-insensitive optical components are the cube-corner (Peck, 1957), and the cat's-eye. The mechanical translation of the tilt-insensitive retroreflector results in a minimum tilt on the returning wave-front. The cat's-eye elements were analyzed in detail by Ezhevskaya and Sinista (1978) and earlier by Beer and Marjaniemi (1966).

The interferometer spectrometer developed by Buijs (1979) uses a real-time adaptive optics technique to remove unwanted wave-front tilts. The interferometer layout is that of the classical Michelson with flat mirrors in both arms. One mirror is translated over a long path, and the other mirror is tilt adjusted during the scan by a servo system to compensate for wave-front tilt errors introduced during the several-centimeter long path travel of the scanning mirror.

Here self-compensating optical configurations are defined and a technique described to analyze them. In general, a tilt of an optical component results in a change in optical path difference (OPD) as well as a wave-front tilt. The change in OPD is self-adjusted by the electronic servo system. The concern here is for wave-front tilts and their compensation. A method to analyze tilts in optical systems, with specific reference to interferometers, was given by Montgomery (1967), who developed tools useful for quantitative analysis of the effect of tilts. Based on the direction cosines of the ray (wave-front normal), Montgomery provides a matrix analysis technique and uses it to analyze the tilt sensitivity of the triangular interferometer configuration. Soares (1978) provides an analysis of the tilt sensitivity and alignment of the triangle path cyclic interferometer. The articles by Montgomery (1967) and by Soares (1978) are not directly concerned with the alignment sensitivities of Fourier interference spectrometers. However, their material provides important tools for the quantitative analysis of tilts in interferometer systems.

Configurations for which the equal-path fringe contrast is not degraded by the tilt of some of the optical elements in the interferometer are compared by Sakai (1977) and are discussed in more detail by their inventors: Scott (1958), Pritchard *et al.* (1967), Buijs (1971), and Rundle (1965). Genzel and Kuhl (1978) describe a tilt-compensated Michelson interferometer for Fourier transform spectroscopy which is built exclusively with spherical and plane mirrors.

A configuration that is compensated for tilts is given by Schindler (1970a). This particular spectrometer was successfully used to record spectra while operating in the vibration environment of aircraft (Toth *et al.,* 1973), and in balloon experiments (Farmer *et al.,* 1976).

B. Analysis Technique

A light beam that strikes a mirror tilted through angle θ leaves the mirror tilted over by angle 2θ relative to the direction the beam would have taken had the tilt misalignment not occurred. However, a beam misaligned by angle θ relative to the direction of its aligned travel will strike a flat mirror and reflect, still misaligned by angle θ, in the sense shown in Fig. 2.21. Figure 2.21a shows a beam that is misaligned in the direction shown by an amount θ. The entering beam is rotated so that it appears to be coming from a point closer to the normal to the mirror surface. The beam reflects, but is rotated so that it appears to be directed toward a point closer to the normal to the mirror surface. Figure 2.21b shows what happens when the entering beam is rotated so that it appears to be coming from a point farther from the normal.

Figure 2.22 shows the cat's-eye afocal system, which has the property of reversing the direction of wave-front propagation by 180°. A pencil of light tilted by angle θ enters the cat's-eye, and the light returns out of the cat's-eye tilted back toward the direction of the original tilt by angle θ in the sense shown in Fig. 2.22.

Figure 2.23 shows the effect of having one flat mirror covering both the entrance and exit of a cat's-eye. Here the flat mirror in front of the cat's-eye is tilted through angle θ, and the reflected beam is tilted through 2θ on entering the cat's-eye. The return beam reflects from the tilted flat mirror and leaves the system parallel to the entering beam. Therefore, the configuration of a flat mirror which covers both the input and output of a cat's-eye is compensated to first order for tilts.

A tilt-compensated interferometer used for Fourier transform spectroscopy contains optical configurations similar to those examined in Figs. 2.21–2.23. Figure 2.24 shows a double-pass interferometer using cat's-eye retroreflectors. Light enters from the left, passes through the glass–air interface of the prism, and strikes the beam-splitter surface. The part of

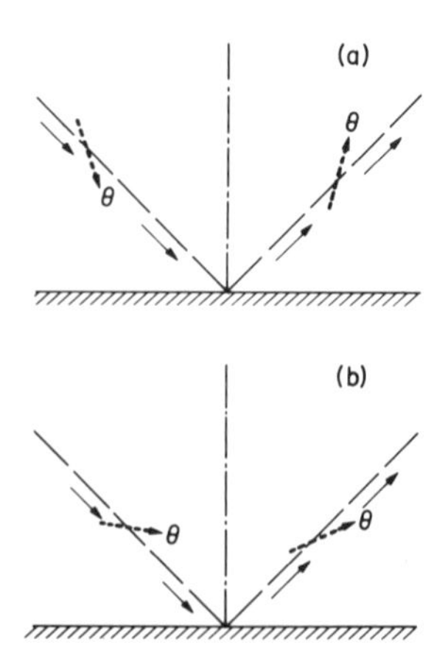

FIG. 2.21. A ray misaligned by θ strikes a flat mirror.

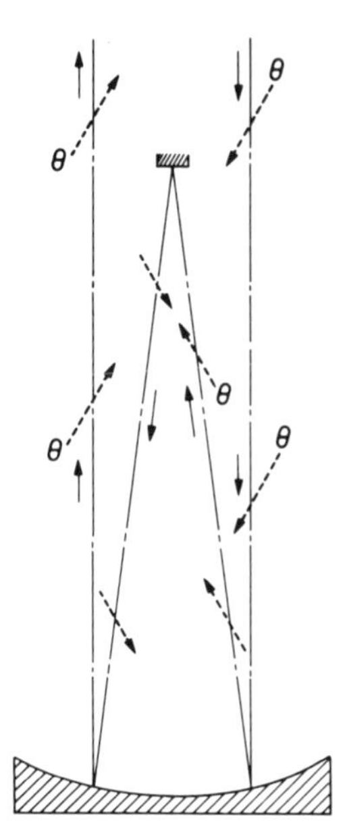

FIG. 2.22. A ray misaligned by θ strikes the tilt-compensated cat's-eye. The misaligned ray leaves the cat's-eye exactly opposite from its entrance.

the light reflected passes through cat's-eye 1, reflects from a fully reflecting coating on the beam splitter–prism block M_1, passes back through the cat's-eye a second time, in a direction opposite to its original passage, and strikes the beam splitter. The part of the light transmitted by the beam splitter passes through cat's-eye 2, reflects from a fully reflecting coating

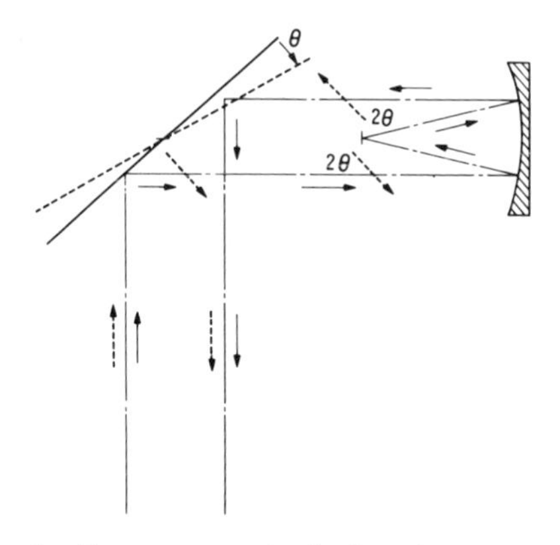

FIG. 2.23. The first order tilt-compensated pair: flat mirror covers both entrance and exit beams of the cat's-eye.

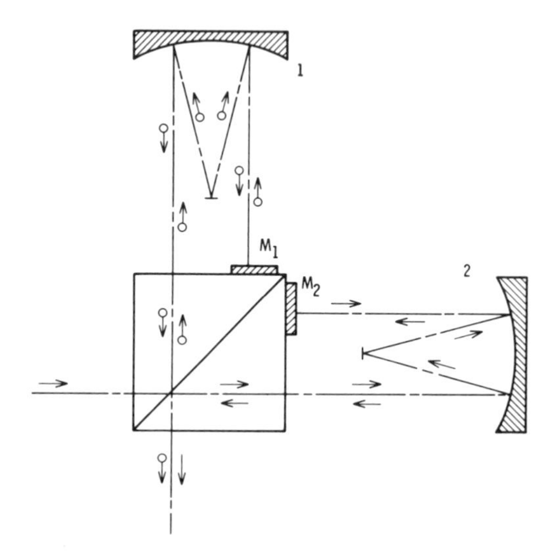

FIG. 2.24. Double-pass interferometer with cube beam splitter.

on the beam splitter–prism block M_2, passes through the cat's-eye a second time, in a direction opposite to its original passage, and strikes the beam splitter. The returning beams are recombined at the beam splitter, and part of the radiation goes back out towards the source; the other goes to a detector. The extended surface of mirror M_1 intersects the extended surface of M_2 by 90°.

In the analysis given below, this beam splitter is systematically rotated about orthogonal axes to investigate this configuration for tilt sensitivity. If we find the interferometer to be tilt-insensitive for component tilts about three orthogonal axes, then we are assured it is tilt compensated for rotation about any axis. Recall that in our definition, *tilt insensitive* permits translation along the optical path difference scan since these are taken up or accounted for by the scan servo system, which moves the retroreflectors to compensate for optical path scanning.

Fig. 2.25 shows the same interferometer as given in Fig. 2.24, but with the beam-splitter cube rotated counterclockwise through angle $\theta/2$ about an axis perpendicular to the paper, passing through the center of the beam splitter. The light that is reflected up into cat's-eye 1 is tilted over by an angle θ shown by the dotted arrow with a circle on its tail. The light exiting from cat's-eye 1 is tilted over by an angle θ shown by the dotted arrow with a circle on its tail. This beam is reflected from the mirrored surface on the cube, which is tilted at angle $\theta/2$. This reflected beam is tilted back so that it is parallel to the path of the beam with the unrotated beam splitter. This

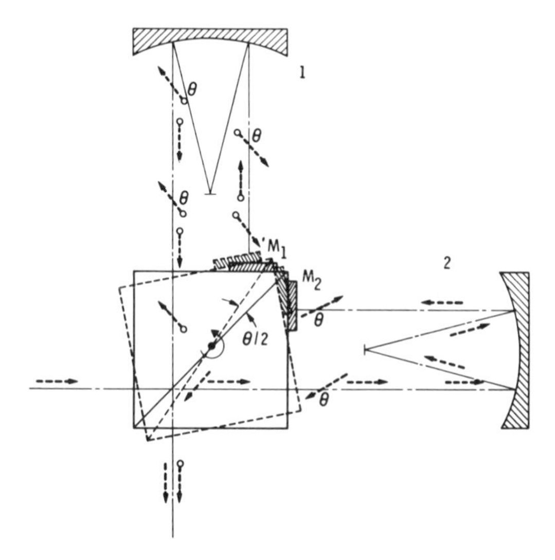

FIG. 2.25. Double-pass interferometer with cube beam splitter rotated through angle $\theta/2$ for analysis of tilt sensitivity.

light returns through the cat's-eye and passes through the beam splitter and out of the system. Returning to the other path, we see that part of the entering light transmits the beam splitter and enters cat's-eye 2. This light is reflected from the flat mirror on the face of the cube, which is rotated through $\theta/2$, and reenters cat's-eye 2, tilted over at an angle of θ. This beam reflects from the beam splitter, is tilted back by angle θ, and leaves the interferometer parallel to the beam from the other arm. Beams from each arm are parallel at the exit of the interferometer, independent, to first order, of the beam-splitter rotation angle.

A rotation of the beam-splitter cube does result in a change in the optical path difference in one interferometer arm relative to that in the other. However, no measurable wave-front tilt error is introduced for beam-splitter rotations of the order of arc-minutes. The importance of second-order effects depends on the F number of the cat's-eye retroreflectors and whether or not the cat's-eye primaries are parabolized. This arises since different portions of the parabola are illuminated as the prism block is rotated and since the parabola has an axis.

This analysis is continued for the configuration shown in Fig. 2.24 to determine tilt sensitivity for rotation about an axis passing once again through a point at the beam-splitter center, but lying in the plane of the paper and normal to the mirror M_2. For purposes of this discussion, the top of the cube is rotated out of the paper, and therefore the lower part has moved into the paper by an angle of $\theta/2$. Light enters from the left, and the portion reflected from the beam splitter enters cat's-eye 1 tilted up, out of the paper by angle θ. It passes into the cat's-eye and leaves it tilted down by angle θ. The light reflects from the tilted mirror M_1 (tilted by $\theta/2$), returns through the cat's-eye, and passes through the beam-splitter block to leave the interferometer with no tilts relative to its path through the unrotated cube. That portion of the light which transmits the beam splitter passes through cat's-eye 2, reflects from mirror M_2 (which is not tilted, only the surface has rotated), and returns through the cat's-eye, back to the beam splitter. At this point the ray is tilted down into the paper by the reflection from the tilted cube.

The wave front that passes through cat's-eye 2 leaves the interferometer tilted relative to that wave front that passes through cat's-eye 1, and this system is *not* tilt compensated for rotations about the axis shown. By symmetry it is clear that rotation about an axis perpendicular to the face of mirror M_1 will also introduce tilt errors.

Return the configuration to that shown in Fig. 2.24 and rotate the prism–beam-splitter assembly about an axis that passes through the beam-splitter center point and is perpendicular to its surface. Rotation about this axis will introduce no tilt errors. Consequently, the interferometer optical configuration shown in Fig. 2.24 is tilt insensitive for rotations about two axes only.

C. Tilt Compensation/Insensitivity

In this part the two interferometer configurations shown in Figs. 2.26 and 2.27, which appear similar, are examined in detail for tilt sensitivity. The optical layouts are almost the same, the distinction being that in one case the light enters the interferometer through a hole in the double-pass retroreflector mirror, and in the other the detector receives light that has passed through the hole in the double-pass retroreflector mirror.

In Fig. 2.26 dotted arrows indicate the ray tilts for tilts of the beam splitter. The light enters from the left and strikes the beam splitter, which has been misaligned by angle $\theta/2$ in the direction shown. The light transmits through the beam splitter into cat's-eye retroreflector 1, returns to strike the tilted beam splitter, and reflects down to the double-pass flat. The beam strikes the flat tilted over by an amount θ. It is reflected from the double-pass flat, travels back to the beam splitter, and reflects again, acquiring another θ tilt for a total of 2θ, and passes into the cat's-eye. Returning from the cat's-eye, the beam reflects from the beam splitter again and leaves the system tilted over by θ. Tracing the other path of the interferometer: the entering beam reflects off the beam splitter and enters

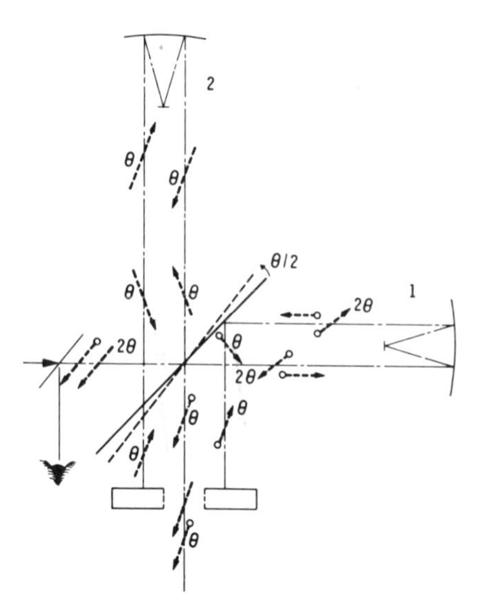

FIG. 2.26. Double-pass interferometer to show tilt sensitivities at both outputs.

cat's-eye 2 tilted over by θ. The light leaves the cat's-eye, reflects from the double-pass flat mirror, and returns to the cat's-eye, passes through the beam splitter, and leaves the interferometer through a hole in the double-pass flat. As shown in Fig. 2.26, if the beam splitter is rotated by angle $\theta/2$, the wave fronts from each are tilted by θ.

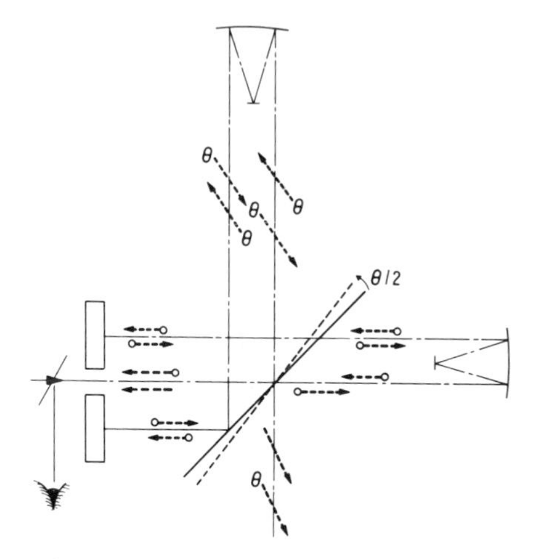

FIG. 2.27. Slight modification to the double-pass interferometer given in Fig. 2.26 showing tilt sensitivites at both outputs.

If the observer sets up a telescope at the output, looking back through the system with the telescope focused at infinity, the two superposed pinpoints from the plane waves in each arm are apparent. As the beam splitter is tilted through angle $\theta/2$, the two superposed pinpoints move through the same angle θ in the field. This configuration is tilt compensated. Vignetting of the signal occurs when the beam splitter is tilted so much that the two superposed pinpoints move out of the field of view.

This interferometer has two outputs. If a beam-splitter mirror, shown in Fig. 2.26, is inserted in the optical path, that light which travels back out of the interferometer in the direction of the incoming beam can be observed. The fringe pattern seen in this direction is tilt sensitive, and for a $\theta/2$ tilt of the beam splitter the wave fronts at this exit tilt by 2θ. As the beam splitter is rotated, the superposed pinpoints of light as seen with a telescope at the output appear to move out of the field of view twice as fast at one output as at the other.

Another configuration is shown in Fig. 2.27, which differs from that shown in Fig. 2.26 by the position of the double-pass retromirror. In Fig. 2.27 the ray enters from the left and strikes the beam splitter, which has been misaligned by angle $\theta/2$ in the direction shown. The ray that passes through the beam splitter passes through the cat's-eye to the double-pass retroreflector, back through the cat's-eye, and to the beam splitter without a tilt. A tilt of θ in the direction shown at the output is introduced at the tilted beam splitter when the ray reflects out of the system. Tracing the other path of the interferometer: the incoming ray is reflected by the beam splitter and is tilted over by angle θ as shown as it enters and leaves the cat's-eye, reflects again from the beam splitter to the double-pass retroreflector, and returns through the system. This ray, like that which passes through the other arm, leaves the interferometer tilted at angle θ. Looking back into the interferometer from the output shown, the configuration is tilt insensitive.

If a beam-splitter mirror, shown in Fig. 2.27, is inserted before the light reaches the interferometer, that light which travels back out of the interferometer in the direction of the incoming beam can be observed. From the figure, one sees that *regardless of the tilt* of the beam splitter, the ray will return directly upon itself. If the observer sets up a telescope at this output and looks back through the system, the superposed points of light will remain stationary during rotation of the beam splitter. If a telescope is set up to observe the pinpoints of light, they appear stationary in the field for beam-splitter rotation.

Both of these configurations are first order tilt compensated for rotations of the beam splitter about an axis normal to the plane of the figure. By inspection one sees that the interferometer is compensated for rota-

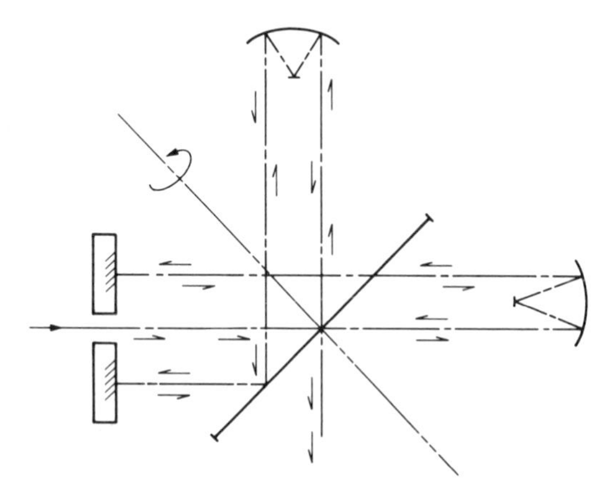

FIG. 2.28. Beam splitter rotated about axis perpendicular to its plane.

tions about an axis normal to the plane of the beam splitter as shown in Fig. 2.28. The remaining axis to examine is that lying in the plane of the figure and passing through the center of the beam splitter as shown in Fig. 2.29. Assume that the rotation is such that the beam-splitter top rotates back.

The light that enters the interferometer and reflects from the beam splitter travels into cat's-eye 1 tilted up out of the paper. The ray leaves the cat's-eye tilted down, and strikes the rotated beam splitter to travel parallel to its original path until it strikes the beam splitter on the return

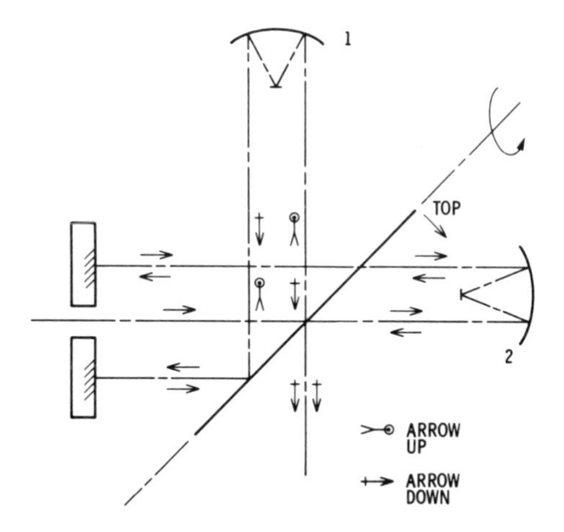

FIG. 2.29. Beam splitter rotated about axis formed by the intersection of the plane of the paper and the beam splitter.

trip, at which point the ray is tilted up on reflection from this surface. The ray returns through the cat's-eye and leaves the interferometer tilted down. That portion of the incoming radiation that transmits through the beam splitter is unaffected by the rotated beam splitter and travels through the system untilted until it returns to reflect off the beam splitter. At this time it is tilted down by the same amount as the light from the other arm. The two wave fronts leaving the interferometer travel in the same direction, and for rotations of the beam splitter the interferometer is tilt compensated.

The configurations in Figs. 2.28 and 2.29 are compensated to first order for small tilts of all elements about all axes.

D. Synthesis of a Particular Configuration

In this part, we use the material given earlier to synthesize a particular interferometer optical configuration whose design was intended to satisfy the needs of a space shuttle high-resolution spectrometer (Abel *et al.*, 1979). The intent of the interferometer spectrometer is to record high-resolution absorption spectra of the earth's stratosphere by observing the sun with an interferometer during the "sunrise" and "sunset" from orbit. A field-widened (Ring and Schofield, 1972) interferometer is not required because the sun is a bright source.

The desired vertical height resolution in the stratosphere and the orbital elements, in particular the height of the spacecraft above the earth, which determines the speed of the spacecraft, dictate that one interferogram be recorded every second. The measurement program requires data of sufficient quality to calculate the abundance of molecular species in the stratosphere (Farmer, 1978). To do this accurately, the absorption features in the spectrum must be nearly resolved.

Figure 2.2 in Section 2.2 shows that at least 4 m of optical path difference are required to resolve CO_2 and CH_4. Jansson *et al.* (1970) show that accurate equivalent width measurements of unblended absorption features are made at spectral resolutions considerably less than the resolution required to resolve the feature. Moving the optical path difference by 4 m/sec is technologically demanding for a spacecraft instrument. These factors, along with the realization that the optical path can be double passed and thus the mechanical stroke shortened, led to the decision to use a 1-m total optical path difference with the equal-path fringe in the scan center, to give 50 cm OPD on either side. A reason for locating the equal-path fringe in the center of the scan was consideration of the mechanical difficulty of turning the scan carriage around quickly and the loss of interferogram data.

The interferometer alignment must survive the harsh vibration environment of a space shuttle launch and the environment brought about by shipping it around the country and to remote locations for ground truth measurements. An optical configuration which leads to the equal-path fringe contrast being first-order insensitive to tilt misalignments was deemed necessary.

One possible configuration is that shown in Fig. 2.30 (Breckinridge *et al.*, 1977). This layout is a modification of the one used by Norton (1978) for the JPL Mark II interferometer (Farmer *et al.*, 1980). In the figure, light enters the system from the lower right through a hole in the double-pass retroflat and strikes the beam splitter. The beam splitter substrate is divided into three areas: a reflecting surface at the top, the beam splitter in the center, and a transparent area at the bottom. The analysis tools developed in the previous section are used to analyze this configuration. Figure 2.31 shows the configuration given in Fig. 2.30, but with the double pass retroflat tilted through angle $\theta/2$. Solid arrows are used to trace the ray through its first pass (until the retromirror), and dotted arrows are used to show the second pass and the effects of the tilted retroflat. Light leaves the interferometer passing to the lower left.

Figure 2.32 shows the configuration given in Fig. 2.30 but with the beam splitter rotated about an axis in the page. The top of the beam splitter is shifted to the left. Light enters the system through a hole in the retroreflector mirror and strikes the beam splitter. The transmitted portion is not tipped over in angle (since the ray traverses a nearly plane parallel plate). The reflected portion is tilted up as shown by the arrow, reflects from the folding flat, and still continues up as it passes through the cat's-

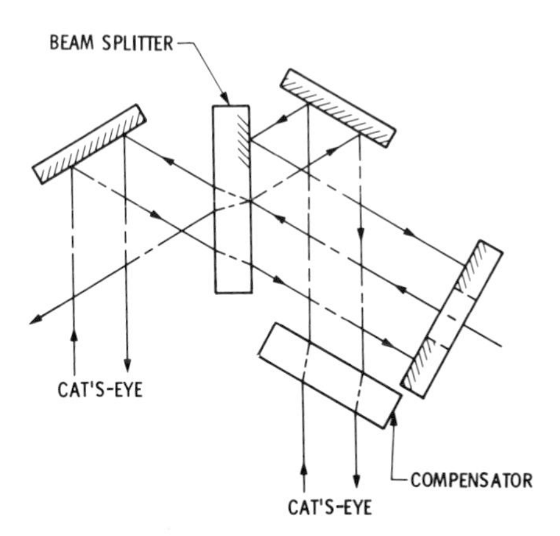

FIG. 2.30. A design approach for an optical configuration of an FTS. [From Breckinridge, Norton, and Schindler (1977).]

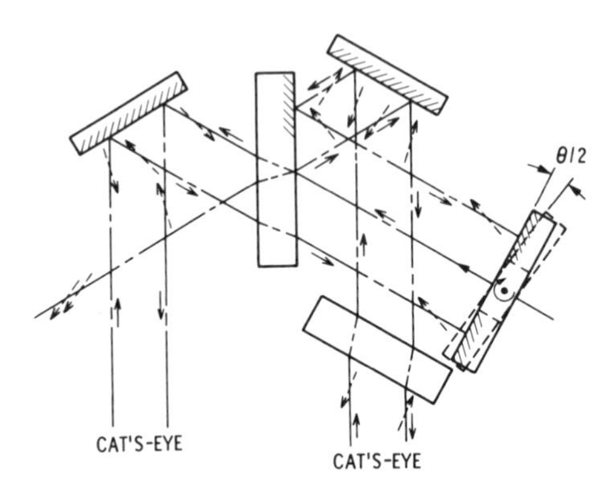

FIG. 2.31. Analysis of the effect of tilting the double-pass retroreflector mirror.

eye. The arrow is down on leaving the cat's-eye and reflects back to the tilted-over beam splitter, which corrects it and sends it to the double-pass retromirror. The light retraces its steps and passes back through the beam splitter, leaving the system in a downward pointing direction. The beam that transmits the beam splitter reflects from it also in a downward pointing direction to match the other beam. The reader can pursue this further to verify that, to first order, this system is tilt insensitive for tilts of all elements about each axis.

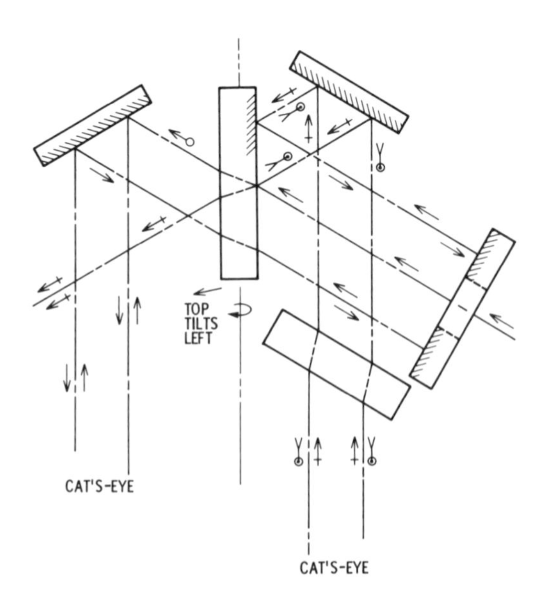

FIG. 2.32. Analysis of the effect of tilting the beam splitter.

Norton (1978) calculated the optical path difference in each of the two arms for the interferometer configuration shown in Fig. 2.30, as a function of the tilts of different components for a point source located on axis in object space, using Eq. (2.19).

Figure 2.33 shows the results of the computer analysis in the form of a plot of the on-axis fringe contrast at 2 μ as a function of tilt in radians and arc-minutes for tilts of the (1) retromirror, (2, 3) bending flats, (3) beam splitter about the Y axis, (4) beam splitter about the X axis, and (5, 6) cat's-eye. The on-axis fringe contrast is most sensitive to tilts of the double-pass retroreflector. A tilt of ten arc-minutes reduces the fringe contrast from 1.00 to 0.95. Were the system not tilt compensated, calculations using the method described above would have shown that a component tilt of one arc-second reduces the fringe contrast from 1.00 to 0.95.

The beam splitter and compensator need to be mounted to minimize vibration and tilts between the two. An approach is to mount them together in a cage assembly which is stiffer than the mounting of that cage assembly to the baseplate.

Spectra recorded with earlier instruments show channeling or ripples in the continuum which may mask shallow absorption features (see Section 2.5,D). To avoid this, the beam splitter and compensator as well as all other transmission optical elements are designed to be wedged.

To produce spectra of the highest signal-to-noise ratio, it was decided to require a symmetric interferogram. The optical path difference should be precisely the same in both arms for *all* wavelengths. The approach to this adjustment is to translate the wedged compensator along a line parallel to the surface of the compensator.

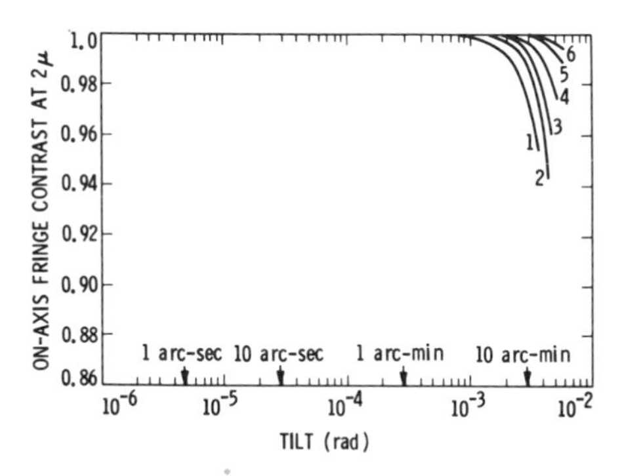

FIG. 2.33. Plot of on-axis fringe contrast at 2 μ as a function of tilt in radians for tilts of (1) the retroflat, (2, 3) bending flats, (3, 4) beam splitter y and x, and (5, 6) cat's-eyes. Graph shows that the interferometer components can be tilted (misaligned) to at least one arc-minute with insignificant effect on the on-axis fringe contrast.

The configuration in Fig. 2.30 shows that the compensator covers both the cat's-eye input and output beams. As a result of this configuration, the thickness and wedge angle on the compensator are not required to be precisely equal. That this is true is shown in Fig. 2.34. Recall that the cat's-eye is double passed. As the wedge is translated, the effect is as though the system has a plane parallel plate whose optical thickness is continuously variable. Figure 2.34 shows a wedge of glass with four path lengths indicated. (The wedge angle is exaggerated.) An arrow indicates the orientation of the light entering the wedge (and thus the cat's-eye) and the light out. Note that the head of the arrow passes through material of thickness a entering the assembly and through material of thickness d upon leaving the assembly. The tail of the arrow passes through a total amount of material $b + c$. By the geometry of the triangle we see that $a + d = b + c$. The plate is, in reality, wedged, and there is not channeling since the value of the wedge angle is such that back reflections are directed out of the optical system.

In Section 2.3 we saw that the maximum equal fringe contrast is obtained if throughout the interferometer reflections are as close to normal incidence as possible. The layout in Fig. 2.30 is shown with a 30° angle of incidence on the mirrors and the beam splitter. This angle can be reduced

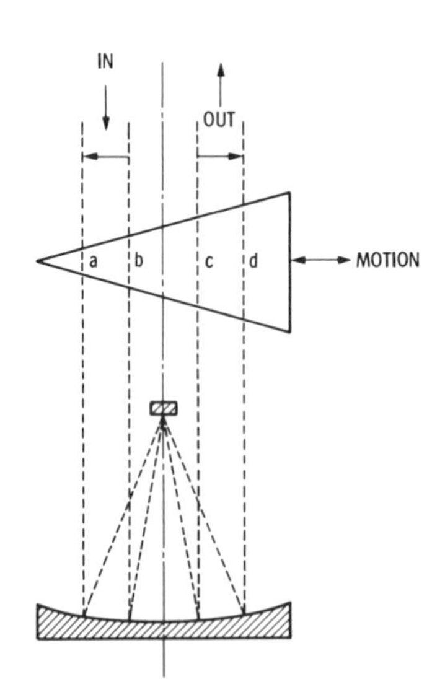

FIG. 2.34. A detail of Fig. 2.30 showing a compensator circular disk with exaggerated wedge angle covering both the inputs to and outputs from the cat's-eye. Since distances $a + d = b + c$, the wedged compensator acts like a plane parallel plate having continuously varying effective optical path thickness as the wedged plate is translated.

to the extent allowed by the finite width of the beams (see Section 2.5). At a 30° angle of incidence the polarization effects on the fringe contrast are negligible.

A disadvantage of this configuration is the relatively large number of passes through the beam splitter/compensator materials and the large number of reflections compared to a standard Connes-type configuration. The very high, broadband reflectivity of gold throughout the red and infrared minimizes the loss from the large number of reflections. High-grade infrared materials such as single crystal KBr have excellent phase transmission properties for a wave front propagating through the bulk material. However, those crystal materials that transmit up to 16 μ such as KBr are relatively soft and very difficult to figure. The system is therefore limited by the quality of surface figure on the compensator and beam splitter, which is typically $\frac{1}{4}$–$\frac{1}{2}$ wave at 632.8 nm over the full aperture. Zinc selenide is not a suitable material, because the very high Fresnel reflection losses reduce the system transmittance to below acceptable values. If zinc selenide were to be used, very large diameter cat's-eyes and beam splitter would be required to overcome the throughput losses due to Fresnel reflections. Recently broadband antireflection coatings have been developed which may lead to the general use of ZnSe for remote sensing FTS.

Complicated wave-front errors can be corrected using a hologram optical element whose design is unique to the particular remote sensing instrument (Chen *et al.*, 1979). Residual wave-front errors caused by aberrations and the figuring errors can be corrected using a reflection holographic optical element on the double-pass retromirror.

Instrument throughput determines the energy density at the output plane, which, if the system is linear, will establish the signal-to-noise ratio. Recall that the experiment for which this instrument is designed is to observe stratosphere absorption spectra at high temporal frequency resolution and with spatial resolution fine enough to measure the vertical scale height or about 2-km vertical height resolution in the atmosphere. A 1 mrad field of view is required (Farmer, 1978).

One-mrad field of view on the sun is approximately $\frac{1}{9}$th the solar diameter. Because of the fast scan time, the 2-μ infrared radiation is modulated at approximately 250 kHz and that at 16 μ is modulated at 31 kHz. The detector HgCdTe is chosen for this approach study based on its sensitivity, its frequency, and its spectral bandwidth coverage. This material, however, has a nonlinear responsivity (Bartoli *et al.*, 1974) at flux levels above 1×10^{-3} W. With a system transmissivity of 0.1, 1×10^{-2} W are required to enter the system, and the entrance aperture must have an area large enough to collect 1×10^{-2} W from approximately $\frac{1}{81}$ the area of the sun, within the wavelength region of interest. Blackbody calculations show that for the sun, only 5% of the energy is at wavelengths greater than 12 μ.

From Allen (1973) the solar constant (integrated over all spectral regions) is 0.136 W/cm^2. The approximate energy collecting area required is given by $(1 \times 10^{-2}$ W$) \times (1/0.136)$ cm^2/W $\times 81 \times (1/0.05) = 119$ cm^2. This area is obtained with a circular aperture about 6 cm in radius. For first-order design purposes we shall assume a circular collecting aperture of about 6.0-cm radius.

Figure 2.17, which defines marginal and chief rays through an optical system, indicates that the diameter of any aperture placed within the system must be greater than the sum of the absolute value of the marginal and chief ray heights to avoid vignetting. Figure 2.20 shows pupil and image plane relationships for the single-pass configuration optical layout of the McMath solar telescope interferometer, given in Fig. 2.19. The center of symmetry is clearly the beam splitter, and images of the system pupil are relayed onto the beam splitter. Locating the pupil at a plane other than at the plane of symmetry will not minimize optical component size for the throughput. If an image of the pupil is not located at the point of symmetry, the sum of the marginal and chief ray heights will be larger than necessary, and some optical elements will require clear apertures larger than otherwise necessary. For the tilt-compensated optical configuration shown in Figure 2.30, the center of symmetry is the double-pass retroreflector, and an image of the optical system pupil should be located here.

Assuming that an image of the pupil of the system is to be placed on the double-pass retroflat, we calculate the size required for the beams on the retroflat. Note that the size of the beams at the beam splitter determines the size of that optical element and that, in general, the beam splitter is the most expensive and difficult element to fabricate. Consequently, it is desirable to locate the retroflat reasonably close to the beam-splitter element. For a high f number system, a calculation of the beam size at the retroflat will approximate the beam size at the beam splitter.

From Eq. (2.33), H $= (6.0$ cm$) \times (0.05$ mrad$) = 0.3$ cm $\times$ mrad. Recall that a resolution of 0.02 cm^{-1} is required. From Eq. (2.4) in Baker (1977), we find the solid angle field of view at the pupil inside the interferometer [see Eq. (2.35)] to be $\Omega_p = \pi/\sigma\delta$, with the maximum wave number $\sigma = 5000$ cm^{-1}, the maximum optical path difference $\sigma = 50$ cm, and $\Omega_p = 1.26 \times 10^{-5}$. From $\Omega = \pi u^2$, $u = 2.0$ mrad. This is the angle that the chief ray makes with the image of the pupil at the retroflat. The angular magnification is given by dividing this angle by the half angle required for the image plane field of view (0.5 mrad). Angular magnification A is then $A = (2.0/0.5) = 4$. The size of the clear aperture (diameter) at the beam splitter is then (12 cm)/4 = 3.00-cm diameter. We are considering the first-order optical design of a system whose configuration is that shown in Fig. 2.30. This design approach requires a beam splitter large enough to have three clear circular apertures, each 3.0 cm in diameter, extending along a diameter. Allowing for workmanship errors near the edge of the piece, a 15-cm-diameter circular beam splitter would be selected for the first-order design.

Another aspect of importance to the first-order design of remote sensing instruments is the packaging. The weight of an instrument based on this approach is less than others of the same capability since the cat's-eyes are double passed, thus shortening the overall instrument. Because of the harsh chemical and dust environment at the launch facility and the contaminant environment reported for the Space Shuttle by Simpson and Witteborn (1977), Bareiss *et al.* (1977), and Scialdone (1979), the interferometer and foreoptics are best placed in an isolated chamber.

In Section 2.4,B it was shown that for a 0.01°C approximate temperature gradient, the change in optical path difference caused by temperature-induced index of refraction fluctuations is on the order of five millifringes. A design approach that requires immersion of the instrument in vacuum both provides contamination control and eliminates path difference variations induced by index of refraction fluctuations. To minimize the weight of this vacuum chamber, and at the same time maximize the volume available inside to locate instrument components, a spherically-shaped chamber is an approach. Unfortunately, the optical configuration of the interferometer as shown in Fig. 2.30, whose aspect is longer than it is wide or high, does not lend itself to being packaged efficiently in a spherical chamber. The next most efficient packaging approach is a cylinder with domed ends. To minimize flexure of the optical component mounts and thus reduce tilt errors and optical surface distortions, the support structure should not be part of the vacuum chamber pressure walls. An approach is to divide the vacuum chamber into two parts: one for foreoptics, the other for the interferometer optics. The support structure should be mounted on the inside walls of the vacuum chamber in such a way that as the chamber flexes, the structure does not.

In this brief section we have shown the development of the first-order optical design of a remote sensing interferometer and how this first-order optical design, which satisfies the desired performance, interacts with the mechanical and electrical systems. System design is, of course, an iterative procedure. The flow to a design approach is not as smooth as implied here. However, we hope that the reader has acquired a feel for a method of tradeoffs and an appreciation for design approach analysis.

E. Another Tilt-Compensated System

An optical configuration which is tilt compensated was described by Woodruff (1977) and is shown in Fig. 2.35.

The characteristics are that one of the cat's-eye apertures is covered by the compensator and that the light that passes into the left-hand cat's-eye reflects directly from the double-pass retroreflector mirror, without passing through compensator materials. This configuration requires that the wedge angle (needed to avoid channel spectra) on the compensator be equal to that of the beam splitter to within approximately one arc-second. Achieving this tolerance and maintaining the needed surface flatness using beam-splitter materials such as BK-7, fused silica, and barium fluoride are not beyond the current state of the art.

A distinct advantage of this configuration over that introduced in Fig. 2.30 is that the wave fronts traveling through the interferometer pass through less beam splitter–compensator material thickness, and they also pass through fewer substrate–air interfaces. Both material thickness and interfaces contribute to wave-front distortions, which decrease fringe contrast.

Consequently, if the materials selected for the beam splitter and compensator are such that matched wedge angles and the required flatness of the faces can be fabricated, then the Woodruff configuration will lead to an interferometer with good fringe contrast.

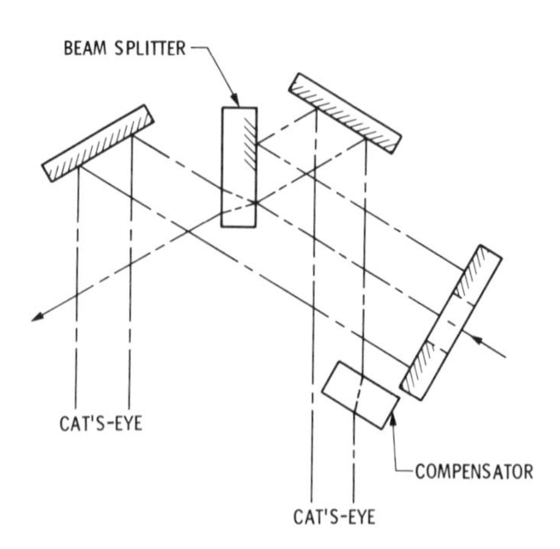

FIG. 2.35. Woodruff tilt-compensated configuration. [Shown with the permission of Mr. Woodruff, who was with Ball Brothers Aerospace Division during the development of this configuration.]

2.7. Cat's-Eye Alignment

The cat's-eye is an afocal optical element that has the property of accepting a plane wave-front traveling in one direction, turning that wave front around precisely 180°, and sending it back out of the system as a plane wave front. Precision alignment is required to assure that a plane wave front into the interferometer will return from the cat's-eye as a plane wave front with minimum aberrations. Wave-front errors in the cat's-eyes degrade the fringe contrast. The cat's-eye is positioned within a Fabry–Perot cavity (Stoner, 1974) as shown in Fig. 2.36. A laser source with pinhole and beam expander is adjusted to give a plane wave-front illumination for the cavity. This plane wave front passes through the beam-splitter element of the first optical flat of the Fabry–Perot cavity, illuminates the parabolized primary mirror, and reflects onto the curved secondary mirror, then off this element, back to the primary, and back out the system to strike the second element of the Fabry–Perot cavity. As with the first plate, this second plate is a partial reflector, and some of the light is reflected to pass through the system again, reflect from the first plate, and retrace its path through the system again. The display on the observation card is one of interference between wave fronts that traverse the cat's-eye one, three, five, seven, . . . times.

If the curved secondary mirror is not positioned precisely, with its center of curvature on the axis of the parabola and at the focus of the parabola, a fringe pattern is seen across the observation card. Tilting of the curved secondary to bring the center of the curvature of the secondary onto the axis of the cat's-eye primary results in a lateral motion of the fringe pattern. As the center of curvature of the secondary approaches the axis of the parabola, a bull's-eye fringe pattern moves onto the observation card.

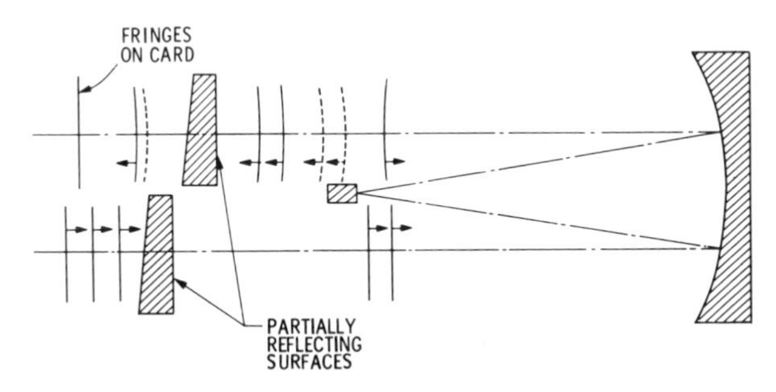

FIG. 2.36. Cat's-eye aligned by positioning it in a Fabry–Perot cavity (Stoner, 1974).

The bull's-eye fringe pattern indicates that the surface of the curved secondary is not positioned at the focus of the parabolized cat's-eye primary. In the presence of defocus, the wave front that makes the first pass through the out-of-focus system is a spherical wave front. The wave front that makes the triple pass is also a spherical wave front but with a much steeper curvature. The same is true of the five-times-pass wave front in relation to the triple-pass wave front, and so on until the brightness of the light from these multiple passes is too insignificant to contribute. The bull's-eye fringe pattern is formed by the intersections of spherical wave fronts of different curvatures, much like the surface intersections of spheres of increasingly smaller radius. As the secondary is translated toward the focus of the primary, the radii of curvature of all the nested spheres become larger and larger, and the paraxial focus is found when the center of the bull's-eye fringe pattern fills the entire field on the observation card.

To make the alignment easy, a mechanical decoupling of the observed fringe translation caused by positioning the secondary's center of curvature and the fringe expansion or contraction caused by focusing is needed. The mechanism to support the secondary is required to rotate about an axis in the surface of the secondary, and the mechanism that focused should move the secondary in and out in relation to the primary without tilting it.

2.8. Conclusion

The Fourier transform spectrometer is a good example of a complex mechanical, electrical, and optical system. All three of these areas of technology are pushed to their limits of understanding to make the interference spectrometer produce high-quality spectra. The performance of an instrument depends on the design approach selected during the first stages of instrument development. In this chapter we presented an introduction to the design of a system based on a particular observational strategy. Fringe contrast, mechanical design, throughput, and interferometer configurations were discussed. Tilt-compensated optical configurations were analyzed and discussed in detail.

ACKNOWLEDGMENTS

The authors would like to express their appreciation to Drs. Robert Norton and Reinhardt Beer for many discussions on interferometry; some of the concepts presented here are theirs. Odell Raper read part of the manuscript and criticized it. Dr. C. B. Farmer, ATMOS Principal Investigator, provided encouragement.

This chapter presents, in part, the results of one phase of research carried out at the Jet Propulsion Laboratory, California Institute of Technology, under contract No. NAS7-100, sponsored by the National Aeronautics and Space Administration.

References

ABEL, I. R., REYNOLDS, B. R., BRECKINRIDGE, J. B., PRITCHARD J. (1979). *Proc. SPIE* **193,** 12–26.

ALLEN, C. W. (1973). "Astrophysical Quantities." The Athlone Press, London.

BAKER, D. (1977). "Spectrometric Techniques" (G. A. Vanasse, ed.), Vol. I, pp. 71–105. Academic Press, New York.

BAREISS, L. E., HETRICK, M. A., and RESS, E. B. (1977). Spacelab Contamination Assessment, Final Tech Rep. NASA-CR-150413, N77-33248.

BARTOLI, F., ALLEN, R., ESTEROWITZ, L., and KRUER, M. (1974). *J. Appl. Phys.* **45,** 2150–2156.

BEER, R., and MARJANIEMI, D. (1966). *Appl. Opt.* **5,** 1191.

BEER, R., NORTON, R. H., and SEAMAN, C. H. (1971). *Rev. Sci. Instrum.* **42,** 1393–1403.

BORN, M., and WOLF, E. (1975). "Principles of Optics," 5th ed. Pergamon, Oxford.

BRACEWELL, R. (1965). "The Fourier Transform and Its Applications." McGraw-Hill, New York.

BRADDICK, H. J. J. (1963). "The Physics of Experimental Method." Chapman and Hall, London.

BRAULT, J. W. (1974). Private communication.

BRECKINRIDGE, J. B. (1971). *Appl. Opt.* **10,** 286–294.

BRECKINRIDGE, J. B., NORTON R. H., and SCHINDLER, R. A. (1977). Wedged Plate, Tilt Compensated, Double-Pass Interferometer Configuration. U.S. Patent Application.

BRECKINRIDGE, J. B., SCHINDLER, R. A., and FARMER, C. B. (1978). October 1978 Optical Society of America Meeting.

BRECKINRIDGE, J. B., MCALISTER, H. A., and ROBINSON, W. (1979a). *Appl. Opt.* **18,** 1034–1040.

BRECKINRIDGE, J. B., NORTON, R. H., SCHINDLER, R. A., and ABELS, I. (1979b). *Proc. SPIE* **191,** Meeting Abstracts.

BUIJS, H. L. (1971). Air Force Cambridge Research Laboratories, Report No. 114 71-0019, p. 163.

BIUJS, H. (1979). A class of high-resolution ruggedized Fourier transform spectrometers, *Proc. SPIE* **191,** 116–119.

Burleigh Instruments, New York (1974).

CHEN, C. W., BRECKINRIDGE, J. B., and WYANT, J. C. (1979). Paper presented at the October 1979 Optical Society of America Meeting, Rochester, New York.

CONNES, J. (1963). Spectroscopic studies using Fourier transformations, *Rev. Opt.* **40,** 45–78, 116–140, 171–190 and 231–265, translated by C. A. Flanagan and published as NAVWEPS Report 8099 NOTS TP 3157, China Lake, California, unclassified.

CONNES, P. (1966). *J. Opt. Soc. Am.* **56,** 896.

CONNES, P. (1970). Astronomical Fourier spectroscopy, *Ann. Rev. Astron. Astrophys.* **8,** 209–230.

CUISENIER, M., and PINARD, J. (1967). *J. Physique Colloq. C2* **28,** C2-97–104.

DELANO, E. (1963). First-order designs and the $y, \bar{y}$ diagram, *Appl. Opt.* **2,** 1251–1256.

EZHEVSKAYA, T. B., and SINITSA, S. P. (1978). *Opt. Spectrosc.* (*USSR*) **45,** 333–336.

FARMER, C. B. (1974). *Can. J. Chem.* **52,** 1544–1559.
FARMER, C. B. (1978). Spacelab Mission 1 Experiment Descriptions, NASA TM-78173, MSFC (P. D. Craven, ed.), pp. 1–5.
FARMER, C. B., RAPER, O. F., and NORTON, R. H. (1976). *Geophys. Res. Lett.* **3,** 13–16.
FARMER, C. B., RAPER, O. F., ROBBINS, B. D., TOTH, R. A., and MULLER, C. (1980). *J. Geophys. Res.* To be published.
FONCK, R. J., HUPPIER, D. A., ROESLER, F. L., TRACY, D. H., and DAEHLER, M. (1978). *Appl. Opt.* **17,** 1739–1747.
FYMAT, A. L. (1972). *Appl. Opt.* **11,** 160–173.
GASKILL, J. D. (1978). "Linear Systems, Fourier Transforms, and Optics." Wiley, New York.
GENZEL, L., and KUHL, J. (1978). *Appl. Opt.* **17,** 3304–3308.
GENZEL, L., and SAKAI, K. (1977). *J. Opt. Soc. Am.* **67,** 871–879.
GOORVITCH, D. (1975). *Appl. Opt.* **14,** 1387–1390.
GRIFFITHS, P. R. (1975). "Chemical Infrared Fourier Transform Spectroscopy." Wiley, New York.
GROSSMAN-DOERTH, U. (1969). *Solar Phys.* **9,** 210–224.
GUELACHVILI, G. (1978). *Appl. Opt.* **17,** 1322–1326.
HANEL, R. A., SCHLACHMAN, B., ROGERS, D., and VANOUS, D. (1971). *Appl. Opt.* **10,** 1376–1382.
HANEL, R. A., and KUNDE, V. G. (1975). *Space Sci. Rev.* **18,** 201–256.
HANEL, R. A. *et al.* (1977). *Space Sci. Rev.* **21,** 103.
HARRIES, J. E. (1977). *J. Opt. Soc. Am.* **67,** 880–893.
HARRIS, F. J. (1978). *Proc. IEEE* **66,** 51–82.
HINKLEY, E. D. (ed.) (1976). "Laser Monitoring of the Atmosphere." Springer-Verlag, Berlin and New York.
HUPPI, R. (1979). Meeting, *Proc. SPIE* **191,** 26–32.
Infrared Phys. (1979). **19,** Submillimeter spectroscopy.
JACQUINOT, P. (1954). *J. Opt. Soc. Am.* **44,** 761.
JACQUINOT, P. (1955). *J. Opt. Soc. Am.* **45,** 996.
JACQUINOT, P., and ROIZEN-DOSSIER, B. (1964). Apodisation, *Prog. Opt.* **3.**
JACQUINOT, P., and DUFOUR, C. (1948). *J. Rech. CNRS* **6,** 91.
JAMES, J. F., and STERNBERG, R. S. (1969). "The Design of Optical Spectrometers." Chapter 11, General Principles of Construction, pp. 183–206. Butler and Tanner, London.
JANSSON, P. A., HUNT, R. H., and PLYLER, E. K. (1970). *J. Opt. Soc. Am.* **60,** 596–599.
JEFFERIES, J. T. (1968). "Spectral Line Formation." Blaisdell, Waltham, Massachusetts.
JONES, R. V. (1961). *J. Sci. Instrum.* **38,** 37–44.
JONES, R. V. (1962). *J. Sci. Instrum.* **39,** 193–203.
JONES, R. V. (1976). *Proc. R. Soc. London Ser. A* **349,** 423–439.
JONES, R. V., and RICHARDS, J. C. S. (1954). *Proc. R. Soc. London Ser. A.* **225,** pp. 122–135.
KATTI, P. K., and SINGH, K. (1966). *Appl. Opt.* **5,** 1962–1963.
LOPEZ-LOPEZ, F. J. (1973). The Application of the Delano y-$\bar{y}$ Diagram to Optical Design, Ph.D. dissertation, Optical Sciences Center, Univ. of Arizona.
MACLEOD, H. A. (1978). *Opt. Act.* **25,** 93–106.
MARTIN, D. H., and PUPLETT, E. F. (1969). *Infrared Phys.* **10,** 105–110.
MATHERS, J. (1977). Goddard Space Flight Center, Private communication.
MICHELSON, A. A. (1927). "Studies in Optics," Phoenix Science Series of the University of Chicago Press, Chicago, Illinois.
MONTGOMERY, A. J. (1967). *J. Opt. Soc. Am.* **57,** 1121–1124.
NEWPORT RESEARCH CO., 1977/78 catalog, Fountain Valley, California.

NORTON, R. H. (1978) Private communication.
NORTON, R. H., and BEER, R. (1976). *J. Opt. Soc. Am.* **66,** 259–264.
PARKER, T. J., LEDSHAM, D. A., and CHAMBERS, W. G. (1978). *Infrared Phys.* **18,** 179–183.
PEARSON, E. T. (1972). F.T.S. Hydrostatic Bearing Design, Engineering Technical Report 46, Kitt Peak National Observatory, Tucson, Arizona.
PECK, E. R. (1957). *J. Opt. Soc. Am.* **47,** 250–255.
PENNER, S. S. (1959). "Quantitative Molecular Spectroscopy and Gas Emissivities." Addison-Wesley, Reading, Massachusetts.
PINARD, J. (1969). *Ann. Phys.* **4,** 147–196.
PRITCHARD, J. L. (1969). U.S. Patent 3,563,617.
PRITCHARD, J. L., SAKAI, H., STEEL, W. H., and VANASSE, G. A. (1967). *J. Physique Colloq. C2* **28,** C2-91–C2-96.
RAINE, K. W., and DOWNS, M. J. (1978). *Opt. Acta* **25,** 549–558.
RICHARDS, P. L. (1977). *Int. Conf. Fourier Spectrosc.* Optical Society of America, Columbia, South Carolina, June.
RING, J., and SCHOFIELD, J. W. (1972). *Appl. Opt.* **11,** 507.
RUNDLE, H. N. (1965) *J. Res. Nat. Bur. Std. U.S.* **69c,** 5.
SAKAI, H. (1977). High Resolving Power Fourier Spectroscopy *in* "Spectrometric Techniques" (G. A. Vanasse, ed.), Vol. I, pp. 2–69. Academic Press, New York.
SCHINDLER, R. A. (1970a). *Appl. Opt.* **9,** 301–306.
SCHINDLER, R. A. (1970b). U.S. Patent 3,535,024.
SCHINDLER, R. A. (1975). U.S. Patent 4,053,231.
SCHRÖDER, B., and GEICK, R. (1978). *Infrared Phys.* **18,** 595–605.
SCIALDONE, J. J. (1979). Assessment of Shuttle Payloads Gaseous Environment, NASA Tech. Memo. 80286, Goddard Space Flight Center, Greenbelt, Maryland.
SCOTT, L. B. (1958). U.S. Patent 2,841,049.
SHACK, R. V. (1975). Class Notes, Optical Science Center, Univ. of Arizona, Tucson, Arizona.
SHURCLIFF, W. A. (1966). "Polarized Light Production and Use." Harvard Univ. Press, Cambridge, Massachusetts.
SIMPSON, J. P., and WITTEBORN, F. C. (1977). *Appl. Opt.* 2051–2072.
SMITH, R. W. (1978). *Opt. Acta* **25,** 721–726.
SOARES, O. D. D. (1978). *J. Phys. E. Sci. Instrum.* **11,** 773–776.
STERNBERG, R. S., and JAMES, J. F. (1964). *J. Sci. Instrum.* **41,** 225–226.
STONER, J. O. (1974). Private communication.
STRAITON, A. W. (1964). Measurement of the radio refractive index of the atmosphere, "Advances in Radio Research" (J. A. Saxon, ed.). Academic Press, New York.
STROHBEHN, J. W. (1971). Optical propagation through the turbulent atmosphere, *Prog. Opt.* **9.**
STRONG, J. D. (1958). "Classical Optics." Freeman, San Francisco, California.
TOTH, R. A., FARMER, C. B., SCHINDLER, R. A., RAPER, O. F., and SHAPER, P. W. (1973). *Nature (London) Phys. Sci.* **244,** 7–8.
WALKER, R. P. and REX, J. D. (1979). *SPIE Proc.* **191.**
WHITEHEAD, T. N. (1954). "The Design and Use of Instruments and Accurate Mechanisms." Dover, New York.
WILLIAMS, C. W. (1966). *Appl. Opt.* **5,** 1084–1085.
WILSON, E. B. (1952). "An Introduction to Scientific Research," pp. 104–108. McGraw-Hill, New York.
WOODRUFF, R. (1977). Private communication.

Chapter 3

Effects of Drive Nonlinearities in Fourier Spectroscopy

ALEXANDER S. ZACHOR

ATMOSPHERIC RADIATION CONSULTANTS, INC.
ACTON, MASSACHUSETTS

ISAIAH COLEMAN

BARTLETT SYSTEMS, INC.
WOODY CREEK, COLORADO

WILLIAM G. MANKIN

NATIONAL CENTER FOR ATMOSPHERIC RESEARCH
BOULDER, COLORADO

ISBN 0-12-710402-X

List of Symbols

- a Abbreviation defined in Eq. (3.11)
- b Abbreviation defined in Eq. (3.11); function defining the apodization/truncation of the interferogram
- c Abbreviation defined in Eq. (3.11)
- f Impulse response of the electrical filter; temporal frequency
- g Interferogram
- $\tilde{g}$ Filtered interferogram
- i $\sqrt{-1}$
- m Index for integer-order Bessel functions
- n Number of poles in the electrical filter
- q Abbreviation defined in Eq. (3.42)
- r Optical retardation
- t Time
- u Arbitrary function of r or x
- v Velocity of moving mirror; $\langle v \rangle$ is average velocity of the mirror
- x $2\langle v \rangle t$, a scaled time variable having units of distance
- y Phase of a random velocity variation; dummy variable
- z Abbreviation defined in Eq. (3.19)
- C Real (symmetric) part of the comlex instrument line shape (ILS)
- E Complex ghost amplitude or spectrum error
- F Complex spectral response of the electrical filter
- G The Fourier transform of g; the true source spectrum except for possible optical distortions in the interferometer
- ILS Instrument line shape or complex instrument line shape
- J Integer-order Bessel function
- P_ξ Power spectrum of random velocity variations
- Q Fourier transform of q
- Re( ) Real part of a complex function or quantity
- S Imaginary (asymmetric) part of the complex instrument line shape (ILS)
- U Arbitrary function of σ; the Fourier transform of u
- X Integration variable with distance units
- Y Dummy variable representing integration limits
- $\mathscr{F}$ Amplitude spectrum of the electrical filter
- α $\sigma_s \phi'(\sigma_s)$ (see Eq. 3.41)
- $\mathscr{S}$ The instrument line shape, a function of σ; $\mathscr{S}(\sigma) = C(\sigma) - iS(\sigma)$
- β Effective (residual) phase spectrum of the electrical filter after introduction of a delay in the reference channel (Eq. 3.33)
- γ Abbreviation defined in Eq. (3.19) and above Eq. (3.41)
- δ The delta or impulse function
- ε Error in the filtered interferogram g resulting from velocity variations
- θ Abbreviation defined in Eq. (3.19)
- ξ The relative velocity variation, either sinusoidal or random
- σ Wavenumber of radiation or temporal frequency divided by $2\langle v \rangle$ (see Section 3.2)
- ϕ Phase spectrum of the electrical filter; effective residual phase after delay compensation
- ξ_0 Zero-to-peak amplitude of a sinusoidal velocity error
- σ_c Bandwidth (3 dB cutoff) of the electrical filter
- σ_{max} Maximum source frequency
- σ_s Wavenumber of a line in the source spectrum
- ψ_0 Phase of the sinusoidal velocity error
- $\langle \ \rangle$ Average over time, retardation, or an ensemble of random functions
- $*$ Convolution operator
- χ $\sigma_s \phi'(\sigma_s) \xi_0$ (see Eq. 3.22)

3.1. Introduction

The scanning Michelson interferometer, commonly known as the Fourier Transform Spectrometer (FTS), is basically a modulator that converts optical radiation to audio frequency signals. Pure "colors" are transformed into pure tones of proportional, but much lower, frequency. An essential component of the FTS is the moving mirror whose velocity determines the ratio of output and input frequencies. Deviation from constant velocity reduces the purity of the output tone corresponding to a monochromatic input.

The FTS output is more properly viewed as a function of the position of the moving mirror rather than as a time signal. In this representation, the instantaneous frequency of the output wave, i.e., the number of cycles per unit mirror displacement produced by a monochromatic input, is independent of the mirror velocity. This frequency (wavenumber) is, in fact, twice the wavenumber of the input radiation for the simplest FTS designs (using additional folding in the interfering beams, it can be made any even multiple of the input wavenumber).

The radiation spectrum of an unknown source is obtained by Fourier analysis of the corresponding FTS output signal (interferogram). The interferogram is reduced to a set of samples that are analyzed in a digital computer, since analog devices do not have the dynamic range required for most FTS applications. Also, the interferogram is passed through an electrical filter prior to sampling to suppress the aliases of out-of-band detector noise. An equivalent statement is that the filter is used to maximize the utilization of the time between samples, which results in maximum signal-to-noise.

Modern FTS designs employ fringe-reference sampling. The reference or sampling channel has a moving mirror slaved to the mirror drive of the main signal channel, and contains an internal laser source that produces high-frequency fringes on a separate detector. These fringes trigger sampling of the main interferogram at equal increments in the displacement of the moving mirror. Thus velocity variations do not compromise the sampling accuracy (purity of tones), at least not directly. In contrast, equal-time samples triggered by a clock reproduce the frequency distortions introduced by velocity variations; this is the same as saying the samples contain errors because they were obtained at unequal increments in mirror displacement.

An effect of the electrical filter is to introduce a time delay in the signal channel. Energy sampled by the reference fringes at some particular mirror position has been modulated by the interferometer at an earlier time. Two successive samples representing a particular *increment* in mirror displacement (the uniform sampling interval) will represent a different increment at the time of modulation, unless the mirror velocity is the same at both times. In other words, the internal fringe reference does not provide for equidisplacement sampling of the true interferogram when there are velocity errors *and* a filter-induced delay in the signal channel. The sampling errors, however, are much smaller than those arising from velocity errors in the clock-sampling case.

The electrical filter generally will have some phase distortion, i.e., its group delay and phase delay are functions of frequency, as are the sampling errors arising from the combination of velocity variations and filter delay. Nishiyama *et al.* (1975) demonstrated that the introduction of a time delay in the sampling channel will partially compensate for the delay introduced in the signal channel by the electrical filter. The fixed sampling delay is, in effect, subtracted from the filter's frequency-dependent group delay, reducing it at most frequencies and eliminating it at the frequency (or frequencies) where the sampling and group delays are equal. The reduction in sampling error can be substantial over the low-frequency portion of a bandpass filter, where phase distortion tends to be small.

The sampling delay could be introduced in the reference channel via an electrical filter. The filter acts as a pure delay to the extent that only the sampling frequency is passed through it (other frequencies are introduced via the velocity variations, but these will not cause appreciable dispersion of the sampling delay if the velocity variations are of low frequency, or if the filter has linear phase near the sampling frequency). In any case, the filter group delay at the sampling frequency should be comparable to the average signal channel group delay. This cannot be achieved by using identical filters in the reference and signal channels (unless they have an unusually wide bandwidth) since the sampling frequency should be at least twice the maximum signal frequency to avoid aliased detector noise at the signal frequencies. The implication is that the delay in the reference channel is most easily implemented using a simple delay circuit.

Another effect of the electrical filter is to introduce amplitude distortion in the signal channel. Amplitude distortion causes nonuniform attenuation over the band (really sidebands) of frequencies corresponding to velocity variations, which means that the interferogram of a monochromatic input is a distorted wave even when viewed as a function of mirror position. Filter amplitude distortion tends to produce sampling errors of roughly the same magnitude as those resulting from filter phase distortion. Stated more accurately, velocity variations manifest their effects through electrical filter distortions when delayed, fringe-reference sampling is used; in assessing the magnitude of these effects, both amplitude and phase distortion must be considered.

Our purpose in this chapter is to quantify the errors in the recovered source spectrum arising from drive velocity variations, when the FTS system employs either conventional fringe-reference sampling or delayed fringe-reference sampling. We shall present a theoretical analysis of the simplest case, consisting of a monochromatic source and a sinusoidal variation in drive mirror velocity. The analytical results are essentially those reported by Zachor and Aaronson (1979) but are developed here (in Sections 3.3–3.6) by a simpler, more direct mathematical approach. We shall also present measurements of spectrum errors and compare them to the predictions of the model (Section 3.8). In Section 3.9 we discuss how the benefits of delayed sampling can be exploited to the fullest by real-time digital filtering of an oversampled interferogram. The effects in the recovered spectrum of random velocity variations are treated in Section 3.10.

From the above discussion it is apparent that the interferogram corresponding to a monochromatic input and a sinusoidal velocity variation is a tone-modulated tone, if the interferogram is represented as a time signal. This type of interferogram has been analyzed by Connes (1961), who showed that the recovered spectrum contains a series of "ghost" lines whose distances from the source line are multiples of the frequency of the velocity variation. These ghosts or FM sideband tones are a well-known phenomenon to communication engineers. Zachor (1977) and Zachor and Aaronson (1979) showed that ghosts are present at exactly the same frequencies when fringe-reference sampling is used, but that the ghosts have different intensities, dependent on the phase and amplitude characteristics of the electrical filter employed in the signal channel and the amount of delay introduced in the sampling channel. The purpose of the mathematical analysis in Sections 3.3–3.7 is merely to derive the complex intensities of these ghosts and their corresponding amplitudes in the recovered spectrum.

3.2. Conventions

Let $\langle v \rangle$ denote the nominal or time-averaged velocity of the scanning mirror, and $\delta \nu(t)$ the velocity error, equal to $\nu(t) - \langle \nu \rangle$, where $\nu(t)$ is the actual velocity at time t. The optical path difference or retardation $r(t)$ in most interferometer designs is twice the physical displacement of the scanning mirror, i.e.,

$$r(t) = 2\langle v \rangle t + 2 \int_0^t \delta v(t_1)\, dt_1 \tag{3.1}$$

if we choose the time origin to coincide with zero path difference. With $x \equiv 2\langle v \rangle t$ as the independent variable, Eq. (3.1) can be written

$$r(x) = x + \int_0^x \xi(x_1)\, dx_1, \tag{3.2}$$

where $\xi(x) \equiv \delta v(x)/\langle v \rangle$ is the relative velocity error.

Note that r and x both have units of length. The interferogram corresponding to $r(x)$ but expressed as a function of retardation r will be denoted $g(r)$. If expressed as a function of time (really $x = 2\langle v \rangle t$), it will be denoted by $g(x)$. Electrically filtered interferograms will be denoted by $\tilde{g}(r)$ or $\tilde{g}(x)$, and the filter impulse response will be denoted by $f(x)$. Fourier transform pairs will be indicated by

$$u(r) \rightleftharpoons U(\sigma). \tag{3.3}$$

The formal relationship will follow the convention

$$U(\sigma) = \int_{-\infty}^{\infty} u(r) \exp(-2\pi i \sigma r)\, dr. \tag{3.4}$$

We shall find in the following analysis that it is usually not necessary to distinguish between the Fourier transforms of $u(r)$ and $u(x)$; thus we can write $F(\sigma) \rightleftharpoons f(x)$. This means that with the adopted conventions σ can be interpreted as either the wavenumber of radiation in the recovered spectrum, or temporal frequency normalized by $2\langle v \rangle$. In other words, the optical spectrum, optical filter, normalized spectral response of the electrical filter $F(\sigma)$, normalized sampling frequency, etc., can all be represented in terms of the variable σ, which we shall call *frequency*.

3.3. The Distorted Interferogram

The interferogram, for reasons stated above, is electrically filtered prior to sampling. The resulting interferogram, considered as a "time" function, is the convolution of $f(x)$ and $g[r(x)]$:

$$\tilde{g}(x) = f(x) * g[r(x)] = \int_{-\infty}^{\infty} f(X) g[r(x - X)]\, dX. \tag{3.5}$$

The function $r(x)$ is given by Eq. (3.2). Its first derivative is

$$dr(x)/dx = 1 + \xi(x) \equiv \nu(x)/\langle \nu \rangle. \tag{3.6}$$

Consider the function $r(x - X)$ which appears in Eq. (3.5). If we expand it in a Taylor series and substitute Eq. (3.6), it becomes

$$\begin{aligned} r(x - X) &= r(x) - X\,dr/dx + (X^2/2!)\,d^2r/dx^2 - \cdots \\ &= r(x) - X[1 + \xi(x)] + (X^2/2!)\,d\xi(x)/dx - \cdots. \end{aligned} \quad (3.7)$$

Note that terms after the second have no contribution if X is restricted to the range $(0, X_{\max})$ and if the relative velocity error $\xi(x')$ can be treated as a constant over any corresponding interval $x - X_{\max} < x' < x$.

The conditions that allow truncation of Eq. (3.7) after the second term are satisfied when the filter bandwidth is large compared to the highest frequency in ξ. The filter, if its impulse response is zero outside of $(0, X_{\max})$, will restrict X to this range in Eq. (3.5). In addition, ξ will vary much more slowly than f, i.e., ξ may be regarded as a constant over any interval of width $X_{\max}$. In particular, for $x - X_{\max} < x' < x$ we can set $\xi(x') \simeq \xi(x)$. Substituting the truncated series [Eq. (3.7)] into Eq. (3.5), and applying this approximation, we obtain

$$\tilde{g}(x) = \int_{-\infty}^{\infty} f(X) g\{r(x) - X[1 + \xi(x)]\}\, dX. \quad (3.8)$$

We note that if the drive mirror does not stop or reverse direction during the scan ($\xi > -1$), r is a monatonic function of x [Eq. (3.6) or (3.2], i.e., r and x are related uniquely, and we can treat ξ as a function of r. Then the distorted interferogram $\tilde{g}$ can be expressed as a function of r:

$$\tilde{g}(r) = \int_{-\infty}^{\infty} f(X) g\{r - X[1 + \xi(r)]\}\, dX. \quad (3.9)$$

This is the desired expression for the distorted interferogram since we are concerned with interferometer systems that sample uniformly in retardation rather than time. Aliasing will have no effect in the recovered spectrum if $\tilde{g}(r)$ is adequately sampled.

Equation (3.9) is the starting point for the following analysis. We emphasize that it rests on two assumptions: that the filter bandwidth is large compared to the highest frequency in $\xi(x)$ and that the mirror velocity is always positive. These are reasonable assumptions for most interferometer designs and for all but the most hostile operating environments.

We acknowledge that it may be generally more appropriate to express the velocity error as a function of time $(2\langle v \rangle t)$ rather than retardation. The distinction becomes unimportant if the maximum value of ξ is much less than unity. In any case, this chapter will treat the effects of errors caused by simple models of $\xi(r)$.

If the filter has infinite bandwidth and zero phase, i.e., if there is no filter in the system, then $f(x) = \delta(x)$ (the delta function), and Eq. (3.9) gives $\tilde{g}(r) = g(r)$, as expected. We note also that if the relative velocity error $\xi(r)$ is identically zero, then $\tilde{g}(r) = \tilde{g}(x)$ is simply f convolved with g.

3.4. Effects of a Sinusoidal Velocity Error in the Spectrum of a Line Source

A change of variable in Eq. (3.9) yields the relation

$$\tilde{g}(r) = \int_{-\infty}^{\infty} \frac{1}{c} f\left(\frac{X}{c}\right) g(r - X)\, dX \tag{3.10a}$$

$$\equiv \int_{-\infty}^{\infty} a(X) b(-X)\, dX = \int_{-\infty}^{\infty} A(\sigma) B(\sigma)\, d\sigma, \tag{3.10b}$$

where

$$\begin{aligned} c \equiv c(r) &\equiv 1 + \xi(r) = v(r)/\langle v \rangle, \\ a(X) &\equiv (1/c) f(X/c) \rightleftharpoons A(\sigma) = F(c\sigma), \\ b(X) &\equiv g(r + X) \rightleftharpoons B(\sigma) = \exp(2\pi i \sigma r) G(\sigma). \end{aligned} \tag{3.11}$$

The equivalence of the two integrals in Eq. (3.10b) is a statement of Parseval's theorem. Note that a, b, and c have simple relationships to functions already identified (Eq. 3.11). The Fourier transforms of a and b given in Eq. (3.11) are expressions of the Fourier time/frequency scaling theorem and the shift theorem, respectively. The spectrum of the undistorted interferogram $g(r)$ is denoted by $G(\sigma)$.

Substituting Eqs. (3.11) into Eq. (3.10b), we obtain $\tilde{g}(r)$ in the form

$$\tilde{g}(r) = \int_{-\infty}^{\infty} \exp(2\pi i \sigma r) F(c\sigma) G(\sigma)\, d\sigma. \tag{3.12}$$

Of course, this result does not imply that $F(c\sigma)G(\sigma)$ is the spectrum of $\tilde{g}(r)$, since c is not a constant but a function of r; nor does Eq. (3.10a) imply that $\tilde{g}(r)$ is expressible as a true mathematical convolution. Equation (3.10a) does show that $\tilde{g}(r)$ can be visualized as the true interferogram "convolved" with an impulse response whose shape varies during the convolution process—its width is proportional to the instantaneous drive velocity at retardation r, and its height is inversely proportional to the velocity. Equation (3.12) states that $\tilde{g}(r)$, at a particular value of r, is equal to the Fourier integral of $F(c\sigma)G(\sigma)$.

If the source spectrum is a single line (delta function) at frequency σ_s, the corresponding undistorted interferogram is

$$g(x) = 2\cos(2\pi\sigma_s x) \rightleftharpoons \delta(\sigma - \sigma_s) + \delta(\sigma + \sigma_s) = G(\sigma). \tag{3.13}$$

Substituting this expression for $G(\sigma)$ into Eq. (3.12), we obtain

$$\tilde{g}(r) = F(c\sigma_s) \exp(2\pi i r \sigma_s) + F(-c\sigma_s) \exp(-2\pi i r \sigma_s). \tag{3.14}$$

We note that $F(-\sigma) = F^*(\sigma)$ since $f(x)$ is real, i.e., the second term of $\tilde{g}(x)$ is the complex conjugate of the first term. Thus we can concentrate on the development of the first term and determine the second by simple deduction.

The filter spectral response can be represented by

$$F(\sigma) = \mathscr{F}(\sigma) \exp[-i\phi(\sigma)] \quad , \tag{3.15}$$

where $\mathscr{F}(\sigma)$ and $\phi(\sigma)$ are the filter amplitude and phase spectra, respectively. We will assume that $\mathscr{F}$ and ϕ can be approximated by linear functions of frequency near $\sigma = \sigma_s$; then

$$\begin{aligned} F(c\sigma_s) &\equiv F[\sigma_s + \sigma_s \xi(r)] \\ &\simeq [\mathscr{F}(\sigma_s) + \sigma_s\xi(r)\mathscr{F}'(\sigma_s)] \exp\{-i[\phi(\sigma_s) + \sigma_s\xi(r)\phi'(\sigma_s)]\}. \end{aligned} \tag{3.16}$$

Let the constants ξ_0, σ_ξ, and ψ_0 denote the amplitude, frequency, and phase, respectively, of the sinusoidal relative velocity error:

$$\xi(r) = \xi_0 \sin(2\pi\sigma_\xi r - \psi_0). \tag{3.17}$$

Substituting Eq. (3.17) into Eq. (3.16) we obtain, for $F(c\sigma_s)$,

$$\begin{aligned} F(c\sigma_s) &= [\mathscr{F} + \sigma_s\xi_0\mathscr{F}' \sin(2\pi\sigma_\xi r - \psi_0)] \\ &\quad \times \exp\{-i[\phi + \sigma_s\xi_0\phi' \sin(2\pi\sigma_\xi r - \psi_0)]\} \\ &= [\mathscr{F} + (\sigma_s\xi_0\mathscr{F}'/2i)(e^{i(\theta-\psi_0)} - e^{-i(\theta-\psi_0)})]e^{-i[\phi - z\sin(\theta-\psi_0)]} \\ &= F(\sigma_s)e^{iz\sin\gamma}\{1 - i\sigma_s\xi_0\mathscr{F}'(\sigma_s)/2\mathscr{F}(\sigma_s)[e^{i\gamma} - e^{-i\gamma}]\}, \end{aligned} \tag{3.18}$$

where we have used Eq. (3.15) and the abbreviations

$$\begin{aligned} \mathscr{F} \equiv \mathscr{F}(\sigma_s), \qquad \phi' \equiv \phi'(\sigma_s), \qquad z = -\sigma_s\phi'(\sigma_s)\xi_0, \\ \theta = 2\pi\sigma_\xi r, \qquad \gamma = \theta - \psi_0. \end{aligned} \tag{3.19}$$

The complex factor $\exp(iz \sin \gamma)$ in Eq. (3.18) is a generating function for Bessel functions; if we apply the Jacobi–Anger formula (Endélyi, 1953)

$$\exp[iz \sin \gamma] = \sum_{m=-\infty}^{\infty} \exp(im\gamma)J_m(z), \tag{3.20}$$

the equation can be written

$$\begin{aligned} F(c\sigma_s) &= F(\sigma_s) \sum_{m=-\infty}^{\infty} J_m(z)e^{im\gamma} \left\{1 - i\frac{\sigma_s\xi_0\mathscr{F}'(\sigma_s)}{2\mathscr{F}(\sigma_s)} [e^{i\gamma} - e^{-i\gamma}]\right\} \\ &= F(\sigma_s) \sum_{m=-\infty}^{\infty} e^{im\gamma} \left\{J_m(z) \vphantom{\frac{1}{1}}\right. \\ &\qquad \left. + i\frac{\sigma_s\xi_0\mathscr{F}'(\sigma_s)}{2\mathscr{F}(\sigma_s)} [J_{m+1}(z) - J_{m-1}(z)]\right\} \cdot \end{aligned} \tag{3.21}$$

The first term of $\tilde{g}(r)$ in Eq. (3.14) then becomes

$$F(c\sigma_s)e^{2\pi i r\sigma_s} = F(\sigma_s) \sum_{m=-\infty}^{\infty} (-1)^m e^{-im\psi_0} e^{i2\pi(\sigma_s + m\sigma_\xi)r} \times \left\{ J_m(\chi) + \frac{i\chi\mathscr{F}'(\sigma_s)}{2\mathscr{F}(\sigma_s)\phi'(\sigma_s)} [J_{m-1}(\chi) - J_{m+1}(\chi)] \right\},$$
$$\chi = \sigma_s\phi'(\sigma_s)\xi_0, \tag{3.22}$$

where we have used Eqs. (3.19), the new definition $\chi = -z = \sigma_s\phi'(\sigma_s)\xi_0$, and the fact that $J_m(-y) = (-1)^m J_m(y)$.

The factor $\exp[i2\pi(\sigma_s + m\sigma_\xi)r]$ in Eq. (3.22) shows that the spectrum of $\tilde{g}(x)$ contains lines (delta functions) at the frequencies $\sigma_s + m\sigma_\xi$, $m = 0$, ± 1, ± 2, ... The source line at $\sigma = \sigma_s$ is reduced from amplitude $\mathscr{F}(\sigma_s)$ to a normally smaller value $[\simeq \mathscr{F}(\sigma_s)J_0(\chi)]$ by the velocity error. Energy removed from the source line appears in the ghost lines, whose intensities will normally decrease rapidly with increasing $|m|$. The second term of $\tilde{g}(r)$ [complex conjugate of Eq. (3.22)] produces a similar line-ghost pattern; in fact, it is the mirror image (about $\sigma = 0$) of the complex pattern corresponding to the first term. Generally, the high-order ghosts will be small enough that the second pattern can be assumed to have no overlap with the first, i.e., no contributions at positive frequencies. Then the spectrum of $\tilde{g}(r)$ at positive frequencies is the Fourier transform of Eq. (3.22). This line spectrum, normalized by the complex constant $F(\sigma_s)$ (the intensity of the source line in the absence of velocity error), is a display of the relative ghost spectrum. The complex relative error in the intensity of the source line is obtained by subtracting unity from its normalized intensity.

We find, using the above prescription, that the spectrum ghosts have complex relative intensities

$$E_m = (-1)^m e^{-im\psi_0} \left\{ J_m(\chi) + \frac{i\chi\mathscr{F}'(\sigma_s)}{2\mathscr{F}(\sigma_s)\phi'(\sigma_s)} [J_{m-1}(\chi) - J_{m+1}(\chi)] \right\},$$
$$|m| = 1, 2, \ldots; \qquad \chi = \sigma_s\phi'(\sigma_s)\xi_0. \tag{3.23}$$

The ghosts are located at frequencies $\sigma_s + m\sigma_\xi$. The complex relative error in intensity of the recovered source line is

$$E_0 = J_0(\chi) - 1 - [i\chi\mathscr{F}'(\sigma_s)/\mathscr{F}(\sigma_s)\phi'(\sigma_s)]J_1(\chi). \tag{3.24}$$

Equations (3.23) and (3.24) are the desired relations describing the first-order effects of the electrical filter, specifically the effects of nonconstant phase and of amplitude distortion $\mathscr{F}'/\mathscr{F}$, when the source is monochromatic and the variation in mirror drive speed is sinusoidal. Note that the source intensity error and ghosts vanish if $\xi_0 = 0$ and/or both ϕ' and $\mathscr{F}'$ are zero (as one would expect and as predicted in Section 3.3).

It should be remarked that Zachor and Aaronson (1979) obtained an expression for E_m which is identical to Eq. (3.23) except for the factor $(-1)^m$. We believe that Eq. (3.23) is the correct version; in any case, both forms correctly predict the numerical results reported in this chapter (Section 3.7) and by Zachor and Aaronson.

3.5. Maximum Relative Height of Spectrum Ghosts

Equation (3.23) predicts that the first-order ghost has complex relative intensity

$$E_1 = -e^{-i\psi_0}\left\{J_1(\chi) + \frac{i\chi\mathscr{F}'(\sigma_s)}{2\mathscr{F}(\sigma_s)\phi'(\sigma_s)}\,[J_0(\chi) - J_2(\chi)]\right\}. \tag{3.25}$$

It is easily demonstrated that the amplitude $|E_1|$ is larger than $|E_0|$ or $|E_2|$, $|E_3|$, . . . when $|\chi| \equiv |\sigma_s\phi'(\sigma_s)\xi_0|$ is sufficiently small. If $|\chi|$ is restricted to even smaller values, the Bessel functions in Eq. (3.25) can be represented with sufficient accuracy by the zeroth through second-order terms in their series expansions. Under this condition, Eq. (3.25) becomes

$$E_1 \simeq -e^{-i\psi_0}\frac{\sigma_s\xi_0}{2}\left[\phi'(\sigma_s)\left(1 - \frac{1}{8}\chi^2\right) + \frac{i\mathscr{F}'(\sigma_s)}{\mathscr{F}(\sigma_s)}\left(1 - \frac{3}{8}\chi^2\right)\right]. \tag{3.26}$$

We note that if $\frac{3}{8}\chi^2 \ll 1$, the second term in brackets in Eq. (3.26) is approximately $i\mathscr{F}'/\mathscr{F} = \exp(i\pi/2)\mathscr{F}'/\mathscr{F}$, while the first is approximately ϕ', i.e., the separate effects of filter amplitude distortion and phase variations are approximately 90° out of phase; they tend to add in quadrature in the amplitude of the first-order ghost. In general, for $\frac{3}{8}\chi^2 \ll 1$, the relative error E_0 and the relative ghost heights $E_{\pm m}$ can be approximated by

$$E_0 \simeq -\left[\frac{\sigma_s\xi_0\phi'(\sigma_s)}{2}\right]^2\left[1 + \frac{i\mathscr{F}'(\sigma_s)}{\phi'(\sigma_s)\mathscr{F}(\sigma_s)}\right], \tag{3.27}$$

$$\begin{aligned} E_m &\simeq (-1)^m\frac{e^{-im\psi_0}}{m!}\left[\frac{\sigma_s\xi_0\phi'(\sigma_s)}{2}\right]^m\left[1 + \frac{im\mathscr{F}'(\sigma_s)}{\sigma'(\sigma_s)\mathscr{F}(\sigma_s)}\right], \\ E_{-m} &\simeq \frac{e^{im\psi_0}}{m!}\left[\frac{\sigma_s\xi_0\phi'(\sigma_s)}{2}\right]^m\left[1 + \frac{im\mathscr{F}'(\sigma_s)}{\phi'(\sigma_s)\mathscr{F}(\sigma_s)}\right], \\ & m = +1, +2, \ldots . \end{aligned} \tag{3.28}$$

It is obvious that the ghosts and source intensity errors in a spectrum containing multiple source lines can be represented by a superposition of arrays of delta functions having relative intensities, or errors in intensity, given by Eqs. (3.23) and (3.24). Of course, the source frequency σ_s will have a different value in these equations for each of the superimposed arrays. Under finite spectral resolution, the complex ghost/error spectrum will consist of the delta function arrays convolved with the complex instrument line shape (ILS), which is the Fourier transform of the truncating/apodizing function.

If the source consists of well-separated emission lines and/or the interferogram is symmetrically truncated, it is valid (Mertz, 1965) to compute the spectrum as the amplitude of the complex Fourier transform of the interferogram. In this case the first-order spectrum ghosts of a line situated at $\sigma = \sigma_s$ will have relative amplitude

$$|E_1| \simeq \frac{\sigma_s \xi_0}{2} \left\{ [\phi'(\sigma_s)]^2 \left(1 - \frac{1}{8}\chi^2\right)^2 + \left[\frac{\mathscr{F}'(\sigma_s)}{\mathscr{F}(\sigma_s)}\right]^2 \left(1 - \frac{3}{8}\chi^2\right)^2 \right\}^{1/2} \tag{3.29}$$

for small $|\chi|$.

If the source lines cannot be resolved (cannot be separated with the available spectral resolution), and the interferogram is nonsymmetrically truncated, some type of correction procedure is necessary to eliminate spurious components in the spectrum arising from the nonsymmetrical (imaginary) part of the complex ILS. We prefer the phase correction method of Mertz (1965), although other techniques can be used to accomplish the same basic purpose. We are interested here not in the details of phase correction or in comparing different methods, but rather in how its use may affect the relative ghost amplitudes in the computed spectrum. We will consider only the Mertz method of phase correction.

Under finite spectral resolution, the first-order ghost of a line will have the normalized spectral shape

$$\mathscr{S}(\sigma) = E_1[C(\sigma) - iS(\sigma)], \tag{3.30}$$

where the complex constant E_1 is given by Eq. (3.26), and $C(\sigma) - iS(\sigma)$ is the complex ILS, or shape of a unit source line; $C(\sigma)$ and $S(\sigma)$ denote the Fourier cosine and sine transforms, respectively, of the truncating/apodizing function. In the actual spectrum the ghost intensity will be modified by the complex filter response [essentially $F(\sigma_s)$ = constant over the width of the ILS], and the ghost will be overlapped by the complex contributions of true source lines and other ghosts. These additive contributions are individually of the form $F(\sigma_s + \sigma_k)\mathscr{S}(\sigma - \sigma_k)E_1(\sigma_s + \sigma_k)/E_1(\sigma_s)$, where σ_k is the distance to the contributing line or ghost; however, $E_1(\sigma_s + \sigma_k)/E_1(\sigma_s)$ would be a real constant (relative line intensity) if the contribution is from an actual source line.

The Mertz method effects the following mathematical operations on the computed complex Fourier spectrum:

(a) It removes from the complex spectrum phase components measurable at low resolution; and

(b) it extracts the real part of the remainder.

The low-resolution phase spectrum removed by the method is called the *reference phase,* and is ordinarily determined from the Fourier transform of a narrow central portion of the interferogram. Thus the method eliminates the filter phase from the complex spectrum. The filter amplitude spectrum can be removed by spectral calibration of the interferometer. We have already eliminated the factors $\exp[-i\phi(\sigma_s)]$ and $\mathscr{F}(\sigma_s)$ by normalizing the complex ghost intensity to the complex intensity of the source line.

Expression (3.30) would represent an isolated, normalized source line in the absence of velocity errors if we set $E_1 = 1$; then the Mertz method, by extracting the real part of the expression, would yield a normalized line shape $\mathscr{S}(\sigma) = C(\sigma)$. Moreover, the contributions from other source lines after phase correction would be of the form $C(\sigma - \sigma_k)$ times the appropriate relative line intensities. In other words, the Mertz method yields the source line spectrum convolved with the symmetric line shape $C(\sigma)$ when there are no velocity errors in the mirror drive [or this spectrum times $\mathscr{F}(\sigma)$ if the instrument is not spectrally calibrated].

Clearly, the Mertz method does not eliminate the asymmetric part of a first-order ghost since E_1 is a complex constant in Eq. (3.30); nor does it eliminate the asymmetric parts of higher-order ghosts or the asymmetric component of error in the shape of a source line, which arises from the complex intensity error E_0 [Eq. (3.24)].

The Mertz correction method will reduce the shape of a first-order ghost to $\mathrm{Re}[\mathscr{S}(\sigma)]$, the real part of Eq. (3.30). The ratio of the ghost height at its center to the line height at its center is then simply

$$\mathrm{Re}[\mathscr{S}(0)]/C(0) = \mathrm{Re}(E_1), \tag{3.31}$$

since $S(0) = 0$. In this sense, the relative ghost height after Mertz correction is

$$|E_1|\ \mathrm{Re}[\exp(-i\phi_{E_1})] = |E_1| \cos(\psi_0 - \psi_1), \tag{3.32}$$

where ϕ_{E_1} is the phase of E_1, and ψ_1 is the phase of the complex expression in braces in Eq. (3.25):

$$\psi_1(\sigma_s) = \tan^{-1}\left[\frac{\chi\mathscr{F}'(\sigma_s)}{2\mathscr{F}(\sigma_s)\phi'(\sigma_s)}\ \frac{J_0(\chi) - J_2(\chi)}{J_1(\chi)}\right]. \tag{3.33}$$

According to Eq. (3.32), the Mertz method of phase correction results in relative first-order ghost heights equal to or smaller than $|E_1|$, the value obtained when the spectrum is computed as the Fourier amplitude spectrum. In fact, if the constant phase angle ψ_0 of the sinusoidal velocity error were exactly $\psi_1(\sigma_s)$, the first-order ghost would vanish. Higher-order ghosts would be present, however, since the equivalent of Eq. (3.32) for the mth-order ghost is [see Eq. (3.23)]

$$|E_m| \cos(m\psi_0 - \psi_m), \tag{3.34}$$

where ψ_m is the phase of the factor in braces in Eq. (3.23).

3.6. Reduction of Ghost Heights by Delayed Sampling

Equation (3.29) explains the motivation for using delayed fringe-reference sampling. Introducing a fixed time delay to the reference fringes or the fringe-triggered sampling commands is equivalent, in the above analysis, to subtracting a linear function from the filter phase spectrum. This tends, at most frequencies, to reduce the quadrature component $[\phi'(\sigma_s)]^2$ in Eq. (3.29). All realizable analog filters have some phase distortion (spectral variations in ϕ'), which makes it impossible to reduce or eliminate this component at all possible source frequencies σ_s. However, the reduction in ghost height can be substantial, particularly at frequencies well below the cutoff of typical low-pass filters, where $\phi'(\sigma)$ and $\mathcal{F}(\sigma)$ are both nearly constant.

In the case of delayed sampling, the effective (residual) phase spectrum of the filter is

$$\beta(\sigma) = \phi(\sigma) - 2\pi x_d \sigma, \tag{3.35}$$

where $x_d = 2\langle v \rangle t_d$, and t_d is the fixed time delay. This equation is merely a statement of the Fourier shift theorem. The definition of x_d assumes that optical retardation is twice the physical displacement of the scanning mirror, as does the definition $x = 2\langle v \rangle t$ adopted in Section 3.2. (More generally, $x \equiv \langle dr/dt \rangle t$ and $x_d \equiv \langle dr/dt \rangle t_d$, where $\langle dr/dt \rangle$ is the average retardation velocity. The value of $\langle dr/dt \rangle$ depends on the design of the interferometer, but generally it is a multiple of $2\langle v \rangle$). According to Eq. (3.35), delayed sampling effects a reduction in local phase spectrum slope equal to $2\pi x_d = 2\pi(2\langle v \rangle t_d)$. If we let $\phi'(\sigma)$ denote either the actual phase spectrum slope or the slope of the residual phase spectrum, the analytical results given above can be used for either the conventional fringe reference case or the delayed fringe reference case.

In frequency regions where the filter group delay is effectively zero as a result of delayed sampling, the ghost heights will depend on second and higher derivatives of the filter phase as well as on filter amplitude distortion. The developed expressions for the relative ghost heights may not give accurate results in this case unless $\sigma_s\phi'' \ll \mathcal{F}'/\mathcal{F}$, since we have neglected the higher derivatives of ϕ and $\mathcal{F}$. In other words, the analytical model is more useful, in the delayed sampling case, for estimating the maximum ghost amplitude over a broad spectral region, than for accurately quantifying the residual error where it is very small.

3.7. Calculated First-Order Ghost Heights for Common Low-Pass Filters

A. Numerical Results

We calculated the first-order relative ghost amplitudes for several common filter designs, for both the conventional and delayed sampling cases. Most of the filters examined are low-pass Butterworth and Bessel, which are characterized by maximally flat amplitude and maximally linear phase, respectively. The spectral response functions of these filters can be written (Weinberg, 1962):

$$F_{\mathrm{Bu}}(\sigma) = \prod_{k=0}^{n-1} [i\sigma/\sigma_c - \exp[i(2k + 1 + n)\pi/(2n)]]^{-1} \tag{3.36}$$

for the Butterworth, and

$$F_{\mathrm{Be}}(\sigma) = \left[\sum_{k=0}^{n} \frac{(2n - k)!(i\alpha_n\sigma/\sigma_c)^k}{2^{n-k}k!(n - k)!}\right]^{-1} \tag{3.37}$$

for the Bessel, where n is the number of poles, σ_c is the selected filter bandwidth (frequency at which the attenuation is -3 dB, corresponding to $\mathcal{F}(\sigma) = 0.707\ldots$), and α_n are the values given in the last column of Table 11-5(b) in Weinberg (1962). Equations (3.36) and (3.37) are convenient forms for computing $F(\sigma)$ [and subsequently its amplitude $\mathcal{F}(\sigma)$ and phase $\phi(\sigma)$] since they are easily programmed for the computer using complex arithmetic coding. Construction of the filters requires knowledge of the s plane pole coordinates, which are tabulated in standard texts on electrical network synthesis. For $n = 1$, Eqs. (3.36) and (3.37) both represent a simple RC filter.

Except where stated, the sampling delay has been matched to the filter delay at zero frequency, i.e., the residual phase $\beta(\sigma)$ is defined as

$$\beta(\sigma) = \phi(\sigma) - \sigma\phi'(0). \tag{3.38}$$

Thus the approximate relative ghost amplitude $|E_1|$ for the case of delayed sampling was evaluated using

$$\phi'(\sigma) \rightarrow \phi'(\sigma) - \phi'(0) \tag{3.39}$$

in Eq. (3.29). This corresponds to the average retardation delay

$$x_{\mathrm{d}} = \phi'(0)/2\pi \tag{3.40a}$$

or time delay

$$t_{\mathrm{d}} = \phi'(0)/(2\pi\langle dr/dt\rangle) = \phi'(0)/[2\pi(2\langle v\rangle)]. \tag{3.40b}$$

For each of the Butterworth and Bessel filters, we calculated $F(\sigma)$ according to Eq. (3.36) or (3.37), and then determined the corresponding $\mathcal{F}(\sigma)$, $\mathcal{F}'(\sigma)$, $\phi(\sigma)$, and $\phi'(\sigma)$. The first-order ghost amplitude $|E_1|$ for the conventional sampling case was obtained using Eq. (3.25), which is a more accurate small-χ representation than Eq. (3.26) or (3.29). The first-order relative ghost amplitudes will be presented as a function of source frequency σ_s normalized to the filter cutoff frequency σ_c. All of the calculated results correspond to a 1% relative velocity error ($\xi_0 = 0.01$).

Figure 3.1 shows the relative ghost amplitude for a simple RC filter.

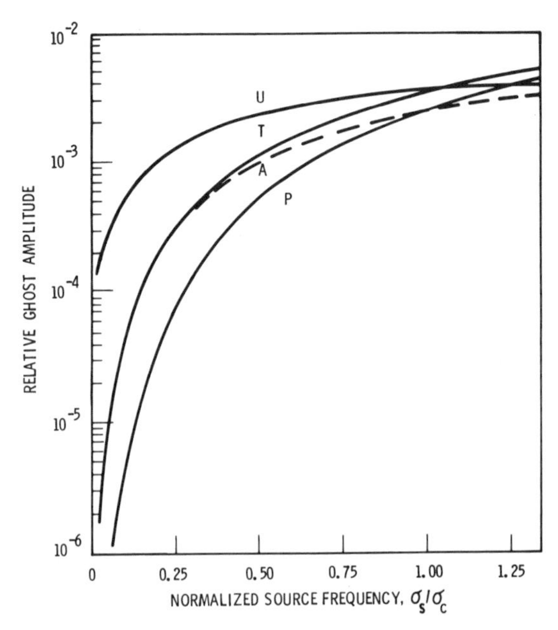

FIG. 3.1. Relative ghost amplitude corresponding to a sinusoidal velocity error of relative amplitude $\xi_0 = 0.01$ for the case of an RC filter. Curve U is the error when conventional fringe-reference sampling is used. Curve T is the error for delayed fringe-reference sampling, equal to the quadrature sum of components A and P due to amplitude and phase distortion (Zachor and Aaronson, 1979).

Curve U is the result for conventional sampling. Curve T is the total ghost amplitude for delayed sampling, which has the quadrature components A and P due to amplitude distortion and phase distortion, respectively [the imaginary and real parts of Eq. (3.26) exclusive of the factor $\exp(-i\psi_0)$]. Note that for the delayed sampling case, the spectrum error is due almost entirely to amplitude distortion when $\sigma_s/\sigma_c \leq 0.5$. Amplitude and phase distortion contribute equally to the first-order ghost of a source line located right at the filter cutoff.

Corresponding results for two-, three- and six-pole Butterworth filters are shown in Figs. 3.2, 3.3, and 3.4, respectively. For the delayed sampling case the spectrum error is dominated by phase distortion at low frequencies and by amplitude distortion at high frequencies; the crossover point moves toward higher frequency as the number of poles is increased. Also, the ghost amplitude relative to that for the conventional sampling case becomes smaller as the number of poles is increased, particularly at frequencies σ_s less than $0.5\sigma_c$. The actual ghost amplitude is quite insensitive to the number of poles when $\sigma_s < 0.5\sigma_c$, and increases with the number of poles at higher frequency.

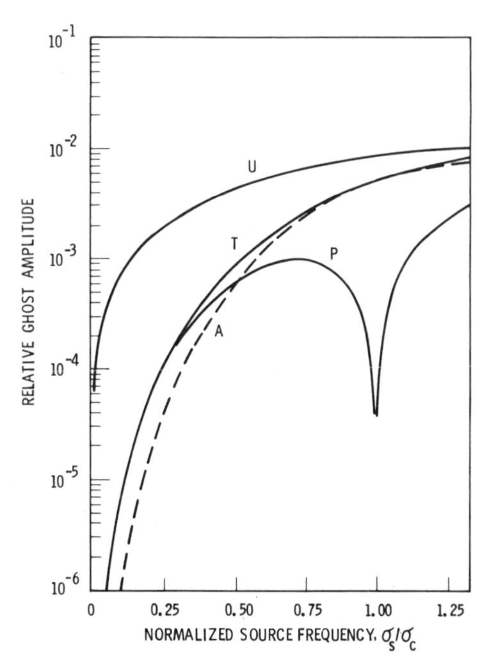

FIG. 3.2. Same as Fig. 3.1, except the filter is a two-pole Butterworth. P is zero at $\sigma_s/\sigma_c = 1$, where β' changes sign (Zachor and Aaronson, 1979).

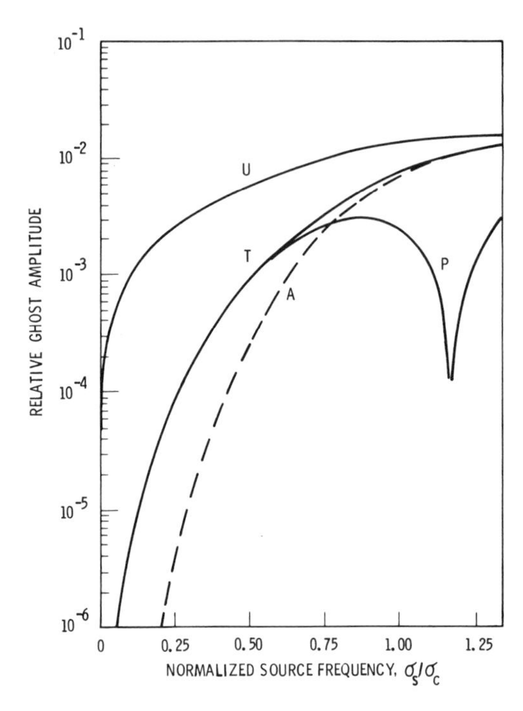

FIG. 3.3. Same as Fig. 3.1, except the filter is a three-pole Butterworth (Zachor and Aaronson, 1979).

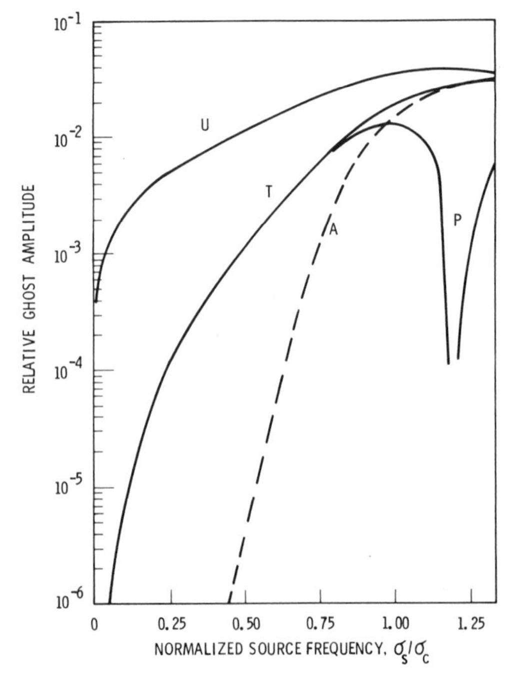

FIG. 3.4. Same as Fig. 3.1, except the filter is a six-pole Butterworth (Zachor and Aaronson, 1979).

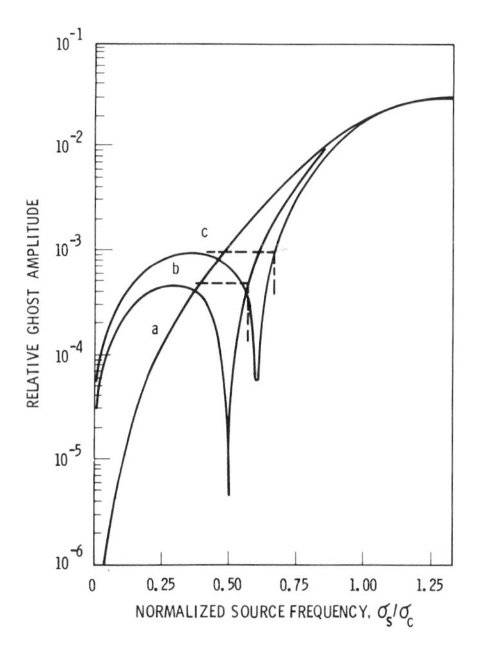

FIG. 3.5. Ghost amplitude for a six-pole Butterworth filter. Curve *a* is the same as curve *T* of Fig. 3.4. Curves *b* and *c* are the corresponding results when the sampling delay equals the filter group delay at $\sigma = 0.5\sigma_c$ and $\sigma = 0.6\sigma_c$, respectively (Zachor and Aaronson, 1979).

Figure 3.5 compares the ghost amplitude distributions for three values of the sampling delay for the case of a six-pole Butterworth filter. Curve *a* is the same as curve *T* of Fig. 3.4; the sampling delay is equal to the filter delay at zero frequency. Curves *b* and *c* correspond to sampling delays equal to the filter group delay (phase spectrum slope divided by 2π) at $0.5\sigma_c$ and $0.6\sigma_c$. Note that if the spectrum contains no lines at frequencies higher than $\sigma_s \simeq 0.57\sigma_c$ (or is optically filtered to this spectral bandwidth), the maximum ghost amplitude for case *b* is four times smaller than that for case *a*. The maximum ghost height would be reduced by approximately the same factor by using the delay of case *c* if the spectral bandwidth is $0.67\sigma_c$. However, within these two optical bands the maximum ghost heights differ by a factor of two between cases *b* and *c*. These examples illustrate that there is an optimum sampling delay for a given ratio of optical to filter bandwidth, and that the maximum ghost amplitude is quite sensitive to both parameters.

Figure 3.6 shows the results for a three-pole Bessel filter. The total spectrum error for the delayed sampling case is overwhelmingly dominated by amplitude distortion at all frequencies. It is higher than the total

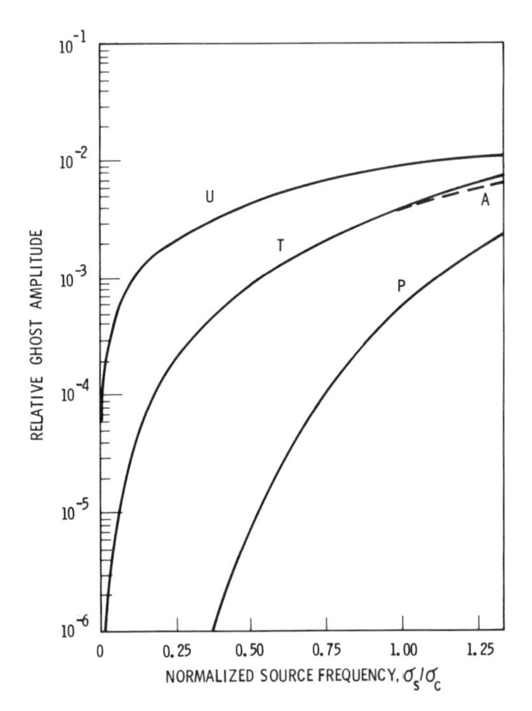

FIG. 3.6. Same as Fig. 3.3, except the filter is a three-pole Bessel (Zachor and Aaronson, 1979).

error for a three-pole Butterworth filter when $\sigma_s/\sigma_c \lesssim 0.5$ and lower then $\sigma_s/\sigma_c \gtrsim 0.5$. We found that the total errors for three- and six-pole Bessel filters are nearly equal at all frequencies [for the delayed sampling case, with β defined by Eq. (3.38)].

Figure 3.7 summarizes the results for total first-order ghost amplitude in the delayed sampling case (the curves T of the previous figures). Included in Fig. 3.7 is the total computed error for a three-pole Chebyshev filter with 0.5 dB gain ripple. This last case is interesting because of the local minimum at $\sigma_s/\sigma_c = 0.5$, which corresponds to a total error approximately one-half that of the other filters. Table 3.I gives the sampling delays used in the computations of relative ghost amplitude. The nondimensional delay values are $x_d\sigma_c$, or, equivalently, the time delay times the temporal-frequency filter bandwidth. The third column of the table lists the figures and specific curves to which the delays apply.

B. Discussion of Results

We have remarked already on how the type of filter and selected sampling delay determines the spectral distribution of first-order ghost amplitude and the relative contributions of filter amplitude and phase distortion.

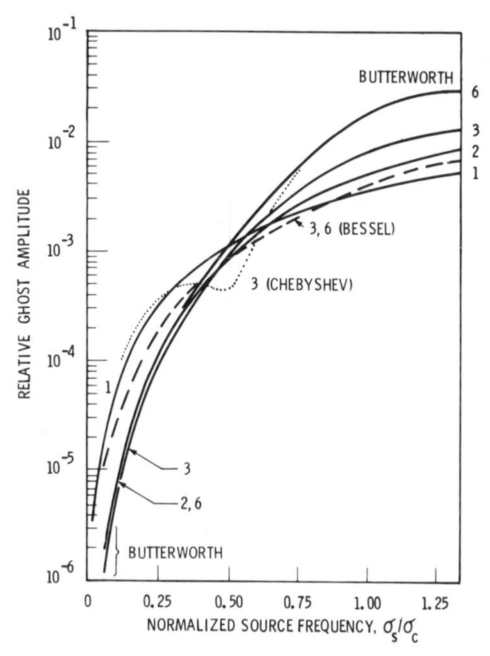

FIG. 3.7. Comparison of the relative ghost amplitudes for different filters, for the delayed sampling case. Except for the three-pole Chebyshev, these are the same as curves *T* in the previous figures. The gain ripple for the Chebyshev is 0.5 dB (Zachor and Aaronson, 1979).

These observations were qualitative, and generally confirm one's expectations based on the well-known characteristics of Bessel and Butterworth filters. The following is a quantitative discussion of the errors and their influence in the selection of the filter under various circumstances.

TABLE 3.I

SAMPLING DELAYS USED IN CALCULATING THE RESIDUAL FILTER PHASE SPECTRUM

Filter	Number of poles	Figures, curve	Delay $x_d\sigma_c$
RC	(1)	1,7	0.158
Butterworth	2	2,7	0.225
	3	3,7	0.318
	6	4,5a,7	0.615
	6	5b	0.686
	6	5c	0.730
Bessel	3	6,7	0.279
	6	7	0.430
Chebyshev	3	7	0.395

To minimize the relative ghost amplitude in the delayed-sampling case, a low-pass Bessel rather than low-pass Butterworth filter should be used only if the source spectrum lines are dispersed throughout the entire filter band. The use of delayed sampling will then reduce the maximum error relative to the conventional sampling case by a factor of 2.3 for a three-pole filter (see Fig. 3.6), or by a factor of 3.7 for a six-pole filter (not shown). The maximum relative ghost amplitude in either case is approximately 4×10^{-3} if the relative amplitude of the velocity error is 0.01. These numbers are not very impressive; the reason is that all filters have significant amplitude or phase distortion near the filter cutoff, which tends to limit the effectiveness of delayed sampling at these frequencies. A much larger benefit accrues from delayed sampling when the maximum spectrum line frequency is one-half the filter cutoff or smaller. When $\sigma_s/\sigma_c \lesssim 0.4$, a Butterworth filter gives smaller ghost amplitudes than a Bessel filter.

It should be noted that the total spectrum error shown in Figs. 3.1–3.7 for the delayed sampling case can be scaled in proportion to the relative amplitude of the sinusoidal velocity error for relative velocity errors up to at least 0.1 (10%). The ghost amplitude scales linearly because for the filters considered, $\chi = \sigma_s \phi'(\sigma_s)\xi_0$ is sufficiently small that Eq. (3.26) is a valid approximation to Eq. (3.25), and in Eq. (3.26) the terms $\frac{3}{8}\chi^2$ and $\frac{1}{8}\chi^2$ are small relative to unity.

Figure 3.7 shows that when the maximum spectrum frequency is approximately one-half the filter cutoff frequency, the maximum first-order relative ghost amplitude (delayed sampling case) is approximately ten times smaller than the relative velocity error, no matter which of the filters is used, excepting the Chebyshev filter. The error may be reduced to $\xi_0/20$ or less by using a six-pole Butterworth and optimal sampling delay (as indicated by Fig. 3.5), or by using a three-pole Chebyshev filter.

Of course, aliasing of detector noise is an important consideration in selecting the filter type and the number of filter poles. Generally, as the filter cutoff frequency is increased the residual ghost amplitudes will be reduced, but at the expense of increased noise aliasing unless the interferogram is significantly oversampled. Oversampling combined with real-time numerical filtering can be used to preserve the output data rate and virtually eliminate aliasing effects. This technique is discussed in Section 3.9.

Finally, we note that if the maximum spectrum frequency is less than $\sigma_c/4$, a Butterworth filter with between two and six poles will result in a maximum relative ghost amplitude equal to approximately $\xi_0/100$, for the delayed sampling case (Fig. 3.7). This ghost amplitude is 40 times smaller than it would be without delayed sampling. An even smaller error might be achieved by using an optimal sampling delay.

Although we have considered only low-pass filters, the general results [Eqs. (3.23)–(3.29)] are applicable to bandpass filters as well. Also, we have considered only standard types of low-pass filters. The method of all-pass network compensation applied to the Butterworth cases may yield higher-order low-pass filters that substantially reduce the spectrum error. The method should also be useful in designing optimal bandpass networks.

3.8. Laboratory Measurements of Spectrum Ghosts

To test the applicability of delay compensation in a practical situation, we studied the output of a Fourier transform spectrometer when the drive velocity was deliberately modulated. The radiation source was a 3.39-μm He–Ne laser which provided sufficient signal intensity that detector noise was negligible. The limiting noise, which determined the precision with which the ghosts could be measured, was due to fluctuations in the laser output and variations in the efficiency of modulation of the interferometer, due presumably to variations in alignment of the moving mirror. The peak-to-peak noise in the signal amplitude was about 0.25% of the peak-to-peak signal. This type of amplitude variation produces noise in the spectrum near the laser line that was typically 0.2–0.3% of the laser signal.

The interferometer is an EOCOM model 7001. This instrument has a maximum path difference of 16 cm. The moving mirror is supported by dual air bearings that provide a smooth, low-vibration, low-friction motion. The motion is controlled by a velocity servo with a linear variable differential transformer (LVDT) velocity sensor. Over the scan range used in this experiment, the servo maintains the velocity constant to better than 0.5%. The servo system has a large bandwidth (~400 Hz) to maintain constant velocity in the presence of environmental vibration. We used this capability to modulate the velocity in a sinusoidal manner.

The velocity was sensed by measuring the frequency of the visible laser reference fringes used for sampling. The reference laser provides a signal that is faster and more linear than the LVDT. It was possible to modulate the mirror velocity up to 30% of its mean velocity with modulation frequencies up to 150 Hz. The system was operated open loop; a sinusoidal error signal was injected, along with the signal from the LVDT, at the summing function of the servo. The velocity resulting from these error signals was monitored, and the sinusoidal error signal adjusted to produce the desired modulation.

In the experiments described, the interferometer was operated at 1 cm maximum optical path difference, and at a velocity such that the unmodulated signal from the 3.39-μm laser source was approximately 4.3 kHz. Each interferogram, after sampling, consisted of 8192 values spaced 2×0.6328 μm apart in retardation. The interferogram was transformed into a spectrum with the assumption that the central sample corresponded to zero path difference.

The signal chain includes an InSb detector, a preamplifier, low-pass and high-pass filters, sample and hold amplifiers, and a 15-bit analog-to-digital (A/D) converter. The low- and high-pass filters are nominally four-pole Butterworth designs. The high-pass filter was set for a 10-Hz cutoff and had a negligible effect on the results. It is the low-pass filter that introduces the phase and amplitude modulation we wish to consider.

The filter cutoff frequency σ_c is selectable to be approximately $4\sigma_s$, $2\sigma_s$, or σ_s, where σ_s is the signal frequency. For a Butterworth low-pass filter of four poles, the theoretical frequency response at $\sigma_c/4$ is extremely flat. When the number of poles is four or more, the accuracy of components required to approach theoretical performance is very high, and we found that the filter amplitude response was not nearly as flat as desired. Adjustment of the response in the vicinity of $\sigma = \sigma_c/4$ for this filter was made on an ad hoc basis to make the amplitude response flat to 0.03% for a frequency deviation of $\pm 10\%$. This produced a transfer function which was not exactly that of a Butterworth filter for any of the three filters, but gave us one filter that produced no amplitude modulation at the source frequency.

The phase shift in the filters was measured directly by comparing the phase of input and output waveforms on an oscilloscope. The phase and amplitude characteristics of the filter in the neighborhood of the signal frequency can be described in terms of the parameters

$$\alpha \equiv \sigma_s \phi'(\sigma_s) \qquad \text{and} \qquad \gamma \equiv \sigma_s \mathcal{F}'(\sigma_s)/\mathcal{F}(\sigma_s).$$

The measured values of amplitude $\mathcal{F}$ and phase ϕ and the corresponding α and γ values for the three filters are given in Table 3.II.

The response of the preamp alone was measured to be flat, but the amplitude response of the detector–preamp combination showed some variation, probably due to detector capacitance. This was compensated in the preamp. However, because the input signal was optical rather than electrical, the compensation could not be done with great accuracy. The preamp–detector combination was measured to be flat to $\pm 0.2\%$.

The first test was trivial. The frequency of the modulation was varied, and it was determined that the ghosts appeared at the proper locations in the spectrum.

The next test compared the amplitude of the $E_{\pm 1}$ ghosts with the prediction of the theory for a fixed modulation frequency and variable modula-

TABLE 3.II

MEASUREMENTS OF CHARACTERISTICS OF LOW-PASS ELECTRICAL FILTERS

	σ(Hz)	$\mathscr{F}$[a]	t_d(μsec)[b]	$\phi = 2\pi f\tau$[c]
Filter 4	3800	0.4883	129 ± 1	3.080
	4300	0.3459	126	3.404
	4800	0.2437	120	3.619
Filter 5	3800	0.9392	54.5 ± 0.5	1.301
	4300	0.9163	55.5	1.500
	4800	0.8887	56.0	1.689
Filter 6	3800	1.0078 ± 0.0003	21.6 ± 0.2	0.5157
	4300	1.0072	21.5	0.5809
	4800	1.0077	21.7	0.6545

[a] Filter 4: $\alpha = \alpha_0 = \sigma_s\phi'(\sigma_s) = 2.32$, $\gamma = \sigma_s\mathscr{F}'/\mathscr{F} = -3.04$; filter 5: $\alpha = 1.66$, $\gamma = -0.24$; Filter 6: $\alpha = 0.59$, $\gamma = 0$.

[b] The effective delay, which includes 2 μsec for delay in the preamplifier.

[c] f is the temporal frequency.

tion amplitude ξ_0. Here we encountered some difficulty. The theory predicts that for values of $\chi = \sigma_s\xi_0\phi'$ up to 0.2, the ghost amplitude normalized to the modulation amplitude, $|E_1|/\xi_0$, should be almost constant, and that the upper and lower ghosts, E_1 and E_{-1}, should have equal amplitude. The observations showed considerable asymmetry in the upper and lower ghosts, and a normalized amplitude which decreased much more rapidly with ξ_0 than calculated. The results are summarized in Table 3.III. One possible source of the discrepancy is the fact that the theory assumes that the modulation is sinusoidal in optical retardation whereas in the actual experiment the modulation is sinusoidal in time. The effect is negligible if the modulation of the velocity is much smaller than

TABLE 3.III

VARIATION OF GHOST HEIGHT WITH MODULATION AMPLITUDE FOR FILTER 5

	Calculated	Observed	
ξ_0	$\lvert E_{\pm1}\rvert/\xi_0$	$\lvert E_{-1}\rvert/\xi_0$	$\lvert E_{+1}\rvert/\xi_0$
0.025	0.820	0.76	0.80
0.050	0.819	0.64	0.70
0.100	0.815	0.51	0.61
0.150	0.809	0.62	0.69
0.300	0.773	0.42	0.45

the mean velocity, but for values of ξ_0 as large as 0.3, the effect probably becomes significant, although its predicted magnitude has not been calculated. We were able to demonstrate theoretically that the first-order ghosts would be asymmetrical when the modulation is sinusoidal in time.

For the largest modulation amplitude ($\xi_0 = 0.3$) and for the filter with cutoff equal to twice the signal frequency, we observed a series of ghosts out to order 3 on each side of the source line. The third-order ghost was less than 0.25% of the parent, and higher order ghosts were lost in the noise. As noted above, the amplitude of the first-order ghost was significantly lower than predicted, but the ratios of the higher order ghosts to the first-order ghost were in satisfactory agreement with the theoretical prediction.

The final test was the most crucial. The amplitude of the first-order ghost, normalized to the modulation amplitude, was measured for each of the three filters, for a range of sampling delays including that corresponding to the propagation delay in the filter. The results are in qualitative agreement with the predictions, but disagree quantitatively.

Using the definitions $\alpha = \sigma_s\phi'(\sigma_s)$ and $\gamma = \sigma_s\mathscr{F}'(\sigma_s)/\mathscr{F}(\sigma_s)$, Eq. (3.23) can be written

$$\frac{|E_m|}{\xi_0} = \left| \frac{J_m(\xi_0\alpha)}{\xi_0\alpha}\,\alpha + \frac{i}{2}\,[J_{m-1}(\xi_0\alpha) - J_{m+1}(\xi_0\alpha)]\gamma \right|. \tag{3.41}$$

Expanding the Bessel functions, and neglecting terms in $(\xi_0\alpha)^4$ and higher powers, we find that the normalized first-order ghost amplitude is

$$\frac{|E_1|}{\xi_0} \simeq \frac{1}{2}\left[\left(1 - \frac{(\xi_0\alpha)^2}{4}\right)\alpha^2 + \left(1 - \frac{3}{4}\,(\xi_0\alpha)^2\right)\gamma^2\right]^{1/2}. \tag{3.42}$$

If α_0 is the value of α for no sampling delay, then with a time delay t_d, $\alpha = \alpha_0 - 2\pi f_s t_d$, where f_s is the temporal frequency corresponding to σ_s.

In Fig. 3.8 we show Eq. (3.42) plotted as a function of α for values of γ corresponding to each of the filters used. The calculation assumes $\xi_0 = 0.15$, the value used in most of the measurements, although the functions vary little with ξ_0. The right-hand end of each curve corresponds to $\alpha = \alpha_0$, i.e., no sampling delay.

In Fig. 3.8 we also show the observed values of $|E_1|/\xi_0$ for the three filters for a variety of values of the sampling delay t_d. For each point, a symbol and a vertical line are plotted. The symbol locates the average intensity of the upper and lower first-order ghosts. The lower and upper ends of the vertical lines correspond to the intensity of the lower and upper ghosts, respectively.

The curve $\gamma = -3.04$ corresponds to the filter operated with the signal frequency approximately equal to the cutoff frequency. The transmittance

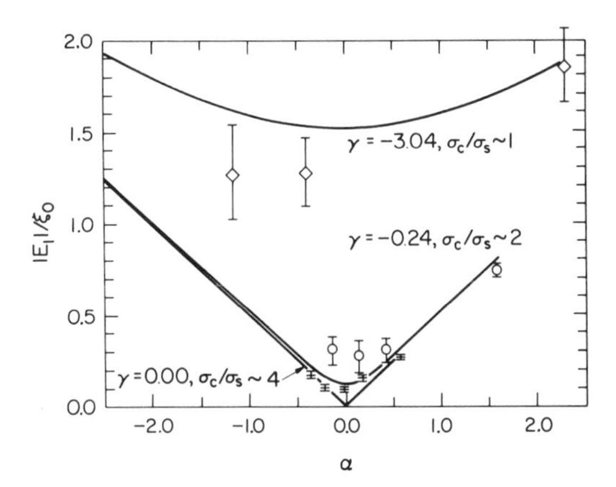

FIG. 3.8. Calculated and measured first-order relative ghost amplitude $|E_1|$ normalized by the amplitude ξ_0 of the sinusoidal relative velocity error. The results are shown for three filters, as a function of $\alpha = \sigma_s\phi'(\sigma_s) - 2\pi f_s t_d$, where f_s is the temporal frequency corresponding to σ_s, and t_d is the delay introduced in the sampling channel. The lower and upper extremities of the vertical bars through the data points indicate the measured values of $|E_{-1}|/\xi_0$ and $|E_1|/\xi_0$, respectively.

of the filter is varying quite rapidly with frequency, and amplitude modulation dominates the ghost intensity. The ghost amplitude is large—almost a quarter of the parent line—and, as noted above, the observed amplitude is significantly less than the calculated amplitude when the ghosts are large. Since amplitude modulation is the dominant factor, delayed sampling has little effect.

The curve $\gamma = -0.24$ is for a low-pass filter operated with the signal frequency near the center of the passband. With no sampling delay ($\alpha = 1.6$), phase modulation dominates amplitude modulation. As α goes to zero (which corresponds to a sampling delay $t_d \simeq 50$ μsec, about half the time between successive samples) the ghost height drops by approximately a factor of five, in reasonable agreement with the calculation. This is a practical case since it requires a bandpass only twice the highest signal frequency, and shows that delayed sampling can give a very significant reduction in spectrum ghosts.

The lowest curve is for the filter with cutoff four times the signal frequency. This represents the most difficult case experimentally because great care is needed to make the system amplitude response absolutely flat. System measurements and careful examination of the interferogram indicate that the amplitude modulation if present at all is extremely small. We can set an upper limit of 0.01 for γ. In this case the ghosts should essentially vanish for the correct sampling delay. In practice, for

$\xi_0 = 0.15$, the minimum ghost amplitude $|E_1|$ was approximately 0.018, a factor of three less than the value with no sampling delay. This illustrates not only the extreme difficulty of making a system that does not produce ghosts, but also the fact that a carefully made system including delayed sampling will perform adequately for most purposes. A good system should have velocity errors at any one frequency substantially less than 1%. With the measured minimum ghost amplitude scaled by ξ_0, the peak-to-peak noise introduced by velocity modulation should be less than 0.1%. In high resolution work it is unusual to have other noise sources any smaller than this in a single scan. Hence, in a practical sense, the reduction in ghost amplitude due to delayed sampling is sufficient for most purposes even if the ghosts are not eliminated completely.

3.9. Use of Digital Filtering to Preserve Signal-to-Noise and Data Rate

In Section 3.7 we showed that the ghosts caused by velocity variations can be made very small by using delayed sampling and by choosing an electrical filter cutoff frequency much higher than the maximum source frequency. However, an increase in filter bandwidth implies increased (aliased) noise at the source frequencies, unless the sampling frequency is increased sufficiently. In other words, it would appear that a reduction in ghost amplitude must be traded for an increase in noise and/or data rate.

In this section we will illustrate, by an example, how the noise/data rate penalties can be reduced or avoided through real-time digital filtering of an oversampled interferogram. A simple version of the technique was applied by Mankin (1978) in an FTS measurement program that obtained high-resolution absorption spectra of the upper atmosphere. The method, in combination with delayed sampling, helped to overcome the effects of 10% relative velocity variations induced by aircraft vibration.

Let us assume that the electrical filter cutoff σ_c is four times the maximum source frequency σ_{max} (which might be determined by an optical filter or the detector). This choice for σ_c would result in relative first-order ghost heights at least 100 times smaller than the relative zero-to-peak velocity error, as shown in Section 3.7. Suppose that we also choose to sample the interferogram at a rate equal to $6\sigma_{max}$, i.e., the retardation sampling interval is $\Delta r = 1/6\sigma_{max}$. Figure 3.9a is a schematic display of the filter spectrum, the optical band (range of measured source frequencies), and their first aliases. Clearly, there is no aliasing of noise into the optical band.

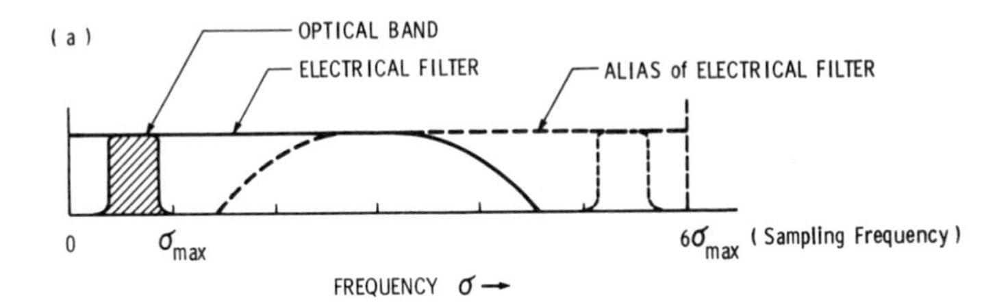

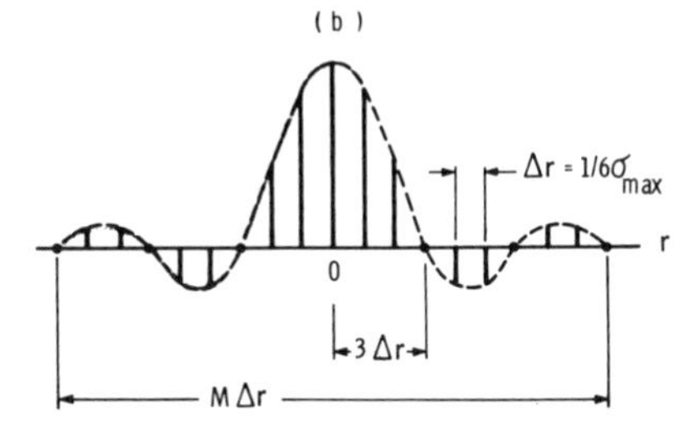

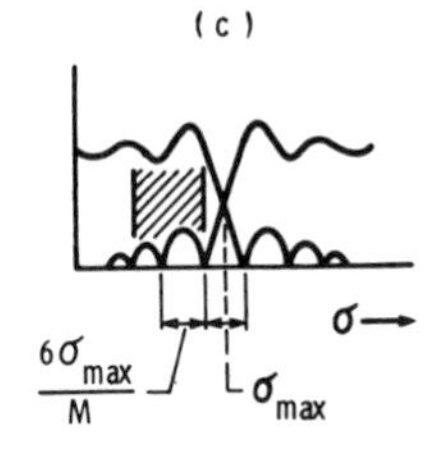

FIG. 3.9. Schematic representation of (a) the filter spectrum, its first alias, and the optical band; (b) the window function weights; and (c) the numerical filter corresponding to the window function. The figure illustrates how over-sampling and real-time digital filtering can be used to preserve both signal-to-noise and data rate when delayed sampling and a wide-band analog filter are used to reduce the effects of velocity errors.

The selected sampling rate in this example is three times greater than would be required under ideal conditions, i.e., if the filter had constant amplitude and linear phase between $\sigma = 0$ and $\sigma = \sigma_c$, there would be no ghosts and no reason to choose σ_c greater than σ_{max}; with $\sigma_c = \sigma_{max}$, and the sampling frequency equal to the Nyquist rate $2\sigma_{max}$, there would also be no aliasing of noise into the optical band.

The output data rate of the system in the actual, nonideal case can be restored to approximately the Nyquist rate by performing some kind of weighted average (numerical filtering) over successive groups of samples. The groups may consist of more than three samples, but the process should effect, in this example, a three-to-one reduction in the total number of samples. A particularly useful "window function", or set of weighting coefficients, is displayed in Fig. 3.9b. It consists of a finite-length "comb" (uniformly spaced set of unit impulses) multiplied by the familiar sinc or sin r/r function. The numerical filtering operation consists of convolving this function with the oversampled data set; however, the function is shifted by 3 Δr rather than Δr (Δr is the spacing in the comb) each time a new output sample is calculated. Note that the first zero in the sinc envelope is at 3 Δr. The equivalent of this operation in the frequency domain is to multiply Fig. 3.9a by the Fourier transform of Fig. 3.9b, whose amplitude is shown schematically in Fig. 3.9c. This transform is an ideal low-pass filter, except for the Gibbs effect, which results from the finite length of the window function. In particular, its cutoff is at σ_{max} (one-sixth the oversampling frequency) and its first alias is centered at $2\sigma_{max}$ (the output data rate).

We see that the method preserves the original data rate and increases the spectrum noise in the optical band only through the sidebands of the first alias of the numerical filter. As indicated in Fig. 3.9c, the width of the Gibbs sidelobes for the sinc window function is $6\sigma_{max}/M$, where M is the number of samples included in the numerical filtering operation; their heights are inversely proportional to this quantity. The sidelobes can be drastically reduced in amplitude (and broadened somewhat) by multiplying the sinc window function by a Hamming, hanning, or Kaiser function (Rabiner and Gold, 1975).

Note that it was sufficient in this example to oversample by a factor of three, even though the electrical filter cutoff is four times the maximum source frequency. Even so, the oversampling rate may stress the capabilities of available A/D converters, especially in FTS applications requiring large dynamic range in the sampled interferogram. The numerical filtering also requires a series of multiplications and additions that must be performed at the oversampling rate. Of course, multiple (parallel) converters and processors can be used to decrease the rate, but this will increase the complexity of the system. These kinds of problems led Mankin (1978) to use a simplified version of the method described here; he used a four-pole Butterworth filter with cutoff frequency three times the maximum source frequency, oversampled by approximately a factor of four, and merely summed four successive A/D conversions to obtain each output sample.

3.10. Some Effects of Random Velocity Variations

Equation (3.12) gives the distorted interferogram in the form

$$\tilde{g}(r) = \int_{-\infty}^{\infty} e^{2\pi i \sigma r} F(c\sigma) G(\sigma)\, d\sigma. \tag{3.43}$$

It will be recalled that $G(\sigma)$ is the true spectrum [really the Fourier transform of the unfiltered interferogram $g(r)$], and $c \equiv 1 + \xi(r) = v(r)/\langle v \rangle$. The error in $\tilde{g}(r)$, after truncation/apodization by the function $b(r) \leftrightarrows \mathscr{S}(\sigma)$, is

$$\varepsilon(r) \equiv b(r)[\tilde{g}(r) - g(r)] = b(r) \int_{-\infty}^{\infty} e^{2\pi i \sigma r}[F(c\sigma) - F(\sigma)]G(\sigma)\, d\sigma. \quad (3.44)$$

We will assume the relative velocity error $\xi(r)$ is small compared to unity so that we can express the difference $F(c\sigma) - F(\sigma)$ in terms of the first derivative of $F(\sigma)$:

$$F(c\sigma) - F(\sigma) \simeq \sigma\xi(r)F'(\sigma). \quad (3.45)$$

Substituting Eq. (3.45) into Eq. (3.44), we obtain

$$\varepsilon(r) = b(r)\xi(r)q(r); \quad (3.46)$$
$$q(r) = \int_{-\infty}^{\infty} \exp(i2\pi\sigma r)F'(\sigma)\sigma G(\sigma)\, d\sigma \rightleftharpoons Q(\sigma) = F'(\sigma)\sigma G(\sigma).$$

Note that $q(r)$ has the form of an inverse Fourier transform [see Eq. (3.4)], a property that we will utilize subsequently. Note also that $\xi \ll 1$ is now a definite requirement. Previously, we assumed $\chi \equiv \sigma_s\phi'(\sigma_s)\xi_0$ was small, but only to obtain a simplified expression for the intensity of the first-order ghost due to a sinusoidal velocity error.

This section deals with the effects of a random relative velocity error $\xi(r)$, which will be regarded as a stationary Gaussian random process with zero mean. We are interested in $\langle|E(\sigma)|^2\rangle$, the mean square of the corresponding spectrum error, when the spectrum is computed as the Fourier amplitude transform of the interferogram. Specifically,

$$\langle|E(\sigma)|^2\rangle \equiv \langle E(\sigma)E^*(\sigma)\rangle, \qquad \varepsilon(r) \rightleftharpoons E(\sigma), \quad (3.47)$$

where the angle brackets denote an average over an ensemble of realizations of $\xi(r)$. The statistical properties ascribed to $\xi(r)$ imply ergodicity, i.e., we may revise the definition (3.47) to

$$\langle|E(\sigma)|^2\rangle = \langle E(\sigma,y)E^*(\sigma, y)\rangle, \qquad \varepsilon(r + y) \rightleftharpoons E(\sigma,y), \quad (3.48)$$

where $\varepsilon(r + y)$ is the interferogram error corresponding to the relative velocity error $\xi(r + y)$, and the angle brackets now denote an average over y, the possible phase values of the velocity error. According to the latter definition and Eq. (3.46),

$$E(\sigma,y) = \int_{-\infty}^{\infty} \exp(-i2\pi\sigma r)b(r)\xi(r + y)q(r)\, dr \quad (3.49)$$

and

$$\langle E^2(\sigma)\rangle = \iint_{-\infty}^{\infty} \left[\lim_{Y\to\infty} \frac{1}{Y}\int_{-Y/2}^{Y/2} \xi(r + y)\xi(r_1 + y)\, dy\right] b(r)b(r_1)q(r)q(r_1) \times e^{-i2\pi\sigma(r-r_1)}\, dr\, dr_1. \quad (3.50)$$

Let $C_\xi(r - r_1)$ denote the quantity in brackets in Eq. (3.50), which is the autocovariance of ξ. Let $P_\xi(\sigma)$ denote the power spectrum of ξ. By the Weiner–Khintchine theorem, $C_\xi(r)$ and $P_\xi(\sigma)$ are Fourier transform pairs:

$$C_\xi(r - r_1) = \int_{-\infty}^{\infty} P_\xi(\sigma_1) \exp[i2\pi\sigma_1(r - r_1)]\, d\sigma_1. \tag{3.51}$$

Substituting Eq. (3.51) for the bracketed quantity in Eq. (3.50) and changing the order of integration, we obtain

$$\begin{aligned}\langle E^2(\sigma)\rangle &= \int_{-\infty}^{\infty} P_\xi(\sigma_1) \int\int_{-\infty}^{\infty} [e^{-i2\pi r(\sigma-\sigma_1)} b(r)q(r)] \\ &\quad \times [e^{i2\pi r_1(\sigma-\sigma_1)} b(r_1)q(r_1)]\, dr\, dr_1\, d\sigma_1 \\ &= \int_{-\infty}^{\infty} P_\xi(\sigma_1) \left| \int_{-\infty}^{\infty} e^{-i2\pi r(\sigma-\sigma_1)} b(r)q(r)\, dr \right|^2 d\sigma_1.\end{aligned} \tag{3.52}$$

The integral whose squared modulus appears in Eq. (3.52) is the Fourier transform of $b(r)q(r)$ evaluated at $\sigma - \sigma_1$. This, in turn, is $\mathscr{S}(\sigma)*Q(\sigma)$ (by the multiplication theorem), also evaluated at $\sigma - \sigma_1$; $\mathscr{S}(\sigma)$ again denotes the complex ILS. Thus

$$\begin{aligned}\langle E^2(\sigma)\rangle &= \int_{-\infty}^{\infty} P_\xi(\sigma_1) |\mathscr{S}(\sigma)*(F'(\sigma)\sigma G(\sigma)]|^2_{\sigma-\sigma_1}\, d\sigma_1 \\ &= P_\xi(\sigma)*|\mathscr{S}(\sigma)*[F'(\sigma)\sigma G(\sigma)]|^2,\end{aligned} \tag{3.53}$$

where we have applied the definition of convolution and used Eq. (3.46) in evaluating $Q(\sigma)$. Equation (3.53) is a generally useful expression for the mean square spectrum error corresponding to small $\xi(r)$. It expresses $\langle E^2(\sigma)\rangle$ in terms of the two-sided power spectrum of the relative velocity error, the spectral line shape, the derivative of the filter spectral response, and the true spectrum.

Suppose that the true spectrum is a single line at $\sigma = \sigma_s$, i.e., $G(\sigma) = \delta(\sigma - \sigma_s) + \delta(\sigma + \sigma_s)$. Then Eq. (3.53) gives

$$\langle E^2(\sigma)\rangle = P_\xi(\sigma)*|\sigma_s F'(\sigma_s)\mathscr{S}(\sigma - \sigma_s)|^2 + P_\xi(\sigma)*|\sigma_s F'(-\sigma_s)\mathscr{S}(\sigma + \sigma_s)^2. \tag{3.54}$$

Unless the resolution is very low, the line shape is effectively a delta function in Eq. (3.54), i.e., $\Delta\sigma|\mathscr{S}(\sigma)|^2$ approaches a delta function as the resolved spectral interval $\Delta\sigma$ approaches zero. If the interferogram truncation/apodization is symmetrical, then $\mathscr{S}(\sigma) = C(\sigma)$ is real and symmetric, and Eq. (3.54) reduces to

$$\langle E^2(\sigma)\rangle = \sigma_s^2[P_\xi(\sigma - \sigma_s) + P_\xi(\sigma + \sigma_s)]|F'(\sigma_s)|^2/\Delta\sigma. \tag{3.55}$$

The corresponding rms spectrum error normalized to the height of the source line is

$$\tilde{E}(\sigma) \equiv \frac{[\langle E^2(\sigma)\rangle]^{1/2}}{C(0)|F(\sigma_s)|} = (\Delta\sigma)^{1/2}\sigma_s[P_\xi(\sigma - \sigma_s) + P_\xi(\sigma + \sigma_s)]^{1/2}\{[\phi'(\sigma_s)]^2 + [\mathscr{F}'(\sigma_s)/\mathscr{F}(\sigma_s)]^2\}^{1/2}, \tag{3.56}$$

where we have used the fact that $C(0) = 1/\Delta\sigma$. The derivative $F'(\sigma_s)$ was evaluated from Eq. (3.15). If the power spectrum P_ξ is flat out to frequency $2\sigma_s$, the relative rms error is constant over $(0,\sigma_s)$ and is equal to

$$\tilde{E} = \sigma_s(2P_\xi\ \Delta\sigma)^{1/2}\{[\phi'(\sigma_s)]^2 + [\mathscr{F}'(\sigma_s)/\mathscr{F}(\sigma_s)]^2\}^{1/2}. \tag{3.57}$$

Equation (3.57) describes the first-order properties of spectrum noise introduced by random velocity errors when the source is monochromatic. The noise is distributed uniformly throughout the spectrum if the power spectrum of the velocity error is flat. The *relative* rms value of the noise is proportional to $(\Delta\sigma)^{1/2}$, to σ_s, and to the rms of the filter phase and amplitude distortions at the source frequency. Note that the noise has approximately the same relationship to σ_s and to the filter distortions as a first-order ghost produced by a sinusoidal velocity error [Eq. (3.29)]. Clearly, delayed sampling is equally effective in suppressing both types of spectrum error.

It should be remembered that the equations developed for the mean square or rms error are based on the assumptions that ξ is small and that the highest frequency in ξ (in P_ξ) is small compared to the bandwidth of the electrical filter. Also, we have defined P_ξ as a symmetrical two-sided function, whereas the customary definition associates the power spectrum only with positive frequencies; Eq. (3.57) would conform to the usual definition if $2P_\xi$ were replaced by P_ξ.

Zachor (1977) describes some additional effects of random velocity errors. It is shown that when the spectrum contains multiple lines and is computed as the Fourier cosine transform, the spectrum noise will include rms ghosts, i.e., local minima or maxima in $\langle E^2(\sigma)\rangle$ will occur midway between pairs of lines and at half the difference frequencies of line pairs. It is shown that when the source spectrum has only low-resolution or "continuum" features, $\langle E^2(\sigma)\rangle$ generally varies over the spectrum and is independent of spectral resolution. In the continuum case the random errors at two widely separated frequencies will be highly correlated, as pointed out by Bell and Sanderson (1972). When the source contains only lines, the random errors are essentially uncorrelated between frequencies separated by more than $\Delta\sigma$.

ACKNOWLEDGMENTS

Figures 3.1–3.7 and much of Section 3.7 were taken from Zachor and Aaronson (1979); this material is based on work performed at Honeywell Inc. for the Jet Propulsion Laboratory,

California Institute of Technology, and sponsored by the National Aeronautics and Space Administration under Contract NAS7-100. Other work reported in this chapter is based on studies supported by the Air Force Geophysics Laboratory (Contract F19628-76-C-0134) and the Defense Nuclear Agency (Contract DNA 001-72-C-0193). Additional manpower and computer resources were contributed by Atmospheric Radiation Consultants, Inc., Bartlett Systems, Inc., and the National Center for Atmospheric Research.

References

Bell, E., and Sanderson, R. (1972). *Appl. Opt.* **11**, 688.

Connes, J. (1961). *Rev. Opt.* **40**, 45, 116, 171, 231.

Endélyi, A. (ed.) (1953). "Higher Transcendental Functions," Vol. II. McGraw-Hill, New York.

Mankin, W. G. (1978). *Opt. Eng.* **17**, 39.

Mertz, L. (1965). "Transformations in Optics." Wiley, New York.

Nishiyama, T., Yamauchi, T., Ohno, M., Morii, M., Ura, N., and Masutani, K. (1975). *Jpn. J. Appl. Phys. Suppl. 14-1* **14**, 67.

Rabiner, L. R., and Gold, B. (1975). "Theory and Application of Digital Signal Processing." Prentice-Hall, Englewood Cliffs, New Jersey.

Weinberg, L. (1962). "Network Analysis and Synthesis." McGraw-Hill, New York.

Zachor, A. S. (1977). *Appl. Opt.* **16**, 1412.

Zachor, A. S., and Aaronson, S. M. (1979). *Appl. Opt.* **18**, 68.

Chapter **4**

Infrared Spectroscopy Using Tunable Lasers

H. R. SCHLOSSBERG

AIR FORCE OFFICE OF SCIENTIFIC RESEARCH
BOLLING AIR FORCE BASE
WASHINGTON, D.C.

P. L. KELLEY†

LINCOLN LABORATORY,
MASSACHUSETTS INSTITUTE OF TECHNOLOGY
LEXINGTON, MASSACHUSETTS

4.1. Introduction

Ultrahigh resolution spectroscopy using tunable lasers has been an active field for some time now. Until recently, however, most of this activity

† The Lincoln Laboratory portion of this work was sponsored by the Department of the Air Force.

ISBN 0-12-710402-X

was dominated by laser specialists demonstrating instrumentation and technique, and communicating through meetings and publications mostly with each other. Recently this situation has been changing, with nonlaser specialists taking up laser spectroscopy because it is the best (and often the only) means of obtaining the information they desire. As usual, such changes are occurring due both to the diffusion of information about the benefits of laser spectroscopy and, perhaps more importantly, the availability of easy-to-use commerical instrumentation.

The purpose of this chapter is to introduce nonlaser specialists to infrared tunable lasers, their capabilities, and their use. We shall discuss spectroscopic techniques which take advantage of the properties of tunable infrared lasers and the physics and practical considerations of using what we see as the most important types of tunable lasers now or in the near future. We shall discuss numerous examples of spectroscopic studies using semiconductor diode lasers, to date the most widely applied tunable IR lasers.

4.2. Properties of and Techniques Using Tunable Infrared Lasers

The attractiveness of tunable lasers for spectroscopic applications stems primarily from two principal attributes of lasers: their narrow linewidth and their high power per unit spectral interval. These attributes are the source of the enormous benefits of laser spectroscopy in terms of resolution and sensitivity. For some purposes, such as obtaining spectra in very small volumes or over very long paths, the spatial coherence of some laser sources is important for focusing to small sizes or for collimation. Rapid tunability is also sometimes an important asset. Together with high spectral intensity, it allows for very rapid spectral scanning.

A. Resolution

Conventional grating spectrometers and interferometers typically have resolution in the infrared of the order of 0.1–0.01 cm^{-1}. At their outer limits of large size, enormous precision, and operating skill, these instruments can be made to yield resolution of the order of 100 MHz. Tunable lasers, on the other hand, can routinely offer resolution of the order of 10 MHz, and with care and skill, resolution of hundreds of kilohertz or less can often be obtained. Along with narrow linewidth, the high spectral intensity of the lasers is important for obtaining high-resolution spectra. In order to obtain an adequate signal-to-noise ratio with a high-resolution spectrometer using a conventional blackbody source, it is necessary to scan wavelengths extremely slowly—in some cases scans may take several days. By contrast to conventional blackbody sources, however, the modest power available from a cw diode laser yields more than 10^{-9} W/Hz, which corresponds to that emitted by a 2000°K blackbody having a diameter of the order of meters. This leads to the possibility of obtaining high-resolution spectra very rapidly, for example, in transient environments such as shock tubes or fast chemical reactions.

Absorption lines of low-pressure gases are normally limited by Doppler broadening, in the infrared to widths of about 30–100 MHz. Therefore, to take advantage of the resolution capability of tunable infrared lasers, a number of techniques have been developed for obtaining resolution beyond the Doppler limit. All of these are possible, directly or indirectly, again owing to the high spectral intensity available.

Three techniques for achieving resolution beyond the Doppler limit, which we shall discuss briefly, are molecular beam spectroscopy, saturated absorption spectroscopy, and Doppler-free, two-photon absorption spectroscopy.

1. *Molecular Beam Spectroscopy*

Laser spectroscopic studies in the visible region using molecular beams to reduce Doppler broadening have been performed for a relatively long time. Direct absorption measurements are difficult due to the small optical density of beams. In the visible, off-axis fluorescence measurements are usually made to indicate absorption. Linewidths and structure in the 100-kHz range have been observed.

In the infrared, long radiative lifetimes generally preclude fluorescence measurements. Direct absorption measurements have been made using multiple parallel beams to obtain long path lengths. Very simple systems consisting essentially of multiple parallel capillary tube arrays attached to a manifold have been successfully used (Chu and Oka, 1975). The capillary arrays are available commercially from Galileo Electro Optics, Sturbridge, Massachusetts.

Figure 4.1 shows a schematic diagram of a molecular beam apparatus employing commercial capillary arrays (Pine and Nill, 1979). A magnified cross section of a capillary array is shown in the inset. An example of the linewidth reduction obtained with a multiple capillary tube collimator is shown in Fig. 4.2. Shown is a derivative transmission spectrum (obtained as discussed below) at 1903 cm^{-1} of the $R(\frac{15}{2})_{1/2}$ doublet of 0.04 Torr of NO. Figure 4.2a was taken with a static cell and shows each doublet component with its Doppler width of 126 MHz. Figure 4.2b shows the

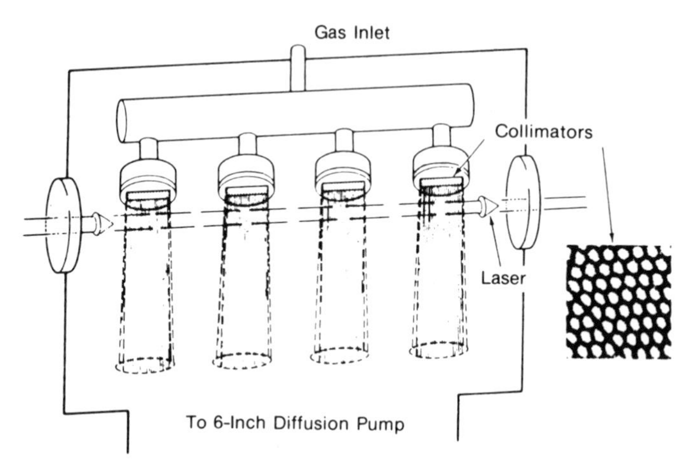

FIG. 4.1. Molecular beam absorption cell. Diameter of capillaries in the multitube collimators is about 5 μm (Pine and Nill, 1979).

molecular beam doublet spectrum with the linewidth of each component reduced to about 6 MHz.

Nozzle beams obtained from a supersonic free jet expansion followed by a collimator ("skimmer") have been developed over some time, primarily for molecular and chemical kinetics work. They offer higher densities than thermal beams, extremely narrow velocity distribution, and low rotational and internal temperatures, which can lead to increased absorption from states of interest, and can considerably help in interpret-

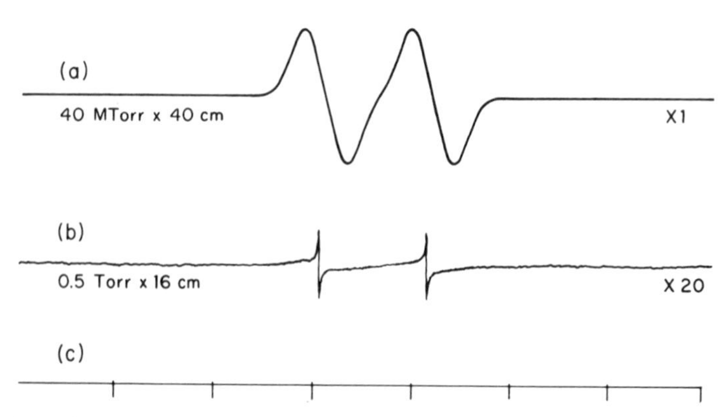

FIG. 4.2. Spectral scan of the $R(15/2)_{1/2}$ doublet in NO under Dppler-limited (a) and sub-Doppler molecular beam (b) conditions. Tick marks (c) are calibration obtained with a Fabry–Perot interferometer of free spectral range 299.8 MHz (Pine and Nill, 1979).

ing the spectra of complex molecules. Dye laser spectroscopy using nozzle beams has recently been reviewed by Levy *et al.* (1977).

Recently pulsed nozzle beams have been developed (Gentry and Giese, 1978). They offer the same benefits as cw beams, but do not require the very high vacuum pumping capability of cw beams. Pulsed beams are very attractive when combined with low repetition rate pulsed lasers. Byer (1979) has recently synchronized a pulsed nozzle beam with pulsed lasers, each operating at 10 Hz.

Elegant experiments using nozzle beams have been performed by Gough *et al.* (1977, 1978). In these experiments the power absorbed from a tunable infrared diode laser beam was detected by directing the molecular beam to a cryogenic bolometer. The vibrationally excited molecules give up their energy to the bolometer, thereby giving a signal proportional to the absorbed power. The technique, like fluorescence measurements in the visible, gives vastly increased sensitivity because all of the signal is due to absorption, as opposed to measurement of a small change in a large signal required in conventional absorption measurements.

The experimental arrangement of Gough *et al.* is shown in Fig. 4.3. Figure 4.4 compares a spectral line of CO taken in a cell and using the apparatus of Fig. 4.3. The narrowing of the line in the beam is evident.

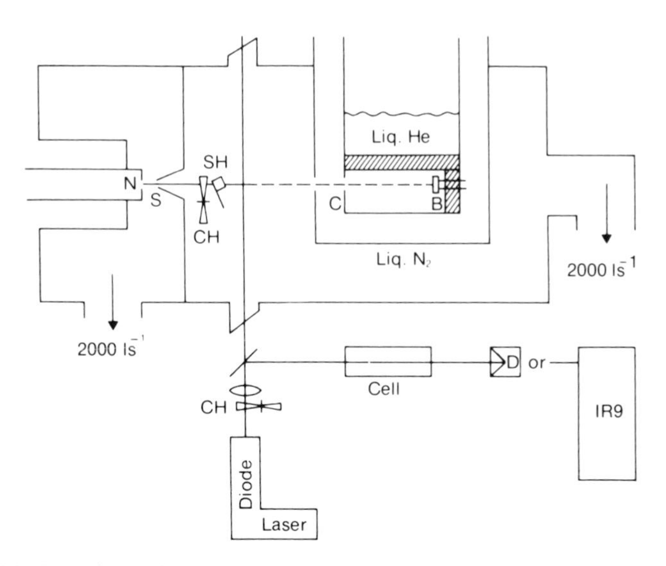

FIG. 4.3. Experimental setup for nozzle beam absorption measurement using liquid helium cooled bolometer detector. N, nozzle; S, skimmer; CH, chopper; SH, shutter; C, collimator, B, silicon bolometer; D, pyroelectric detector; IR9, conventional IR spectrometer (Gough *et al.*, 1977).

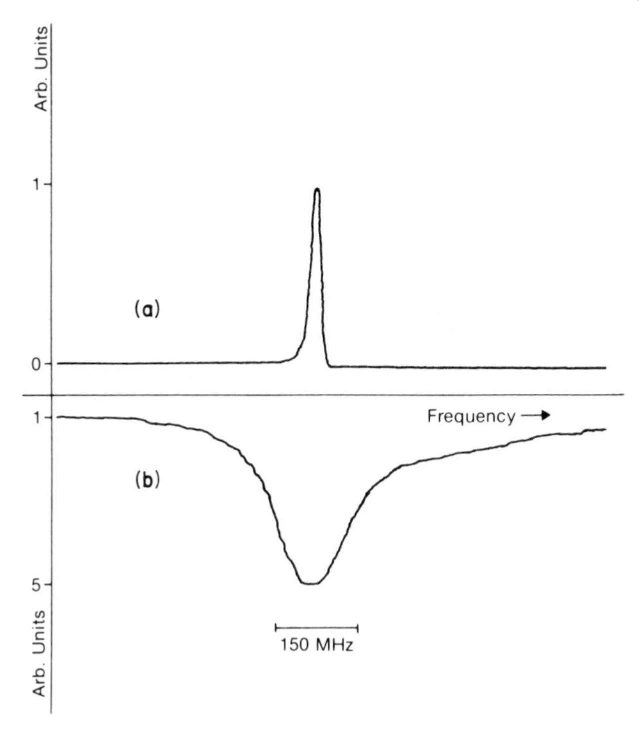

FIG. 4.4. Here $v = 0 \rightarrow 1$, $J = 1 \rightarrow 2$ transition of CO as measured (a) on the nozzle beam of Fig. 4.3 and (b) in an absorption cell (Gough *et al.*, 1977).

The width of the narrow line is about 20 MHz and is due to frequency jitter of the diode laser.

As pointed out by Gough *et al.*, their experiment shows the potential of tunable infrared laser spectroscopy to perform the analysis of internal states in typical nozzle beam studies—for example, crossed beam kinetics experiments or surface kinetic studies.

2. *Saturation Spectroscopy*

Saturation spectroscopy essentially uses a moderately intense laser beam to change the population distribution (i.e., saturate) of a small group of molecules in the thermal velocity distribution and then probes the change in the absorption line due to the saturation.

The earliest, and simplest method of saturation spectroscopy is the so-called Lamb dip. A laser beam is made to traverse a cell, is reflected back on itself, and then detected as shown schematically in Fig. 4.5a. If the laser is tuned off line center, different molecules in the velocity dis-

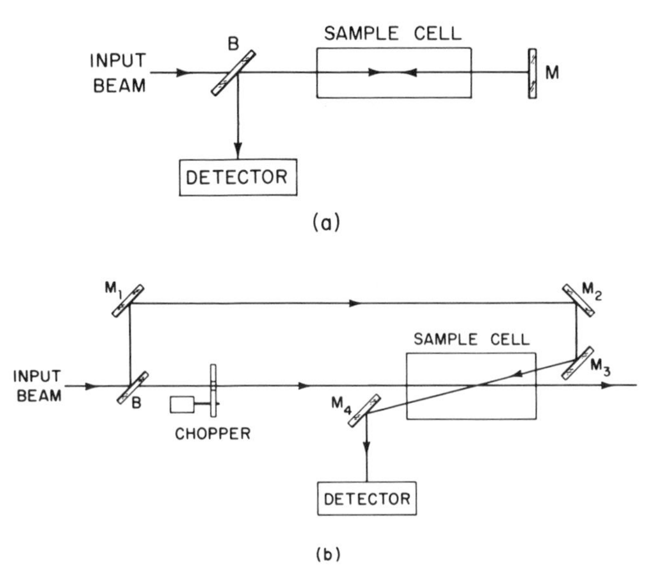

FIG. 4.5. Schematic setup for Lamb dip experiment (a) with reflected wave, (b) with weak probe wave. B, beam splitter. Probe beam can also be obtained from a separate source.

tribution interact with the forward and reflected waves, whereas near line center it is the same group of molecules, with velocity component in the beam direction near zero, which interacts with both waves. Hence a feature is seen in the absorption line at line center which is characterized by the non-Doppler-broadened (homogeneous) width of the line.

An important variant of the Lamb dip technique (Letokhov and Chebotayev, 1969) shown in Fig. 4.5b sends most of the laser beam, chopped, in the forward direction. A small part of the beam, the probe, is split off before the chopper and sent in the backward direction. As the laser frequency is scanned, the probe beam is detected synchronously at the chopping frequency of the main beam. In this arrangement, there is no signal except that due to saturated absorption.

Lamb dip experiments give the homogeneous linewidth for molecules with near zero velocity component in the beam direction and can, therefore, be used to obtain information about collision processes for these molecules. An interesting variant of the Lamb dip technique (Mattick *et al.*, 1973) can be used to obtain linewidths of molecules with nonzero velocities in the beam direction, and such data can be unfolded to obtain collision broadened linewidth versus molecular velocity.

In this technique, again a strong chopped saturating field is incident on the sample in one direction, and a probe field traversing the sample in the other direction is synchronously detected at the chopping frequency. In this case, however, the saturating and probe fields have different frequencies, on opposite sides of the peak of the Doppler-broadened line. As the probe is tuned, a saturation resonance is observed when it is off line center by the same amount that the saturating field is off in the other direction, i.e., when the saturating and probe fields are symmetrically located about line center. The resonance again comes about because at this tuning the same velocity group interacts with both saturating and probe fields, their Doppler shifts being equal and opposite. The velocity along the beam direction which is probed is given by $v = c\Delta/\nu_0$, where Δ is the detuning of the frequency from the line center frequency ν_0. Saturation resonances can be observed for detunings of many Doppler widths, corresponding to velocities many times the thermal velocity.

To observe saturation resonances in molecular absorption, sufficient intensity must be used for saturation to occur. Naturally, exactly how much intensity is needed depends on molecular and experimental parameters, but roughly the change in population of a state must be at least a few percent of the original population. Thus $\sigma\tau I \simeq 0.1$, where σ is the transition cross section, τ the population relaxation time (which can be radiative, collisional, or determined by diffusion or beam transit time), and I the beam intensity in photons per cm^2/sec. This implies that tens of milliwatts per square centimeter are required for typical transitions at low pressure. Most of the tunable lasers can reach these intensities, in some cases with beam shaping required. The low-power tunable lasers could also be used advantageously as the probe sources in many interesting saturation spectroscopy measurements.

Many different kinds of saturation resonances have been observed in the infrared using fixed frequency lasers and in many cases Stark or Zeeman tuning of the molecule. Good reviews have been given, among others, by Brewer (1972), Shimizu and Shimoda (1972), Letokhov (1975), and Letokhov and Chebotayev (1977). An important class of saturation resonances involves two closely spaced levels with allowed transitions to a common third level, the two frequencies of which cannot be resolved in ordinary spectroscopy because of Doppler broadening. We refer to the reviews for details.

3. *Doppler-Free Two-Photon Spectroscopy*

It was first pointed out by Vasilenko *et al.* (1970) that in a two-photon absorption process, there will be a substantial reduction of the Doppler linewidth if the two photons travel in opposite directions. If two waves in opposite directions of frequencies ν_1 and ν_2 are incident on a molecule with velocity v and transition frequency ν_0, then for a resonant two-photon absorption $\nu_1 + \nu_2 + (\nu_1 - \nu_2)v/c = \nu_0$, and hence the Doppler width will be given by $(\nu_1 - \nu_2)v_{\mathrm{Th}}/c$, where v_{Th} is the thermal velocity. If ν_1 and ν_2 are close enough so that this is less than the homogeneous width, then the homogeneous width will be seen. Further, the cross section will be substantially increased because virtually all of the molecules in the velocity distribution will participate in the absorption.

In the infrared, the intensities required for Doppler-free two-photon absorption are impractical, unless there is a near resonant intermediate level, i.e., an allowed transition near, but not exactly at ν_1 or ν_2. This is a common occurrence when the two-photon transition is an overtone transition and each frequency is nearly resonant with a transition at the fundamental of the same vibrational mode.

Doppler-free two-photon absorption in the infrared has been observed by Bischel *et al.* (1975) on a $2\nu_3$ transition in CH_3F using two CO_2 laser frequencies. The $2\nu_3$ transition was Stark tuned into resonance with the sum of the CO_2 laser frequencies, one of which was off resonance with a ν_3 transition by 7426 MHz. Figure 4.6 shows the relevant energy levels and laser frequencies and the Doppler-free two-photon spectrum obtained by Stark tuning.

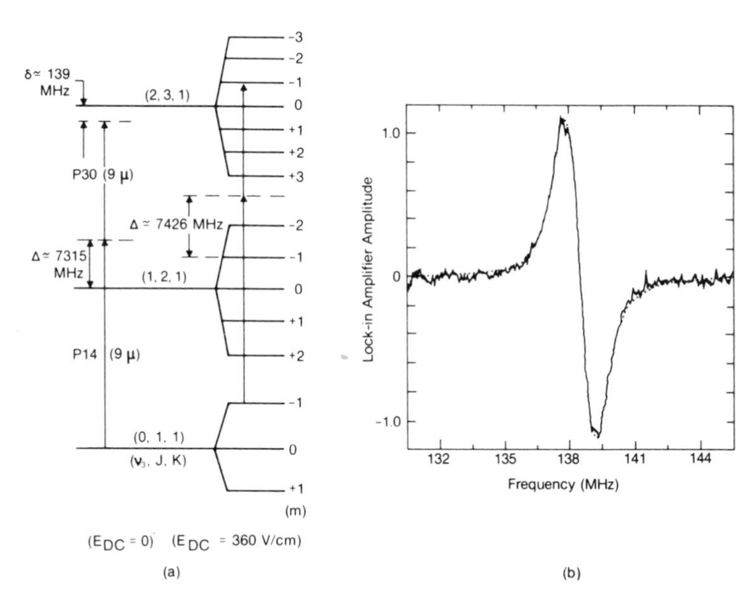

FIG. 4.6. (a) Energy levels and (b) absorption line of Doppler-free two-photon absorption experiment in CH_3F. (After Bischel *et al.*, 1975.)

The experimental arrangement for Doppler-free two-photon absorption is similar to that shown in Fig. 4.5b for saturated absorption. It is only necessary again for one beam, say ν_1, to be intense, and to observe the absorption of a probe beam at ν_2 traveling in the opposite direction. The intensity needed for the strong beam is somewhat greater than that for a saturation resonance because of the off-resonant intermediate state, but this is compensated, in part, by the fact that all of the molecules participate. The absorption $A(\nu_2)$ of the probe at frequency ν_2 can be conveniently estimated by the following formula:

$$A(\nu_2) = (A_0/4)|\mu_{12}E_1/h(\nu_{01} - \nu_1)|^2(\Delta\nu_D/\Delta\nu_H),$$

where A_0 is the peak absorption coefficient (at resonance) for a transition between levels 0 and 1; $\Delta\nu_D$ and $\Delta\nu_H$ are the Doppler and homogeneous widths, respectively, of such a transition; $h\nu_{01}$ is the energy spacing between levels 0 and 1, μ_{12}, the dipole matrix element between levels 1 and 2 (2 is the highest level), and E_1 the field strength of the field at frequency ν_1 [$I_1 = (c/8\pi)E_1{}^2$].

As an example of the required intensity, the signal of Fig. 4.6 was obtained with ~375 W/cm². If the intermediate state detuning were less by a factor of 20, then less than 1 W/cm² would have been required. A tunable probe would have offered the flexibility to Stark tune the intermediate transition as closely as desired.

B. Sensitivity

Lasers make possible, or much easier, a number of techniques that can greatly increase the sensitivity of spectral measurements. These are in addition to the basic increase due to the high spectral brightness of tunable lasers as sources.

1. *Derivative Spectroscopy*

Derivative spectroscopy is a technique that is well known in microwave spectroscopy (Townes and Schawlow, 1954) and is valuable for the same reasons in tunable laser spectroscopy. In derivative spectroscopy, a small ac frequency modulation is impressed on the laser output, so that its frequency is given by $\nu_1 + \delta \cos \omega_M t$, where ν_1 is the frequency which is slowly scanned. If the response of a sample gas to be measured, e.g., its transmission has a frequency dependence $T(\nu)$, then by a Taylor expansion, the response to the laser input is $T = T(\nu_1) + (dT/d\nu)_{\nu_1} \delta \cos \omega_M t + \ldots$. If the response signal (transmitted power) is synchronously deleted at the modulation frequency ω_M, then the $(dT/d\nu)_{\nu_1}$ term is detected; hence the name *derivative spectroscopy*. If the second harmonic of the modulation is detected, then, from further expansion of the Taylor series, it is apparent that the second derivative $(d^2T/d\nu^2)_{\nu_1}$ is detected.

The Taylor series expansion is accurate, and real derivatives versus frequency ν_1 are obtained, only if the frequency excursion δ is very small compared to the linewidth of the line to be measured. For accurate line-shape measurements it is essential that this be the case. If, on the other hand, it is only desired to observe a weak line and measure its center frequency, larger modulation, typically $\frac{1}{3}$ of the linewidth, gives best signal-to-noise ratio results. In this case the actual derivative is not measured (the line shape is distorted), but the name *derivative spectroscopy* is still commonly used. Also, in this case, if the actual line shape is known (e.g., Lorentzian), then parameters (e.g., linewidth) can be computed from the measured shape.

Derivative spectra are important because they emphasize signals that change significantly with frequency over a linewidth compared to signals that change more slowly, such as laser power and material parameters.

The NO spectrum shown in Fig. 4.2 is a derivative spectrum. An example of the improvements in sensitivity obtainable with derivative spectroscopy is illustrated by Fig. 4.7, which compares a conventional absorp-

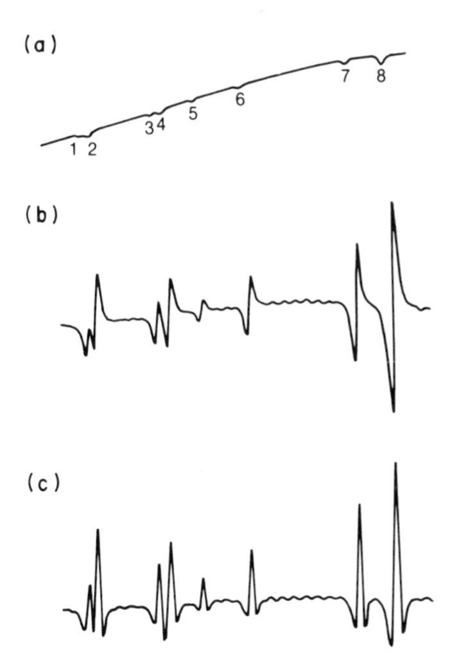

FIG. 4.7. Transmission spectrum—ν_2 band of NH_3 comparing (a) direct absorption measurement, (b) first derivative scan, and (c) second derivative scan. Pressure, 5.4 Torr; path length, 16.4 cm. (From Laser Analytics Inc. advertisement, with permission.)

tion spectral scan of a tunable diode laser with first and second derivative scans for weak NH_3 lines in the region of 1248 cm^{-1}. The sloping background in the straight absorption spectrum is due to source power variation. The total scan is about 0.4 cm^{-1}.

Derivative spectra can also be obtained by applying frequency modulation to the position of the absorption line, while scanning the input frequency. The classic example of this technique is Stark modulation spectroscopy in the microwave region. There are also numerous cases of relatively fixed frequency laser spectra, both linear and nonlinear, obtained by Zeeman or Stark tuning the molecule with a small ac field also applied to obtain a derivative spectral line. The Stark tuned line shown in Fig. 4.6 was obtained in this manner.

2. *Multipass Cells*

Multipass ("White") cells have been used in conventional IR spectroscopy. However, the spectral brightness and spatial coherence of tunable laser sources make it possible to make more passes over greater distances and to measure thereby absorption over enormously long path lengths. For example, a White cell has been demonstrated (Kim *et al.,* 1978) with a tunable diode laser which can be used to measure absorption over path lengths up to 1.5 km. The cell had a total length of 11 m, making it compact enough to allow temperature control from 160 to 300°K.

3. *Optoacoustic Spectroscopy*

Optoacoustic spectroscopy is another technique for observing a signal which is directly proportional to the power absorbed by a gas sample, rather than to the power transmitted through the sample. Although the technique was suggested for spectroscopy by Alexander Graham Bell (McClelland, 1979), the high power per unit spectral interval available from lasers has led to a large resurgence of interest in optoacoustic spectroscopy of gases. A detailed review of important aspects of optoacoustic spectroscopy has recently been published (Pao, 1977). The optoacoustic technique is illustrated in Fig. 4.8. The laser light is chopped before entering a sample cell. Absorption by the gas causes heating, which is periodic because of the chopping. This periodic heating drives a sound wave, which is converted to a voltage by a sensitive microphone to provide the output signal. The microphone voltage is normally synchronously detected with the chopper frequency as a reference.

The optoacoustic signal can be enhanced by using a chopping frequency at an acoustic resonance of the cell (Dewey, 1977). The sensitivity of

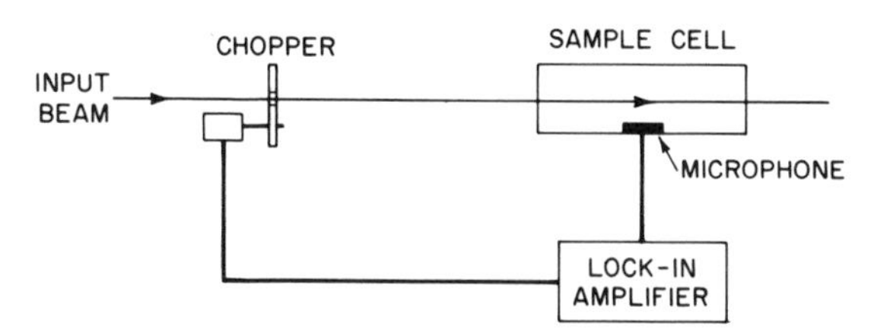

FIG. 4.8. Schematic experimental setup for optoacoustic spectroscopy.

optoacoustic spectroscopy generally is limited by absorption in the cell windows and walls, and by environmental noise. A detection sensitivity of $9 \times 10^{-8}\,\text{cm}^{-1}$ has been achieved using 280 mW of power in an acoustically resonant cell (Max and Rosengren, 1974). Multipass systems can be used to increase the absorption. These are especially sensitive to window absorption. In addition, the passes must be close together for good acoustic coupling. Using a multipass, acoustically resonant system, an equivalent absorption sensitivity of $7.5 \times 10^{-9}\,\text{cm}^{-1}$ has been achieved with only 1 mW of input power (Koch and Lahmann, 1978).

Derivative spectroscopy can also be used with optoacoustic detection (Dewey, 1977). The source frequency is modulated instead of a chopper being used. Periodic heating is caused by the periodic change in absorption with frequency. Derivative spectroscopy should alleviate problems of window and wall absorption. In a high Q acoustically resonant system, stability of the acoustic resonator would be a requirement for very sensitive measurements.

4. *Heterodyne Spectroscopy*

In heterodyne spectroscopy high resolution and sensitivity are achieved in emission spectra of gases, or in absorption spectra using a high temperature background. Important cases are those where the gas or the thermal source are remote, e.g., looking at hot plumes or exhausts or looking at atmospheric absorption using the sun as a source (Hinkley and Kelley, 1971; Menzies, 1971).

In heterodyne spectroscopy the laser is used as a local oscillator in a system shown schematically in Fig. 4.9. The laser output is mixed with the chopped thermal radiation to be examined on a fast infrared detector. The IR detector signal feeds an IF amplifier with center frequency Ω and bandwidth B. A spectrum is obtained by sweeping the local oscillator frequency or by tuning the IF frequency Ω. The amount of tuning is limited by the detector response. For faster data collection, a bank of rf filters with different center frequencies Ω can be used to collect the spec-

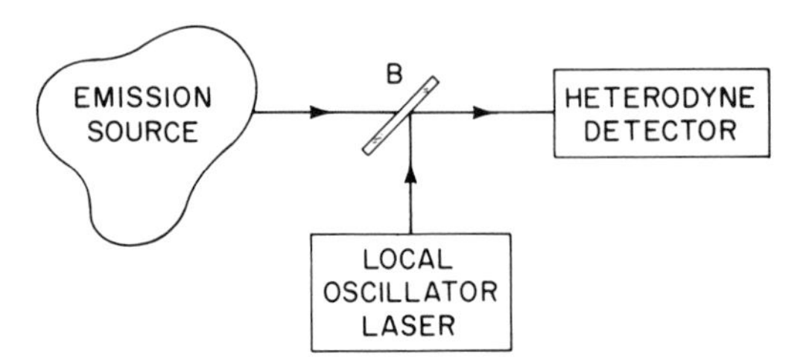

FIG. 4.9. Schematic experimental setup for heterodyne spectroscopy.

trum all at once (Hillman *et al.*, 1977). The resolution of a heterodyne spectrum is $2B$. Thus it is determined completely by electronics and is limited only by the stability of the local oscillator.

If the local oscillator power can be made large enough, a condition is reached where the dominant noise in heterodyne detection is due to the shot noise associated with the local oscillator power. If, in addition, the collection optics for the thermal radiation are large enough to just resolve the source, then the signal-to-noise ratio of a heterodyne system is given (Hinkley and Kelley, 1971) for a photodiode detector by

$$\mathrm{S/N} = 2\eta T_{\mathrm{p}} T_{\mathrm{o}} (B\tau)^{1/2} [\varepsilon_{\mathrm{s}} (e^{h\nu/kT_{\mathrm{s}}} - 1)^{-1} - \varepsilon_{\mathrm{b}} (e^{h\nu/kT_{\mathrm{b}}} - 1)^{-1}],$$

where T_{p} and T_{o} are the transmission of the path to the source and of the optical system, respectively; T_{s}, ε_{s} and T_{b}, ε_{b} are the source and background temperature and emissivity, respectively; τ is the postdetection integration time, and η the effective quantum efficiency of the detector. For a representative high-resolution spectrum at 10 μm with a 1200°K blackbody, the background is negligible, and for $\tau \sim 1$ sec, $B = 5$ MHz, $T_{\mathrm{o}} = 0.1$, $\eta = 0.4$, and $T_{\mathrm{p}} = 1$, $\mathrm{S/N} \simeq 150$. The spectrum is observed, of course, by the decrease in T_{p} as a line is scanned.

For remote sensing, an important feature of the signal-to-noise ratio equation is that it is independent of range providing the receiving telescope meets the resolution criteria. For example, at 10 μm, and 1-km range, a 1-m source would require 2-cm optics.

The most commonly used detectors for infrared heterodyne detection are $\mathrm{Hg}_{1-x}\mathrm{Cd}_x\mathrm{Te}$ photodiodes. State-of-the-art $\mathrm{Hg}_{1-x}\mathrm{Cd}_x\mathrm{Te}$ detectors have bandwidths of close to 2 GHz and effective quantum efficiencies of about 45% (Spears *et al.*, 1975; Ku and Spears, 1977). Shot noise limited performance of these detectors can be obtained with local oscillator power of less than 0.5 mW with a 500-MHz bandwith (Ku and Spears, 1977).

Laser heterodyne detection was reviewed by Menzies (1976) and is extensively discussed by Kingston (1978).

4.3. Semiconductor Diode Lasers

A. Description

Recombination semiconductor lasers operate by stimulated emission on transitions of electrons between states in the conduction and valence bands. Population inversion and lasing have been achieved by electron beam and optical pumping, but by far the most common means of pumping such lasers is with current in a forward biased diode. Stimulated emission occurs in the region of the p–n junction as electrons in the conduction band in or transported from the n region combine with holes in the valence band in or transported from the p region.

Infrared semiconductor lasers have operated to wavelengths as long as about 30 μm. GaAs and its ternary alloys operating in the near infrared in the region of 0.8 μm have received intense research and development effort because of their importance to optical fiber communications. For the same reason, more recently, intense effort has been directed toward quarternary alloys of $In_xGa_{1-x}As_yP_{1-y}$ operating in the region of approximately 1.0–1.6 μm. We shall not discuss these lasers here, but shall limit ourselves to longer wavelength types.

The tunable diode lasers of primary use in infrared spectroscopy are made from lead salts and operate in the 2–30-μm range. Binary compounds including PbS, PbSe, and PbTe, and pseudobinary alloys such as $Pb_{1-x}Sn_xTe$ and $PbS_{1-x}Se_x$ are used, depending on wavelength range desired.

Figure 4.10a shows a photograph of a lead salt diode laser mounted in a header which provides heat sinking and electrical current. Figure 4.10b is a schematic diagram of a diode laser.

The nominal center frequency of a tunable diode laser is most usually obtained by choosing the appropriate lead salt alloy and its composition. Figure 4.11 shows the wavelength range available from different compositions of some of the commonly used alloys. Coarse tuning of a diode laser can be achieved by varying its temperature, pressure, or the magnetic field applied to it.

Fine tuning of diode lasers and wavelength scanning for recording spectra are most commonly accomplished by varying the diode current, which changes its temperature due to resistive heating. The output frequency of the laser is determined primarily by the mode frequencies of the Fabry–Perot resonator formed by the end reflectors, since the resonator linewidth is much less than the gain linewidth of the active medium. The small temperature variation with current results in an index of refraction

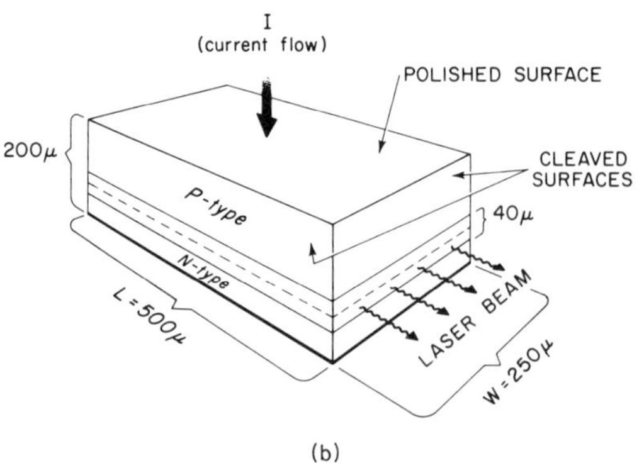

FIG. 4.10. (a) Photograph of a lead salt diode laser. The diode is the small, rectangular crystal sandwiched between the bottom support stud and the upper contact bar. The support stud and the contact bar provide both electrical contact and heat sinking (Kelley, 1977). (b) Schematic drawing of a lead salt diode laser.

variation which changes the optical path length between the end reflectors and hence the resonator mode frequencies. The effect of real length changes due to thermal expansion is negligible by comparison. Usually diode lasers oscillate in several cavity modes, spaced by about 1 cm^{-1}, the number increasing with increasing current. For high-resolution spectroscopy, it is necessary to isolate one mode and tune it. The isolation is usually done with a monochromator.

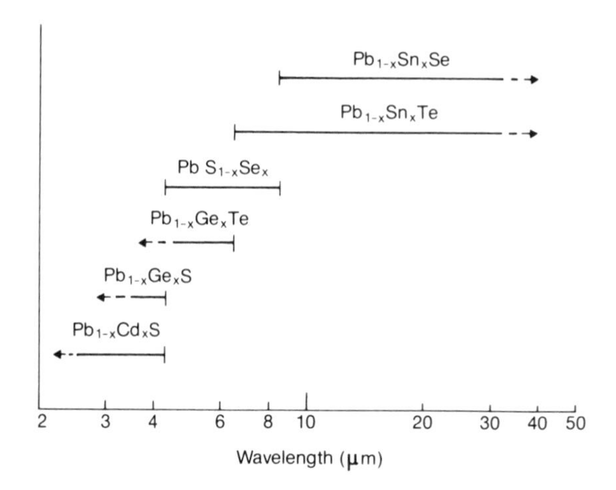

FIG. 4.11. Wavelength range of lead salt alloys (Kelley, 1977).

A single mode can be continuously tuned by changing diode current over a range of 0.5 to several wavenumbers, at which point a mode jump occurs. Figure 4.12 illustrates mode tuning and jumping characteristics. Mode jumping results from a shift of the gain spectrum relative to the mode frequencies as the band gap of the material changes with tempera-

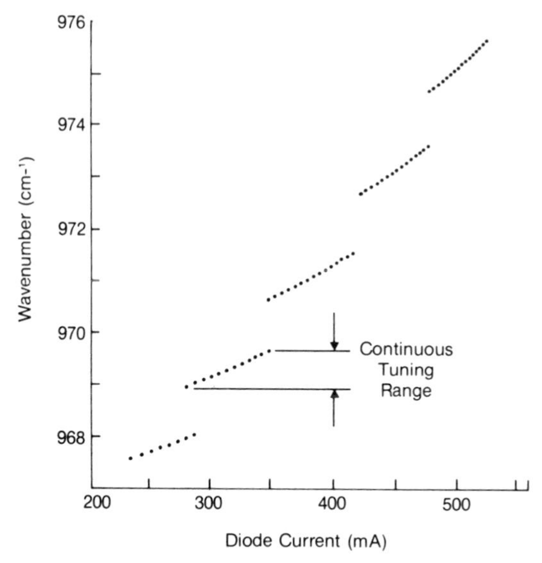

FIG. 4.12. Fine tuning and mode jumping from a lead salt laser. Tuning is continuous within a mode (Hinkley *et al.*, 1976a).

ture. The laser cavity tuning rate with temperature is typically about one-third the band gap tuning rate.

A number of important advances were made in the technology of lead salt diode lasers in the early to mid-nineteen seventies. Increases in device yield and internal quantum efficiency resulted, in part, from improvements in the size and quality of vapor grown crystals (Harmon and McVittie, 1974). Direct current carrying capacity was greatly increased by new, low-resistance contacting methods and improved heat-sinking techniques. The output efficiency and optical beam quality have been improved by the use of a stripe geometry to suppress parasitic oscillation modes. In a stripe geometry device, the active region is confined to the central portion (~50 of the 250 μm dimension in Fig. 4.10) of the junction. Other improvements included improved optical quality and application of optical coatings to the end mirrors.

Single mode power levels available from lead salt diode lasers are typically in the $\frac{1}{10}$–1-mW range depending on wavelength region and other factors. The highest reported single mode powers are 50 mW from a PbS diode at 4.3 μm (Ralston *et al.*, 1974) and 6 mW from a $Pb_xSn_{1-x}Te$ laser in the 10–12 μm region.

An accurate measurement of the linewidth of a $Pb_xSn_{1-x}Te$ laser was made sometime ago by measuring the beat frequency spectrum of the diode laser with a very stable CO_2 laser (Hinkley and Freed, 1969). A linewidth of 54 kHz was measured for a mode with 240-μW power, and the linewidth varied inversely with mode power as predicted by fundamental theory of quantum phase fluctuations. In practice, unless suitable care is taken, linewidths are greater than this fundamental limit and are determined by environmental factors.

Lasers such as that shown in Fig. 4.10 are limited to operation near liquid helium temperatures. Major advances were made through the development of heterostructure lasers, in which a different composition of material is in contact with the active region. Previous work on GaAs–$Ga_{1-x}Al_xAs$ lasers had shown that such structures can serve to confine both the light and the injected carriers to the active region, thereby greatly reducing the threshold current and improving the performance of the laser.

A single heterostructure laser was reported (Walpole *et al.*, 1973) in which a layer of PbTe was deposited by vacuum evaporation on vapor-grown $Pb_{0.88}Sn_{0.12}Te$. Continuous-wave laser operation was achieved up to 65°K. A double heterostructure laser that could operate at liquid nitrogen temperature was demonstrated in 1974 (Groves *et al.*, 1974). A liquid phase epitaxy technique was used to grow an active layer of *n*-type $Pb_{0.88}Sn_{0.12}Te$ on a PbTe substrate, and to cover it with a PbTe layer. A stripe geometry device was made by growing the active layer in a 50-μm opening in a MgF_2 coating on the substrate.

A stripe geometry, double-heterostructure PbSnTe laser grown by molecular beam epitaxy has been operated up to 114°K (Walpole *et al.*, 1976). Within its cw operating range of 12–114°K heat-sink temperatures, the laser was tuned from 15.9 to 8.54 μm. The temperature tuning characteristics of this laser as well as the threshold currents are illustrated in Fig. 4.13.

Varying temperature is today by far the most common means for coarse tuning of diode lasers. Compact closed cycle coolers are commercially available which can provide heat-sink temperatures variable from about 10°K to room temperature. Temperature stability and vibration isolation of laser diodes in these coolers limit short-term laser frequency stability to about 5–10 MHz (Jennings and Hillman, 1977). By varying heat-sink temperature, a typical diode laser can be tuned several hundred wavenumbers to about 16 μm and greater than 50 cm^{-1} at longer wavelengths.

Hydrostatic pressure applied to a diode laser can yield a very broad tunability. In an early demonstration, a pulsed lead selenide diode laser at

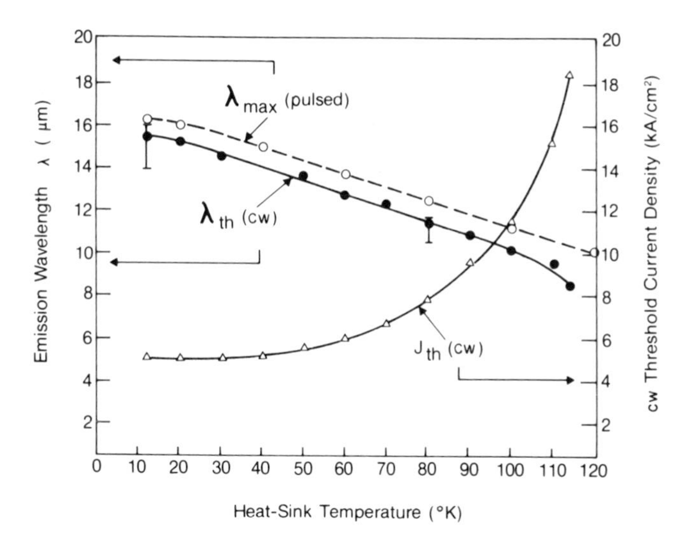

FIG. 4.13. The solid curve through the experimental points (solid dots) gives the wavelength of emission of the threshold mode versus heat-sink temperature. The range of emission at heat-sink temperatures of 12 and 80°K is indicated by the vertical lines for current densities of several kiloamperes per square centimeter above threshold. The dashed curve through the experimental points (open circles) gives the longest wavelength mode seen under low-duty-cycle pulsed operation as a function of heat-sink temperature. The curve through the triangular data points gives the cw threshold current density as a function of heat-sink temperature (Walpole *et al.*, 1976).

77°K was tuned from 8 to 22 μm using hydrostatic pressure up to 14 kbar (Besson *et al.*, 1965). At very low temperatures, helium gas freezes at high pressure. Uniaxial pressure in principle could be used, but it is difficult to avoid crushing the diodes. The advent of diode lasers operating at 77°K and above, therefore, allows the possibility of broad pressure tuning. Figure 4.14, showing the variation of band gap of binary semiconductors at 77°K with pressure, indicates the possibility of covering the entire wavelength region between 2 and 35 μm using only binary compounds and one or two alloy compositions with pressure tuning. In practice, pressure would probably be used for coarse tuning, with current being used for fine tuning (Pine *et al.*, 1973).

It is also possible to coarse or fine tune the frequency of a diode laser with magnetic field. This results from the quantizing of energy states in both the conduction and valence bands into Landau levels in the magnetic field, with each Landau level split into two spin states. Because of the large mass anisotropy in the lead salts, the levels and hence the coarse tuning are strong functions of magnetic field orientation. The fine tuning of

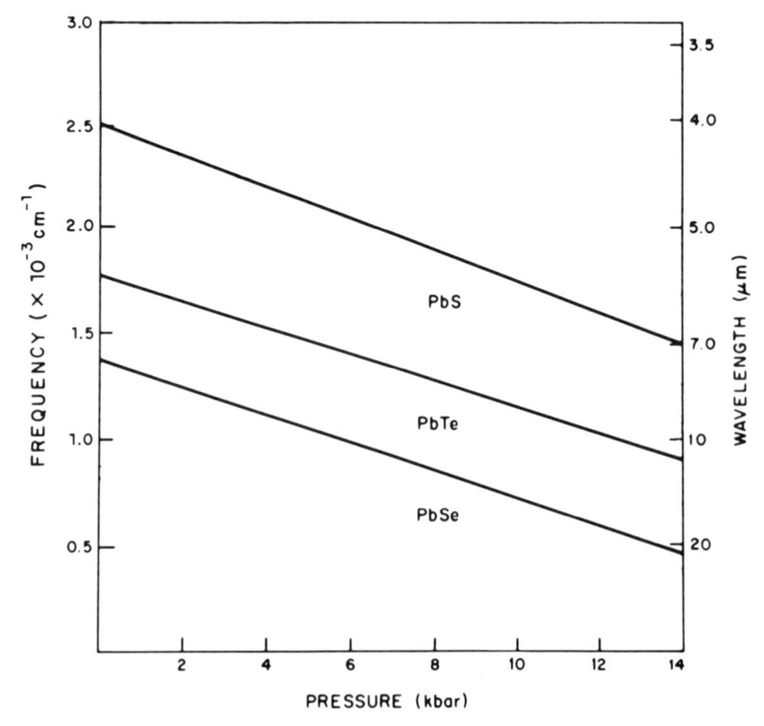

FIG. 4.14. Band-gap variation with hydrostatic pressure for some binary semiconductor materials at 77°K.

a diode laser mode with magnetic field occurs in a manner similar to fine tuning with current. The tuning rate within a mode and for different modes can vary with magnetic field. For a $PbS_{0.82}Se_{0.18}$ diode laser, it has been measured to vary between 0.4 and 2.0 MHz/G (Nill *et al.*, 1973).

B. Commercial Availability

The only commercial supplier of tunable lead salt lasers in the United States at present is Laser Analytics Incorporated, Bedford, Massachusetts, a subsidiary of Spectra Physics Incorporated. The Telefunken Company, in Germany, is also a supplier. The Laser Analytics devices are stripe geometry, single heterojunction diodes made by a compositional interdiffusion process (Hinkley *et al.*, 1976b). They have been operated as high as 100°K. Figure 4.15 shows a temperature tuning curve for one such laser. Commercial lasers are available with alloy compositions selectable for operation anywhere within the range of 2.7–30 μm. Attainable power is quoted as greater than 0.25 mW. The company sells a variety of accessories as well as a complete system, including a closed cycle refrigerator, laser power supply, monochromator, optics, sample cells, and detectors. The laser power supply is capable of performing a number of functions,

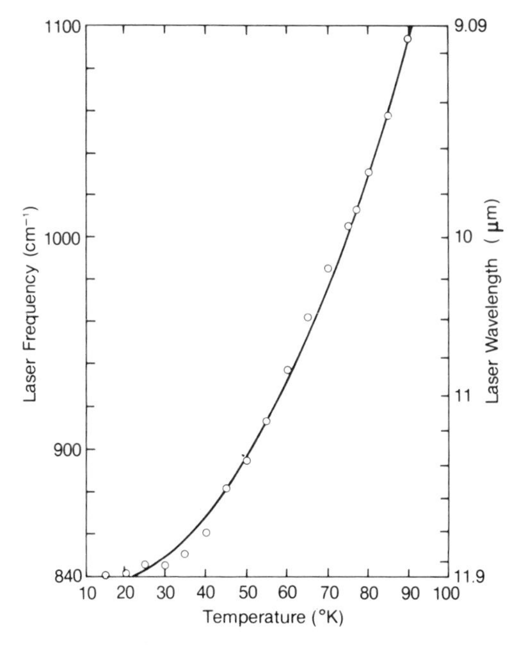

FIG. 4.15. Temperature tuning curve for a compositional interdiffusion lead salt diode laser (Hinkley *et al.*, 1976b).

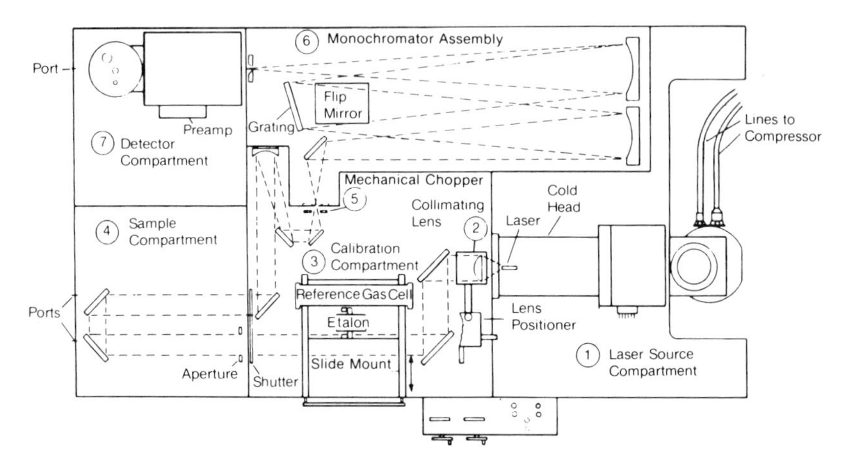

FIG. 4.16. Optical schematic of Laser Analytics Inc. diode laser source spectrometer. (From Laser Analytics Inc. brochure, with permission.)

including scanning at variable rates, supplying a variable amplitude ac current for derivative spectroscopy, and providing burnout protection for the laser diode. A schematic of the complete system is shown in Fig. 4.16, and its quoted specifications are given in Table 4.I.

TABLE 4.I

SPECIFICATIONS OF LASER ANALYTICS INC. STANDARD MODEL LS-3 LASER SOURCE SPECTROMETER[a]

Resolution	$<3 \times 10^{-4}$ cm^{-1}
Stability	
Frequency	
short term (<1 sec)	3×10^{-4} cm^{-1}
long term (30 min)	10^{-3} cm^{-1}
Amplitude	
short term (<1 sec)	<0.3%
long term (30 min)	<2%
Reproducibility	
Rapid scan (scan to scan)	
frequency	3×10^{-4} cm^{-1}
amplitude	0.3%
Sweep	
frequency	3×10^{-4} cm^{-1}
amplitude	0.5%
Spectral Frequency Calibration	
Absolute	
gas reference line	<0.01 cm^{-1}
monochromator dial	<1 cm^{-1}

(Continued)

TABLE 4.I (*Continued*)

Relative	
étalon	<2% of scan
Tunability	
Sweep mode of power supply	
tuning rate	0.0001–0.03 cm^{-1}/sec
tuning range	0.6 cm^{-1}
Modulation mode of power supply	
tuning rate	<0.2–600 cm^{-1}/sec
tuning range	0.5 cm^{-1}
repetition rate	
standard	50–1000 Hz
optional	500–10,000 Hz
Frequency coverage (all modes)	>50 cm^{-1} for SDL-10,20,30,40[b] >200 cm^{-1} for SDL-20 W, 30 W
Resetability (T and I changes)	
For $T < 20°K$	$\Delta\nu \leq 0.01$ cm^{-1}
For $T > 20°K$	$\Delta\nu \leq 0.1$ cm^{-1}
Sensitivity	
Source intensity	>0.1 mW (SDL-40)
all modes (max attainable	>0.2 mW (SDL-30)
within laser tuning range)	>0.5 mW (SDL-20)
	>0.25 mW (SDL-10)
Detector preamplifier output	
responsivity (peak λ)	0.3 V/μW typical
noise	—
broadband (10 Hz–500 kHz)	<1 mV
narrowband (1 Hz bandwidth)	<1 μV
System signal-to-noise ratio	
broadband (modulation scan mode)	>30 dB
narrowband (sweep mode with lock-in)	>60 dB
Optical Performance	
Collimation	
$f/1$ IR-2 lens	<2 mrad
$f/1$ KBr lens	2 mrad
Laser Characteristics	
Optical size	<50 × 50 $(\mu m)^2$
Current tuning rate	1–30 cm^{-1}/A
Approximate temperature tuning rates	
single mode	<1 cm^{-1}/°K
all modes	≤4 cm^{-1}/°K
Mode spacing	<0.5 cm^{-1}
Beam divergence	$f/1$ typical

[a] From Laser Analytics Inc. Brochure, with permission.

[b] SDL-10 within 3500–2320 cm^{-1}; SDL-20, 20 W within 2320–1180 cm^{-1}; SDL-30, 30 W within 1180–650 cm^{-1}; SDL-40 within 650–380 cm^{-1}.

C. Practical Considerations

1. *Laser Lifetime*

There has been a variety of experiences with respect to both shelf life and operating lifetime of diode lasers. Some diodes that have operated well have been found not to operate at all after being unused for several months. Apparently, keeping diodes at liquid nitrogen temperature alleviates this problem, indicating that some diffusion or solid reaction is probably occurring. In many cases such failure will occur in commercial diodes very soon after manufacture, often during shipping. This problem is being alleviated by lengthier and better testing procedures (Mantz, 1979).

Repeated temperature cycling of diodes can cause failures, usually in the heat-sink contact. Improved contacting and testing have somewhat alleviated these problems. It is, of course, possible to maintain the diodes at low temperature. The difficulty with this in a closed cycle cooler comes from the finite life of the compressor; typically it is rated at about 7000 hr.

2. *Mode Jumping*

Mode jumping has been discussed above; it limits the continuous tuning range of a single mode to somewhere between 0.5 and about 3.0 cm^{-1}. Modes can usually be separated with a monochromator, and complete coverage over a more extensive range is often obtained by choosing modes whose tuning ranges overlap. However, this is not always possible, and depending on the application, it may be necessary to have several diodes to observe a particular spectrum of interest. In a few cases the multimode behavior of the diode can be so severe that abnormally small regions will be scannable continuously.

3. *Calibration*

The monochromator used to isolate modes provides a laser frequency calibration to better than 1 cm^{-1} accuracy. The most common means of obtaining a fine relative calibration of diode laser output frequency in a scanned spectrum is by means of a solid or air spaced, moderate finesse, Fabry–Perot étalon. As the laser frequency passes through Fabry–Perot orders, a sequence of frequency markers is generated, spaced by the free spectral range of the étalon. Figure 4.16 shows an étalon that can be inserted into the optical path in place of the sample gas cell by means of an external control. The diode frequency is scanned twice, once to record the gas spectrum and once to record the étalon frequency calibration. An alternate arrangement is to beam split the single mode output from the monochromator, pass one beam through the cell, the other through the reference étalon, and detect each beam individually. This method eliminates possible drifts between successive scans; but it requires two detectors, and it is subject to errors (Flicker *et al.*, 1978) because of the varying angle of incidence of the light emerging from the monochromator exit slit and entering the étalon as the laser frequency is scanned, unless the monochromator grating is synchronously scanned.

An arrangement for obtaining an étalon calibration with one detector without the above-mentioned difficulties is shown in Fig. 4.17a (Chraplyvy, 1978). It is a modification of a procedure previously used to normalize a spectral scan to the input intensity (see below). The diode laser beam is split into two beams, one beam passing through the étalon, the other through the sample cell. The two beams are chopped, each with a different chopping frequency. The two beams are then recombined; a single mode is isolated with the monochromator and then detected. The reference étalon and sample spectrum signals are separated from the total detected signal by passing it through two lock-in amplifiers, each tuned to the appropriate chopping frequency. Figure 4.17b shows an étalon calibration and spectrum obtained by this procedure (Chraplyvy, 1978).

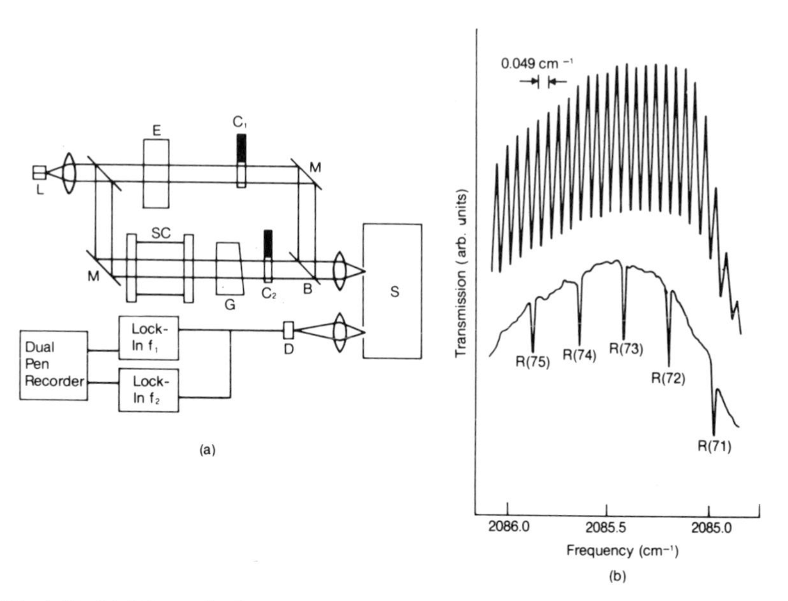

FIG. 4.17. (a) Schematic diagram of single detector, single scan technique for recording spectrum and Fabry–Perot calibration. (b) Recording of calibration and CO spectrum using setup in (a) (Chraplyvy, 1978).

A convenient and commonly used étalon is a germanium solid 2.54 cm long, which gives a free spectral range of about 0.05 cm^{-1}. Because of the high index of refraction of germanium, this étalon need not be coated. For very high precision spectra, a smaller free spectral range is needed, and an air-spaced étalon must be used. Reddy *et al.* (1979a) have described the need for and the design of several air-spaced étalons. In addition, they have shown the substantial (more than 10^2) advantage of an air-spaced étalon in terms of environmental stability.

Absolute frequency calibration is most commonly performed using one or a few spectral lines of a low-pressure gas with accurately known frequencies. This calibration gas can be used for absolute calibration the same way the étalon is used for relative calibration, or it can be placed in series with the sample gas if it will not interfere. Low-pressure discharges of CO_2 in the 9–11-μm region and CO in the 4.5–8-μm region are convenient since the lines are well spaced and their frequencies are known to very high accuracy from stable CO_2 and CO laser measurements. Absolute accuracies to better than 10 MHz are possible. Recently Wells *et al.* (1979) have proposed the OCS ν_2 overtone band as a good calibration standard in the 9.5-μm region. They have derived accurate molecular constants for the 02^00–00^00 transitions of $^{16}O^{12}C^{32}S$ by locking diode laser frequencies to the peak of a number of lines in this band, and measuring the diode laser frequencies by heterodyning with a CO_2 laser. Several hot band and isotopic OCS frequencies were also measured this way. A table of frequencies, generated from the derived constants, spaced by about 0.4 cm^{-1} from 1025 cm^{-1} to 1074 cm^{-1} was presented. Wells *et al.* (1979) have indicated they are planning to generate OCS calibration spectra in the region of 11 μm using the ν_1 band by heterodyning with a $^{13}CO_2$ laser, and in the 6-μm region using the $2\nu_1$ band.

In the 12–17-μm range, CO_2 and HCN should be convenient standards. Devi *et al.* (1979) have used a diode laser spectrometer with a 0.15-cm^{-1} free spectral range, air-spaced étalon for relative calibration to measure several lines of the CO_2 ν_2 band relative to nearby lines of HCN as standards. The values obtained were compared with values obtained from state-of-the-art Fourier spectroscopy quoted to have a precision better than 0.001 cm^{-1} (Kauppinen, 1979). The results proved consistent to within 15 MHz (0.0005 cm^{-1}). The reproducibility of diode laser measurements was at least as good.

The procedure described above of heterodyning gas laser lines with diode lasers tuned to line center can, of course, be directly used in appropriate wavelength ranges. For example, Worchesky *et al.* (1978) have measured D_2O frequencies this way. Where suitable reference gas lines are available, however, there would appear to be little advantage to this procedure except under very unusual circumstances.

4. *Absorption Measurements*

Figure 4.7 illustrates a difficulty in obtaining accurate absorption data from a diode laser scanned spectrum, namely, the variation of single mode laser power with frequency (current). A classic spectroscopic technique for dealing with this problem and obtaining direct readout of transmission is the dual beam spectrophotometer. The design and construction of such an instrument using a diode laser have been discussed by Dubs and Gunthard (1978). Figure 4.18 gives a result obtained by them.

Diode intensity variation can be compensated for by ratioing transmitted intensity signals to input intensity measurement. If the very small drift errors involved are tolerable, separate scans can be made with gas cell, étalon, and with empty cell. The results can be digitally recorded and ratioed, then manipulated or plotted (Eng and Mantz, 1979a). Beam splitting can, of course, also be used. It should be possible to add a third channel to the arrangement of Fig. 4.17 to measure reference intensity. The output of each lock-in shown in the figure would then be ratioed with the output of a lock-in tuned to this third channel frequency in an analog ratio circuit before plotting, to obtain intensity normalized spectra and étalon frequency calibration.

5. *Noise Characteristics*

The low frequency amplitude noise characteristics of infrared diode lasers that would affect spectroscopic or gas analysis applications have

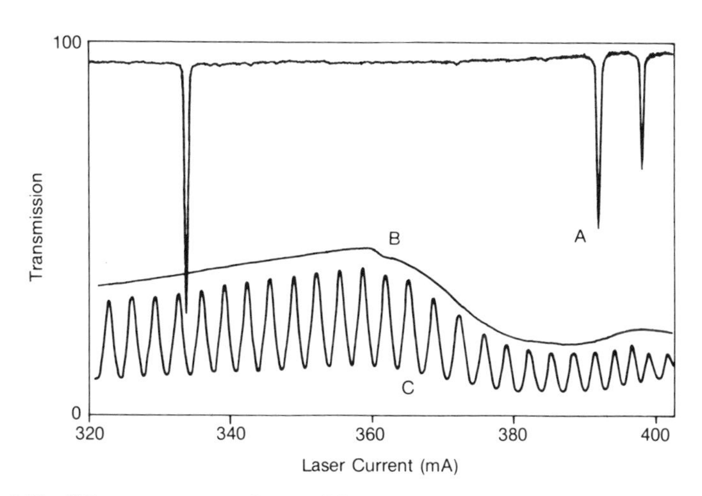

FIG. 4.18. CH_4 gas spectrum, 10-cm cell length, 0.8-Torr pressure, room temperature. *A*, Double-beam spectrum, *B*, transmitted laser intensity, *C*, transmitted laser intensity through étalon. Diode laser temperature is 37°K, laser current sweep rate 0.5 mA/sec; étalon fringe spacing is 0.05-cm^{-1} (Dubs and Gunthard, 1978).

been measured by Eng *et al.* (1979a). They have distinguished vibration noise from their closed cycle cooler, excess laser noise, and mechanical noise in the laser contacts. The latter can be eliminated. A result of this study is a recommendation that laser modulation frequencies be between 4 and 10 kHz.

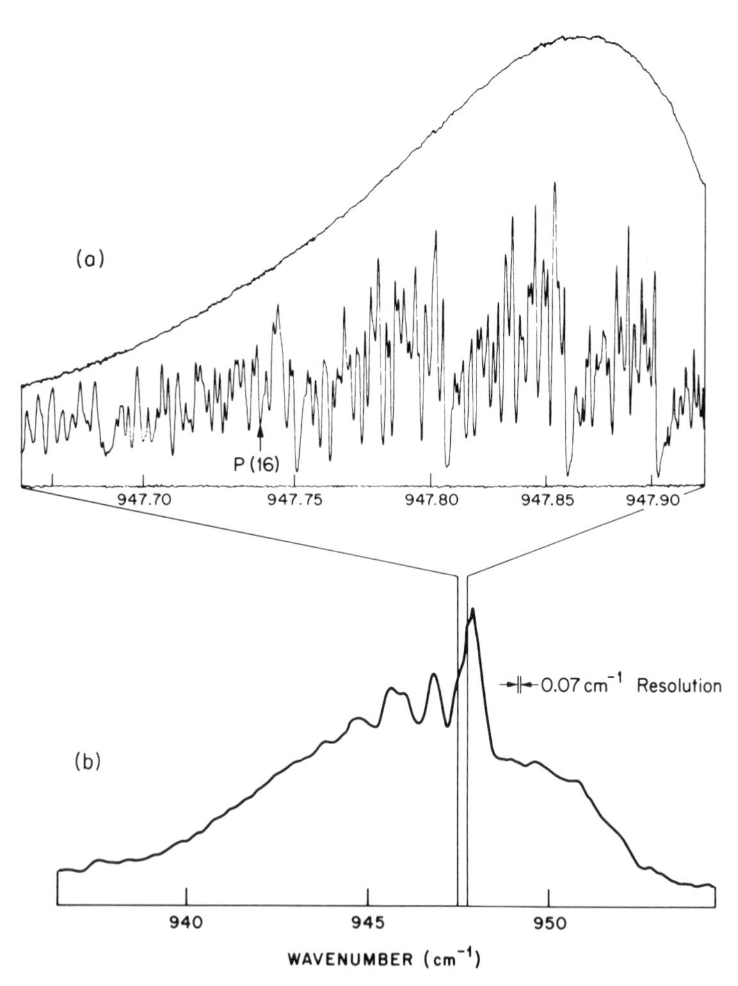

FIG. 4.19. Spectra of sulfur hexafluoride. (a) Diode-laser scan (SF_6 pressure, 0.1 Torr, cell length, 10 cm, resolution, 3×10^{-6} cm^{-1}) is of band segment near the P(16) CO_2 laser line. (b) Grating spectrometer scan of the ν_3 band, taken in a 25-cm cell at a pressure of 0.1 Torr SF_6 has 0.07 cm^{-1} resolution. Grating-spectrometer scan from Brunet and Perez (1969). (Hinkley and Kelley, 1971. Copyright 1971 by the American Association for the Advancement of Science.)

D. EXAMPLES OF DIODE LASER SPECTROSCOPY

1. *Spectra*

A dramatic example of the capability of tunable diode lasers in molecular spectroscopy is demonstrated in the case of sulfur hexafluoride SF_6. Figure 4.19 is a widely referenced spectrum taken by Hinkley (Hinkley and Kelley, 1971). Shown is a high-resolution grating spectrometer spectrum (the best available at the time) of the ν_3 band and a small wavenumber region of that spectrum scanned with a tunable diode laser. The enormously rich and complex structure completely unrevealed by the grating spectrometer is evident.

Even more impressive than revealing the complex structure of SF_6 is the role of diode laser spectroscopy used by the group at Los Alamos Scientific Laboratory in understanding that complex spectrum. With ob-

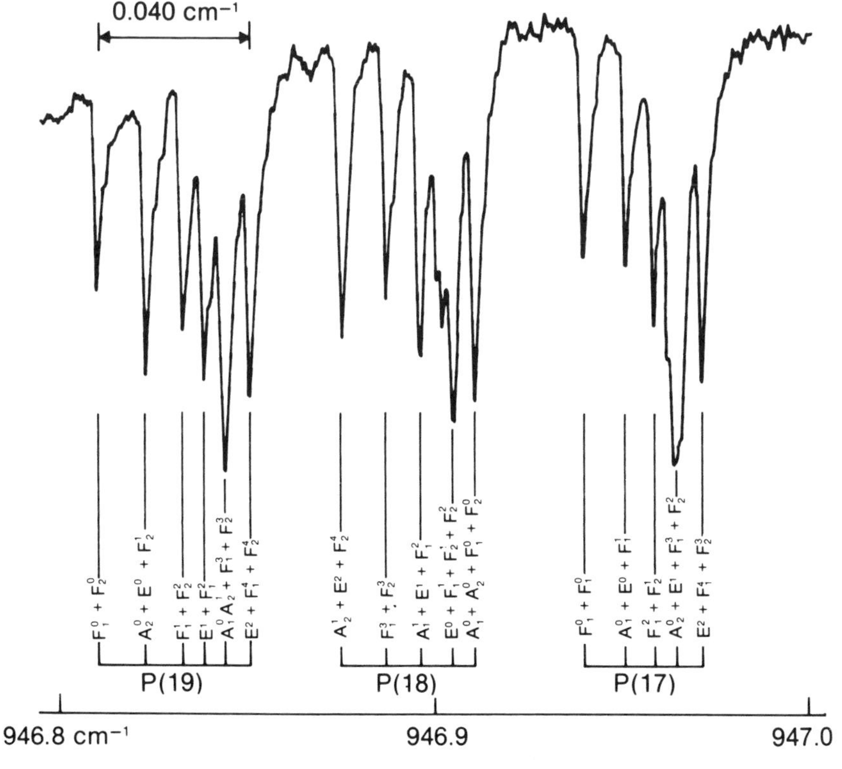

FIG. 4.20. Portion of the low-temperature ν_3 spectrum of SF_6 with line identifications (Aldridge *et al.*, 1975).

servation of large portions of the SF_6 spectrum with tunable diode laser spectroscopy at 130–140°K to reduce interference from hot bands, over 5000 lines in the room temperature spectrum of SF_6 have been assigned, and their line positions can be calculated to an accuracy of 0.002 cm^{-1} (Aldridge *et al.*, 1975; McDowell *et al.*, 1976, 1977). Figure 4.20 shows a low-temperature spectrum of SF_6 with line assignments.

The 16-μm ν_3 band of UF_6 is important for laser isotope separation investigations. Figure 4.21 shows a theoretical spectrum and a laser diode spectrum taken with a nozzle beam. The nozzle beam is crucial to obtaining low temperatures (to avoid hot band interferences) while still maintaining sufficient density for taking spectra (as well as for isotope separation). Spectra such as Fig. 4.21 were important in showing distinct absorption features of $^{235}UF_6$ on which isotope separation could be based.

Ultrahigh-resolution spectra of molecules using tunable diode lasers are being routinely obtained today by some 200 diode laser spectrometers (Butler, 1979). Table 4.II shows molecules whose spectra have been published after 1976 in studies not discussed in the text. For earlier work we refer to a similar table in Hinkley *et al.* (1976a).

As another example of tunable diode laser spectroscopy, Fig. 4.22 shows a spectrum of the $(\nu_4 + \nu_5)^{0+} - \nu_4$ hot band Q branch of $^{12}C^{13}CH_2$ observed in a natural sample of acetylene (Reddy *et al.*, 1979b). The spectrum illustrates Doppler-limited resolution and also provides a good example of the sensitivity of tunable laser spectroscopy. In previous grating spectroscopy with a conventional source, there was no indication of the illustrated spectrum.

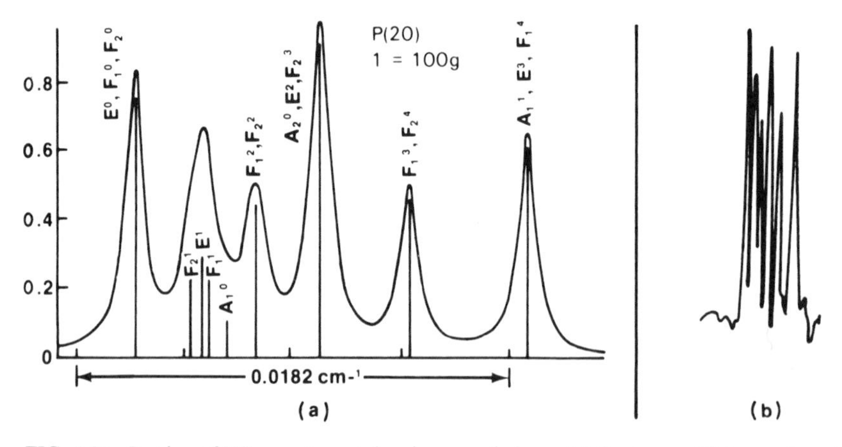

FIG. 4.21. Portion of UF_6 spectrum taken in a nozzle beam (a) theory, and (b) experiment (Nill, 1977).

TABLE 4.II

SOME TUNABLE DIODE LASER SPECTROSCOPY STUDIES PUBLISHED AFTER 1976 WHICH ARE NOT DISCUSSED IN THE TEXT

Molecule	Wavenumber region	Reference	Footnote
H_2O	600, 1000	Eng and Mantz (1979a)	a
CO_2	618, 940	Eng and Mantz (1979b)	a
	1040, 667	Travis *et al.* (1977)	b
C_3O_2	860	Jennings *et al.* (1978)	b
NH_3	800–1200	Capellani and Restelli (1979)	—
	1080	Sattler and Ritter (1978)	—
	945	Nereson (1978)	—
NO^+	2285	Bien (1978)	a,c
CH_3OH	1034	Sattler *et al.* (1978)	—
CH_3F	1190	Hirota (1979)	—
	1010	Sattler and Simonis (1977)	—
CS	1270	Yamada and Hirota (1979)	—
	1225	Todd and Olson (1979)	—
HNO_3	895	Brockman *et al.* (1978)	—
D_2O	1060	Worchesky *et al.* (1978)	—
OsO_4	960	McDowell *et al.* (1978)	—
SF_6	615	Person and Kim (1978)	—
		Travis *et al.* (1977)	b
CF_4	1073	Radziemski *et al.* (1978)	—
CF_2Cl_2	918	Jennings (1978a)	—
		Restelli *et al.* (1978)	c
		Cappellani *et al.* (1979)	—
$CFCl_3$	850	Cappellani *et al.* (1979)	—
ClO	853	Margolis *et al.* (1978)	—
UF_6	625	Travis *et al.* (1977)	b
WF_6	713	Travis *et al.* (1977)	b
H_2	1035	Reid and McKellar (1978)	d

[a] Intensities, pressure broadening coefficients.
[b] Nozzle beam spectroscopy at ~50°K.
[c] Vibrational quenching rates.
[d] Rotational quadrapole transition.

2. *Linewidths, Shapes, and Shifts*

The very narrow linewidth of a tunable diode laser makes it particularly well suited for studying molecular line shapes, particularly with self or foreign gas pressure effects. Since, at atmospheric pressure, typical linewidths are of the order of 0.1 cm^{-1}, the limited continuous tuning range of a laser mode is generally not a problem, and for these measurements absolute calibration is not necessary. Figure 4.23, showing a hot

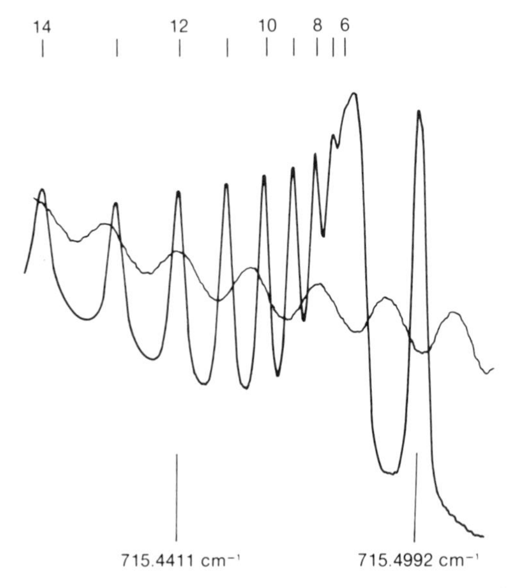

FIG. 4.22. A portion of the observed Q branch of the $(\nu_4 + \nu_5)^{0+} - \nu^{1f}$ band of $^{12}C^{13}CH_2$ and the $Q(26)$ line of the corresponding band of $^{12}C_2H_2$. Sample path length 6 m, pressure of gas 160 μTorr (Reddy *et al.*, 1979b).

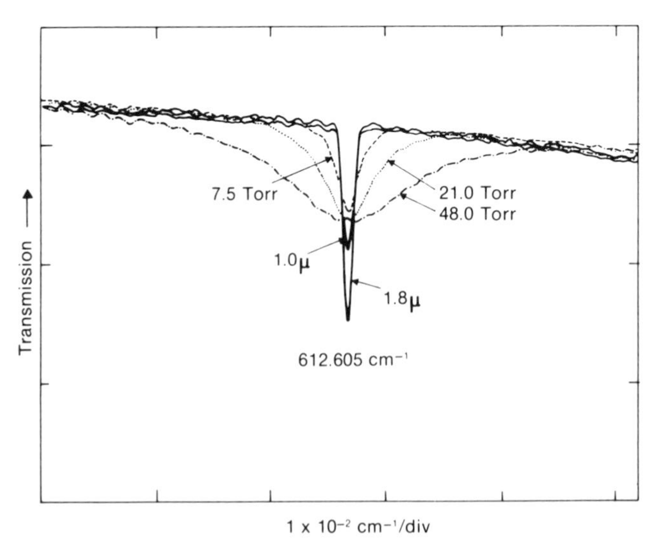

FIG. 4.23. The $R(27)$ line of the $01^{1d}0$ to $02^{2}0$ hot band transition in N_2O measured at room temperature in a 7-cm cell (dotted and dashed curves) and in the long path cell at 200 m (solid curves) (Kim *et al.*, 1978).

band ν_2 transition of N_2O in the 16-μm region, is an almost textbook illustration of absorption line shapes (Kim *et al.*, 1978). It was taken with the aid of a very long path multipass cell described in Section 4.2,B. Clearly illustrated are the Doppler regime, where the peak absorption varies with pressure while the linewidth is constant, and the pressure broadened regime, where the peak absorption stays constant while the linewidth grows with pressure.

Figure 4.24 shows measurements taken of the linewidth versus pressure of H_2O ν_2 band degenerate transitions $16_{0,16}-15_{1,16}$ and $16_{1,16}-15_{0,15}$. The curves illustrate the phenomenon of collisional or "Dicke" narrowing. As the pressure is increased, the Doppler width is narrowed by the effect of velocity averaging collisions. This effect becomes significant when the mean free path for such collisions becomes less than the infrared wavelength (actually $\lambda/2\pi$). If pressure broadening has not become dominant by the time this condition is reached, then collisional narrowing is observed. For H_2O, pressure broadening effects of collisions are small enough to observe the narrowing effect for states of high rotational energy (~2000 cm^{-1}).

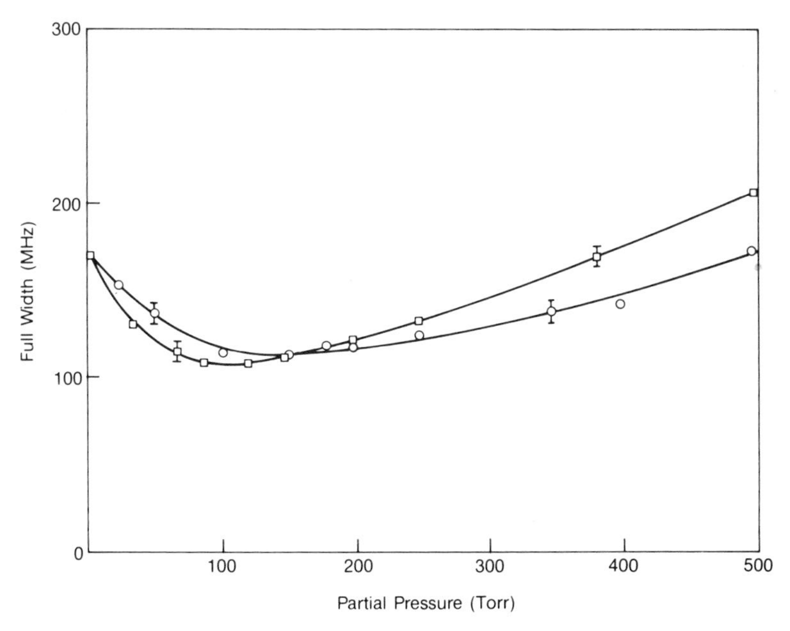

FIG. 4.24. Full width at half maximum of absorption constant for the water vapor line at 1879.01 cm^{-1} versus buffer gas pressure, showing the phenomenon of collisional narrowing. ○, H_2O–Ar; □, H_2O–Xe. Water vapor pressure is 2.0 Torr (Eng *et al.*, 1972).

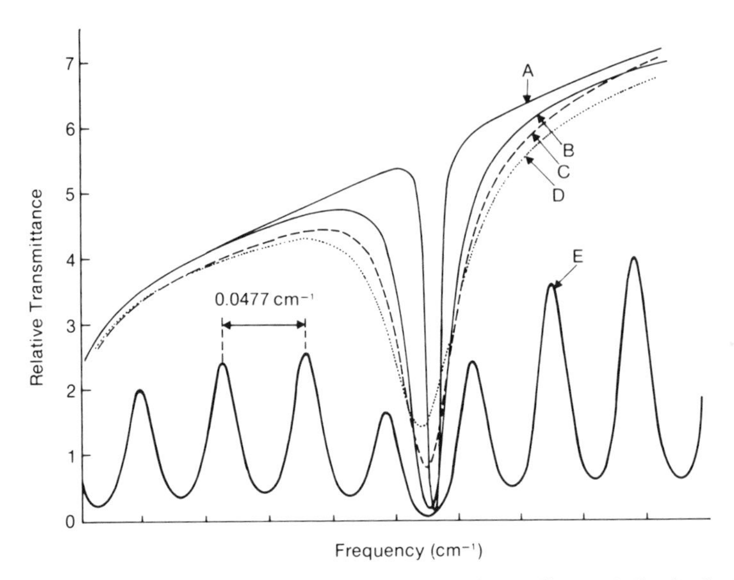

FIG. 4.25. Tunable diode laser absorption spectrum of the $H^{79}Br$ $P(7)$ line in the fundamental region. Curve A is a spectrum of 2.9-Torr HBr. Curves B, C, and D also have 2.9 Torr in addition to Ar with total pressures of 200.6, 400.1, and 600.2 Torr, respectively. Curve E is an étalon trace for wavelength scale determination. Frequency increases from left to right.

Figure 4.25 shows spectra of the $P(7)$ line of the $H^{79}Br$ fundamental at 4.3 μm taken at 2.9 Torr of HBr and various added Ar pressures (Mantz and Eng, 1978). Table 4.III summarizes data obtained on various HBr lines in terms of line intensity, and Ar pressure broadening and shift coefficients. The ratios of the measured intensities were within 10% of what they should be for HBr at room temperature. These HBr:Ar measure-

TABLE 4.III

SUMMARY OF LINE INTENSITIES AND PRESSURE-BROADENING AND SHIFT COEFFICIENTS MEASURED WITH A DIODE LASER FOR HBr WITH Ar[a]

Line	Intensity (cm^{-1} molecule cm^2)	Broadening (cm^{-1}/atm.)	Shift (cm^{-1}/atm.)
$P(7)$	$(3.08 + 0.74) \times 10^{-20}$	$(2.23 + 0.08) \times 10^{-2}$	$(8.95 + 0.55) \times 10^{-3}$
$R(3)$	$(12.29 + 1.07) \times 10^{-20}$	$(3.75 + 0.23) \times 10^{-2}$	$(10.63 + 1.48) \times 10^{-3}$
$R(4)$	$(9.90 + 0.65) \times 10^{-20}$	$(2.75 + 0.10) \times 10^{-1}$	$(6.58 + 1.07) \times 10^{-3}$
$R(6)$	$(5.76 + 1.01) \times 10^{-20}$	$(2.52 + 0.30) \times 10^{-2}$	$(8.97 + 1.91) \times 10^{-3}$
$R(8)$	$(2.11 + 0.46) \times 10^{-20}$	$(2.05 + 0.38) \times 10^{-2}$	$(4.75 + 0.16) \times 10^{-3}$

[a] Mantz and Eng, 1978.

ments are interesting in that they were made to assess the possibility of on-line determination of fill pressures in quartz halogen lamps. These lamps use mixtures of HBr and Ar at pressures up to 5 atm and are subject to large variations in processing (Butler, 1976).

3. *Gas Analysis*

Spectral lines observed with tunable diode lasers can be used to detect the presence of molecular gases and measure their concentrations at extremely low levels. This capability has a large number of potential monitoring and diagnostic applications.

As an example, Reid *et al.* (1978a,b) have developed an extremely sensitive instrument and have analyzed its application in air pollution measurements. The instrument (their second, improved version) is shown in Fig. 4.26. Air samples are slowly flowed through the sample 200-m path length White cell at reduced pressures. The tunable diode laser output traverses the White cell and is detected at detector *A*. Derivative spectroscopy at a modulation frequency of 1 kHz is used. Figure 4.27 shows a reference spectrum of 1 Torr of NH_3 near 1140 cm^{-1} and a second derivative spectrum of 20-Torr laboratory air in the White cell in the same wavelength region. The NH_3 lines in the air spectrum are apparent and correspond to an NH_3 concentration of 20 ppb (parts per billion). In operation, the diode laser frequency is locked to the center of the absorption line of the gas to be detected by placing a sample of that gas in the cell labeled 2 and detecting with detector *B*. The first derivative signal is used

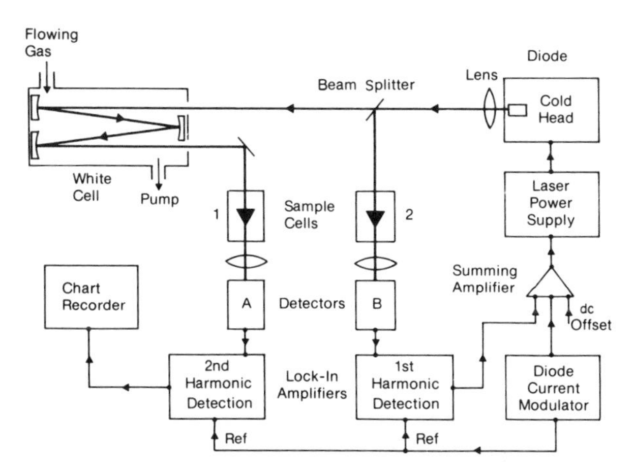

FIG. 4.26. Schematic diagram of high-sensitivity laser absorption spectrometer (Reid *et al.*, 1978b).

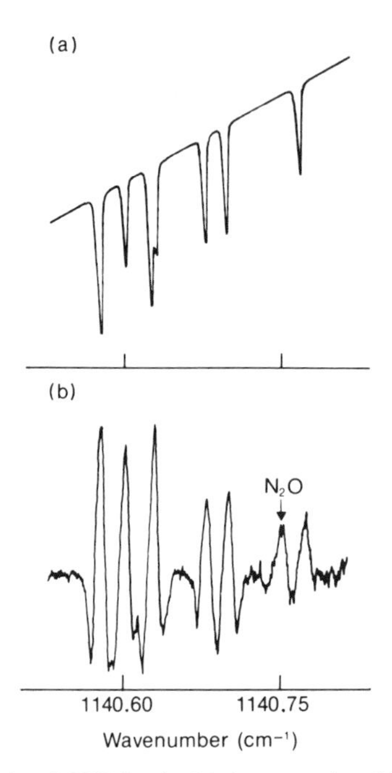

FIG. 4.27. Detection of 10-ppb NH_3 in air. (a) A conventional high-resolution scan of NH_3 lines near 1140 cm^{-1}. Pressure is ~1 Torr, cell length, 10 cm. (b) A laser absorption scan over the same wavelength region in air. Path length is 200 m in air at 20 Torr, with a time constant of 3 sec. The air scan contains two additional lines due to N_2O. One is indicated at 1140.747 cm^{-1}. A second, stronger N_2O line is coincident with the NH_3 line at 1140.606 cm^{-1} (Reid *et al.*, 1978b).

in a feedback loop to the laser power supply to perform the lock in a conventional manner. The dc offset can be used, if desired, to scan the laser about line center.

Table 4.IV lists the minimum concentration of various pollutants measurable at the given wavelengths with the instrument of Fig. 4.26. It should be pointed out that the first seven entries in Table 4.IV can all be observed with the one diode, temperature tuned. The concentration levels given in Table 4.IV correspond to an absorption sensitivity of 3×10^{-8} m^{-1}. A critical factor in achieving this sensitivity to absorption is illustrated in Fig. 4.28. Shown is the second derivative signal with no sample gas, with the beam blocked (trace A), and with the laser frequency modulation off (trace B). But with the beam unblocked and modulated, trace C was observed, a fringe pattern with amplitude much greater than the detector noise. This and other patterns were traced to accidental étalon effects, in the case shown, between two antireflection coated lenses

TABLE 4.IV

MINIMUM DETECTABLE CONCENTRATION FOR MOLECULES OF ATMOSPHERIC IMPORTANCE[a]

Molecule	Absorption coefficient of strong lines ($m^{-1} \cdot ppb^{-1}$)	Approximate wavenumber of lines (cm^{-1})	Sensitivity (ppb)
SO_2	3.5×10^{-8}	1140	0.9
O_3	2×10^{-7}	1050	0.015
N_2O	5×10^{-8}	1150	0.6
CO_2	3.5×10^{-10}	1075	90
H_2O	2×10^{-9}	1135	15
NH_3	2×10^{-6}	1050	0.015
PAN	3×10^{-7}	1150	0.09
CH_4	3×10^{-6}	1300	0.009
SO_2	3.5×10^{-7}	1370	0.09
NO_2	5×10^{-6}	1600	0.006
NO	3×10^{-6}	1880	0.009
CO	10^{-5}	2120	0.003
CO_2	10^{-4}	2350	0.0003

[a] From Reid *et al.* (1978a), modified for absorption sensitivity of 3×10^{-8}/m achieved by Reid *et al.* (1978b) with the system of Fig. 4.26.

spaced by 38 cm and having an effective reflectivity $\sim 10^{-3}$. With care, these effects were eliminated. It is suggested by Reid *et al.* (1978a) that previous efforts which could not measure total power absorption less than 10^{-3} of incident were also limited by étalon effects as in Fig. 4.28.

To investigate the viability of diode laser absorption spectroscopy for isotopic analysis of gaseous compounds, Whelan *et al.* (1977) have analyzed the 4.5-μm ν_3 band of $^{14}CO_2$ at Doppler-limited resolution. Figure 4.29 is a diode laser scan of three low-pressure gas cells in series enriched, respectively, in $^{14}CO_2$, $^{13}CO_2$, and normal $^{12}CO_2$, showing the noninterference of one $^{14}CO_2$ line with more abundant isotope lines (Eng *et al.*, 1978). It was found that the optimum $^{14}CO_2$ line for detection is $P(18)$ and that the only possible interfering species in air would be N_2O. Using a system with sensitivities similar to those of Reid *et al.* (1978a), Butler *et al.* (1978) have estimated that in a background 7.6-Torr air the minimum concentration of $^{14}CO_2$ detectable would be 2.6×10^4 molecules/cm^3 or about 10^{-2} pg in a six-liter White cell. This corresponds to about 1.7×10^{-2} pCi in equivalent radioactivity units.

In some cases diode laser spectroscopy can be useful for atom concentration measurement. Diode laser spectroscopy of the ground state fine structure splitting at 404 cm^{-1} has been proposed for measuring fluorine

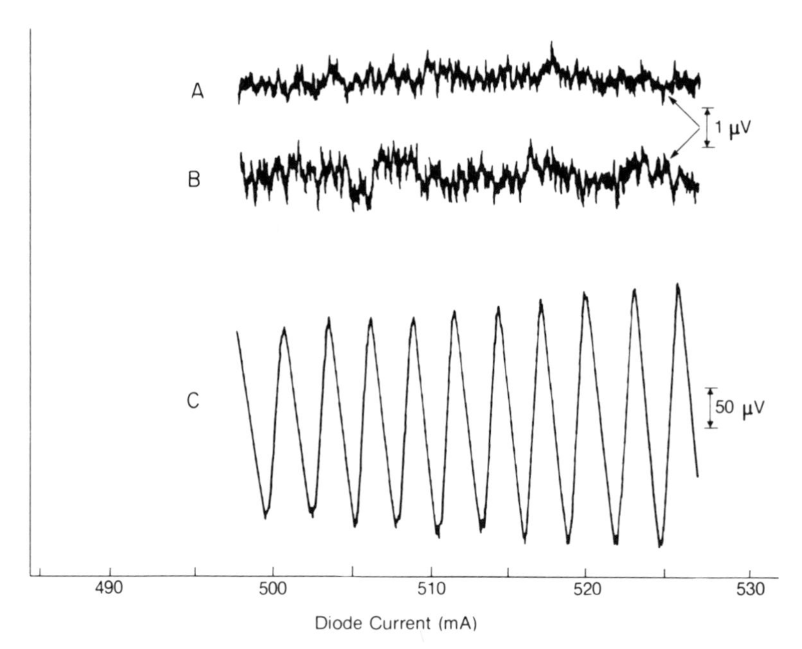

FIG. 4.28. Lock-in amplifier output using second harmonic detection. In the absence of modulation, sensitivity is limited by detector noise. When a 1-kHz modulation is applied to the diode current, the interference fringes in trace C appear. These fringes result from an étalon formed by two antireflection coated lenses in the beam path. The étalon spacing is 38 cm, with an effective reflectivity of $\sim 10^{-3}$ (Reid *et al.*, 1978a). Trace A, beam-blocked, detection noise; B, beam unblocked, modulation off; C, beam unblocked, modulation on.

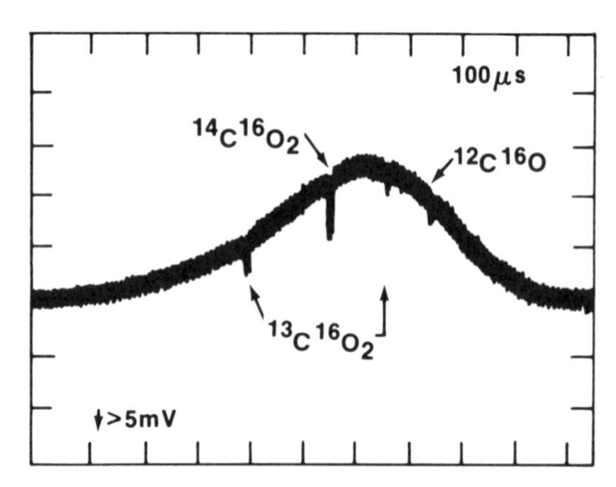

FIG. 4.29. Transmission scan showing the $^{14}C^{16}O_2$ ν_3 band P_2 line at 2224.257 cm^{-1} and nearby $^{13}C^{16}O_2$ and $^{12}C^{16}O_2$ lines. Absorption was measured with three cells in series, each cell containing one CO_2 isotope (Eng *et al.*, 1977).

atom concentrations (Schlossberg, 1976), in particular for analyzing chemical laser dynamics. The feasibility of fluorine atom probing has recently been demonstrated by Stanton and Kolb (1980). Figure 4.30 shows the fluorine atom spectrum obtained by them. Approximately 10^{15} fluorine atoms/cm^3 in the 10-cm path length cell were produced by flowing an F_2–Ar mixture through a 2450-MHz microwave discharge. The signal in Fig. 4.30 corresponds to about 1.5×10^{-3}/cm absorption. Stanton and Kolb estimate that using multipass optics and derivative detection, a sensitivity of 10^{12}–10^{13} atoms/cm^3 could easily be achieved, more than adequate for chemical laser diagnostics.

Other atoms with ground state fine structure splitting in the 2.7–30 μm range could also be probed this way, e.g., Br and Cl. Probing atoms in excited states, where energy spacings are smaller, is an interesting possibility, with the absorption sensitivities that have been demonstrated.

Diode laser spectroscopy has been shown to be a viable technique for measuring concentrations of species in shock tubes and flames (Sulzmann *et al.*, 1973; Hanson, 1977; Hanson *et al.*, 1977). Temperature measurements can also be made by comparing the intensities of closely spaced absorption lines originating from different levels of a molecular species. Hanson and Falcone (1978) have demonstrated this technique in a flame. Figure 4.31 shows one of their results using the $v = 3 \leftarrow 2$, $R(2)$ and $v = 2 \leftarrow 1$, $P(4)$ lines of CO near 2101 cm^{-1} to measure a temperature of 1790°K. This compared to a thermocouple measurement of 1770°K. It is significant that the measurement (Fig. 4.31) was made within 50 μsec and could be made faster, and at high repetition rates. A modification of the

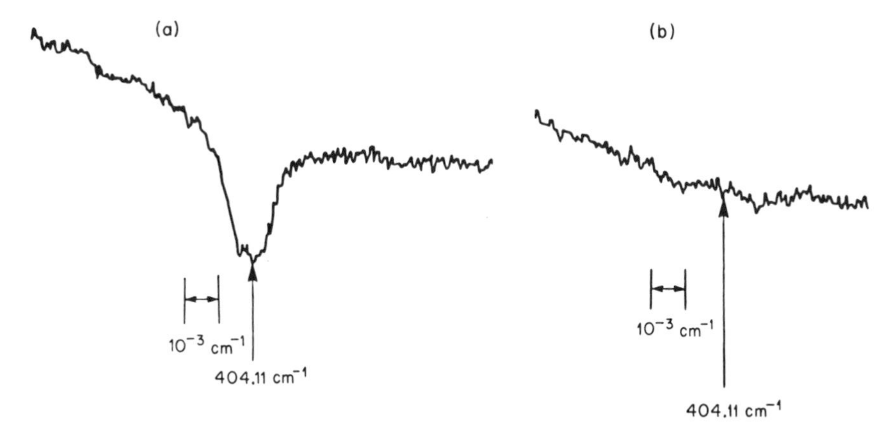

FIG. 4.30. (a) Absorption line of $^2P_{3/2}$–$^2P_{1/2}$ transition at 404.11 cm^{-1} in atomic fluorine measured in an F_2 discharge. (b) Spectral scan of same region with discharge off to verify identification of absorber (Stanton and Kolb, 1980).

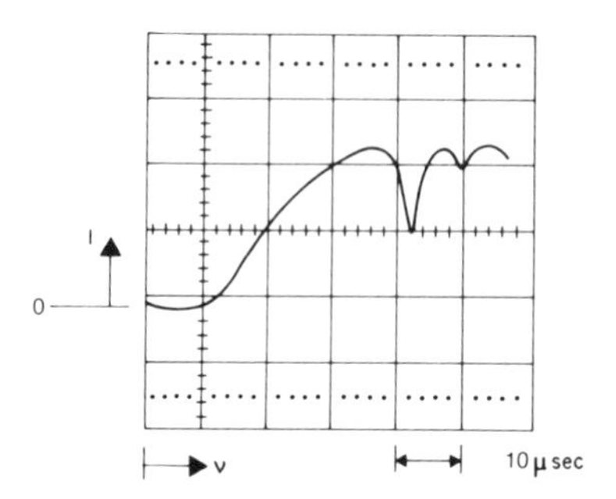

FIG. 4.31. Oscillogram data for temperature measurement in fuel-rich atmospheric pressure, propane–air flat flames showing fast single sweep (10 μsec/division) of CO lines $v = 3 \leftarrow 2$, $R(2)$ and $v = 2 \leftarrow 1$, $P(4)$ at 2101 cm^{-1}; the temperature by absorption is 1790°K, and the temperature by thermocouple measurement is 1770°K (Hanson and Falcone, 1978).

technique is to use closely spaced lines of different CO isotopes (Kolb, 1979), e.g., ^{13}CO and ^{12}CO, which would have a large difference in v for transitions within a given diode operating interval (e.g., $v = 5$ for ^{12}CO and $v = 0$ for ^{13}CO). The large decrease in relative population of the high v is largely compensated by the natural isotope abundance, making for similar absorptions and a relatively simple measurement.

Partial pressures of gases in a mixture can be analyzed nonperturbatively using diode laser spectroscopy. An example is a study of the partial pressures of H_2O, SO_3, and H_2SO_4 over azeotropic aqueous solution of H_2SO_4 (Eng *et al.*, 1978). Figure 4.32 shows a diode scan near 1416 cm^{-1} of the vapor over an H_2SO_4 solution at 165°C. The broad dip in the center is due to atmospheric water vapor in the unpurged optical path outside the sample cell. The narrow line in the dip at 1416.13 cm^{-1} is a line of the low-pressure water vapor in the cell. The other lines in Fig. 4.32 are SO_3 lines. The partial pressures of H_2O and SO_3 were measured at a variety of H_2SO_4 bath temperatures by comparing line strengths to those measured in reference cells under known conditions. The total pressure was measured in a specially constructed U-tube manometer, that measurement then inferring the H_2SO_4 partial pressure. Partial pressures measured by Eng *et al.* (1978) are listed in Table 4.V.

Diode laser monitoring of pollutants over long atmospheric paths has received considerable attention (Ku *et al.*, 1975; Hinkley, 1976; Hinkley *et al.*, 1976b; Chaney *et al.*, 1979). Figure 4.33a shows an experimental long-path monitoring arrangement, and Fig. 4.33b shows CO concentrations measured with this system over a 2-km path in Saint Louis. Table 4.VI shows the predicted measurement sensitivity for a number of other pollutants based on a 0.3% sensitivity to total absorbed power over a 1-km

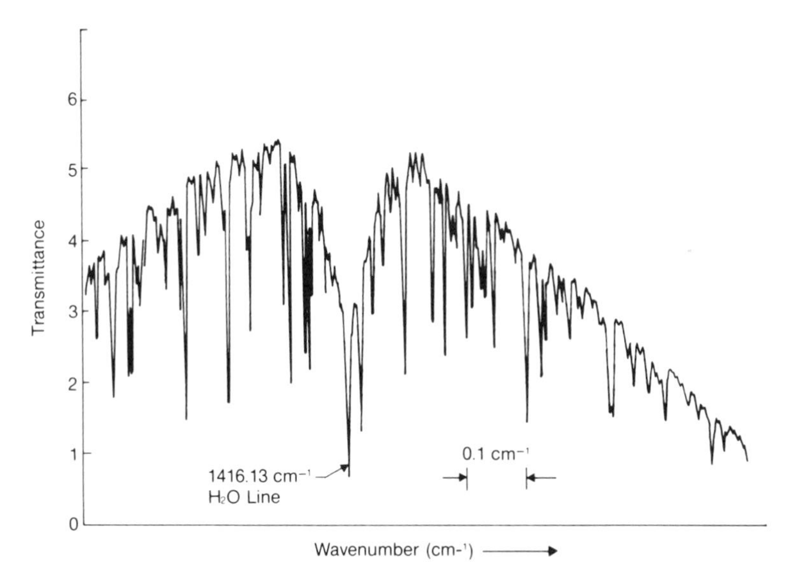

FIG. 4.32. Absorption spectra of the vapor above a hot H_2SO_4 solution showing the H_2O line at 1416.13 cm^{-1} and numerous SO_3 lines. H_2SO_4 reservoir temperature was about 165°C. With the reservoir at room temperature, the background has a trapezoidal shape as well as a broad dip due to the absorption at atmospheric pressure by the H_2O line at 1416 cm^{-1} (Eng *et al.*, 1978).

path. This is about the present limit for a long open path, as discussed by Eng *et al.* (1979b).

4. *Heterodyne Spectroscopy*

As described in Section 4.2, heterodyne spectroscopy is a passive means of obtaining very high resolution spectra using gas emission or absorption of radiation from a thermal source (e.g., the sun). Diode lasers

TABLE 4.V

PARTIAL VAPOR PRESSURES ABOVE A HOT H_2SO_4 AZEOTROPIC SOLUTION[a]

Reservoir temp. (°C)	Vapor pressure (Torr)				Dissociation constant
	H_2O	SO_3	Total	H_2SO_4	
107	0.024	0.022	0.08	0.034	0.0155
150	0.23	0.21	0.75	0.32	0.151
200	2.3	2.0	7.8	3.5	1.31

[a] Eng *et al.*, 1978.

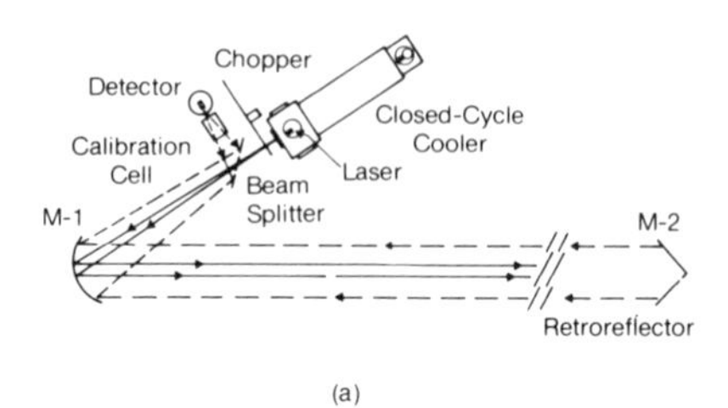

(a)

(b)

FIG. 4.33. (a) Schematic of laser and optical system for long-path monitoring of atmospheric pollutant gases. *M*-1 is an off-axis paraboloidal mirror, *M*-2 a remote retroreflector whose diameter is approximately 12 cm. (b) Monitoring of ambient CO in site 105 (Inner City Site in St. Louis, Missouri) over a total (round-trip) path of 2 km on 2 October 1974. Comparative bag samples were analyzed off-line by a chromatographic technique, as indicated. Averaging time for the laser measurements was 10 min for this presentation (Hinkley, 1976. Copyright by Chapman and Hall, Ltd.).

TABLE 4.VI

PREDICTED SENSITIVITY FOR SOME POLLUTANTS IN LONG-PATH DIODE LASER MONITORING BASED ON 0.3% ABSORPTION SENSITIVITY[a]

Molecule	Formula	(cm^{-1})	k(atm^{-1} cm^{-1})	Sensitivity (ppb)
Freon-11	CCl_3F	847	110	0.27
Freon-12	CCl_2F_2	921	275	0.11
Vinyl chloride	C_2H_3Cl	940	11	2.8
Ethylene	C_2H_4	950	42	0.71
Ozone	O_3	1052	22	1.3
Ammonia	NH_3	1085	93	0.33

[a] Hinkley *et al.*, 1976b.

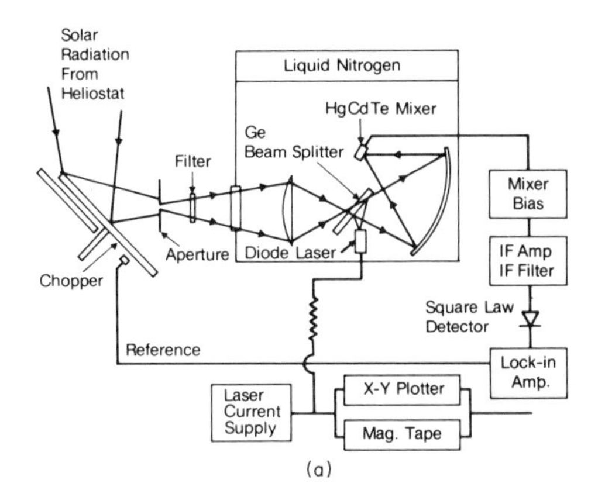

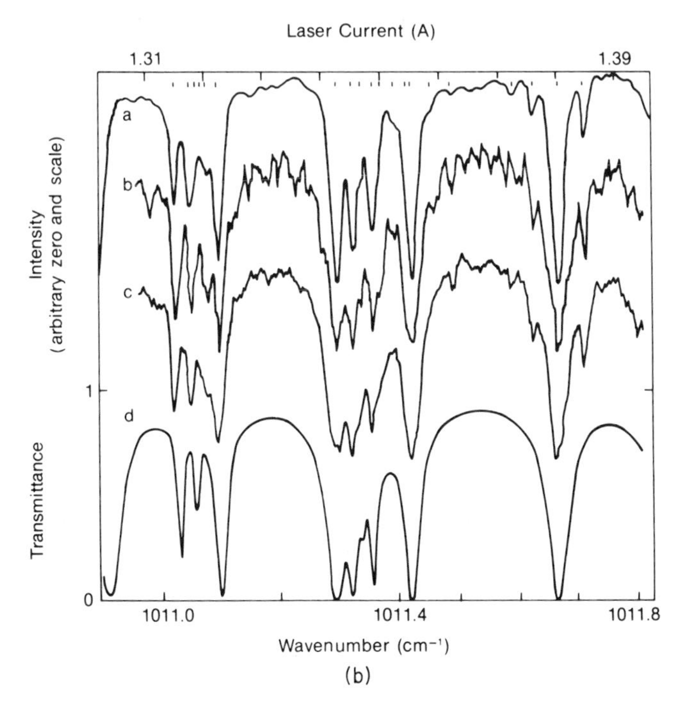

FIG. 4.34. (a) Apparatus used for heterodyne spectroscopy of upper atmospheric ozone. (b) Ozone absorption spectra. Trace *a* was obtained by passing tunable diode laser radiation through an absorption cell. Traces *b* and *c* are heterodyne spectra of atmospheric ozone, and *d* is a calculated atmospheric spectrum (Frerking and Muehlner, 1977).

as local oscillators are attractive because they can be tuned close to the line to be observed.

Menzies (1976) suggested the value of heterodyne spectrometers for studying ozone in the atmosphere. Atmospheric ozone spectra, using the sun as source and a diode local oscillator, have been published by Frerking and Muehlner. Figure 4.34a shows their experimental arrangement. Chopped solar radiation filtered to a 380-cm^{-1} bandwidth is mixed with diode laser radiation (in a single mode) near 1011 cm^{-1} by a fast HgCdTe detector. An IF cutoff frequency of 100 MHz was used, giving a 200-MHz resolution. In Fig. 4.34b, trace *b* shows the spectrum obtained with a 1-sec integration time. For comparison, trace *a* is a diode laser absorption spectrum, and trace *d* is a computed spectrum based on best available data for ozone molecular parameters and atmospheric concentration profiles. The resolution in the computed trace was made the same as the experimental ones. Trace *c* is the average of *b* and five other, similar scans, each full scan taking 10 min.

The signal-to-noise ratio of the heterodyne spectrum in Fig. 4.34b is less than the shot noise limited value by a factor of about 100. Of this about a factor of 6 is attributable to low diode local oscillator power, such that its shot noise did not exceed the IF amplifier noise. More recently Ku and Spears (1977) have demonstrated heterodyne radiometry with a signal-to-noise ratio with a diode local oscillator within a factor of 2.5 of the value obtained with a CO_2 local oscillator. About a factor of 2 is due to low single mode diode power—0.15 mW, whereas 0.4 mW of CO_2 power was required to reach quantum noise limited operation. The detector used had an effective heterodyne quantum efficiency η of 0.45, implying that the signal-to-noise ratio with the diode local oscillator was within a factor of 6 of *ideal* ($\eta = 1$) quantum limited heterodyne detection. Much of this success is attributable to observing, and avoiding in use, diode laser regions of current with large excess noise. Figure 4.35 shows a heterodyne absorption spectrum of C_2H_4 near 942 cm^{-1} taken by Ku and Spears (1977) with a 900°C blackbody source, an IF bandwidth of 300 MHz, and a 0.4-sec postintegration time.

5. *Saturation Spectroscopy*

A saturated absorption spectrum (inverted Lamb dip) using a diode laser has been demonstrated in NH_3 (Jennings, 1978b). The experimental arrangement is shown in Fig. 4.36a. A standing wave cavity around the NH_3 sample is formed by the mirror S_1 and the output face of the diode. The absorption is monitored by detecting the power from the zinc selenide beam splitter as the laser is tuned. A monochromator (not shown) was

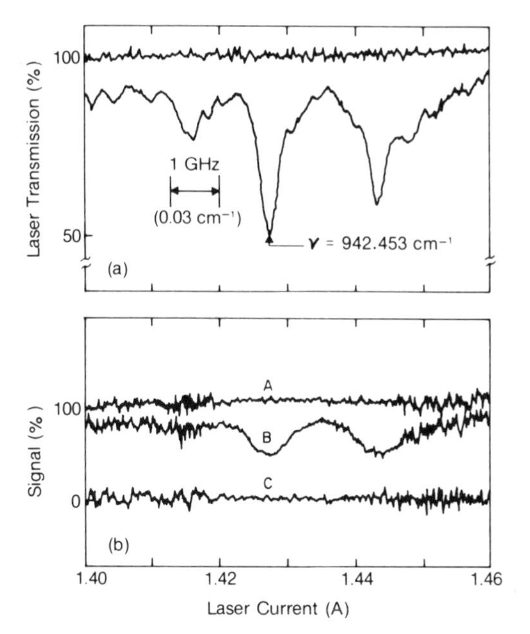

FIG. 4.35. (a) Direct diode-laser absorption scan of three C_2H_4 lines near 942 cm^{-1}; P = 10 Torr C_2H_4. (b) Heterodyne absorption scan of the above lines with spectral resolution of $2B$ = 600 MHz; T_{RC} = 0.4 sec. Trace A: P = 0; B: P = 10 Torr C_2H_4; C: shutter closed (Ku and Spears, 1977).

placed between mirror S_2 and the detector to isolate a single mode with approximately 0.75-mW power. Intensity at the focus in the sample is estimated as about 1 W/cm^2. The observed spectrum is shown in Fig. 4.36b. The NH_3 line is the ν_2, $sP(4,3)$ at 887.878 cm^{-1}. At a pressure of 1 Torr, the Doppler width (FWHM) is 80 MHz and the pressure broadened (homogeneous) width 46 MHz, giving an overall Voigt linewidth of 92 MHz. The measured dip width and overall width are in rough agreement. The homogeneous linewidth measurable was limited by the approximately 10-MHz spectral width of the laser.

The NH_3 experiment is important in demonstrating the capability of diode lasers for saturation spectroscopy. Refinements in the arrangement of Fig. 4.36 would include eliminating reflection from the diode face and using the technique of detecting the strong saturating wave modulation frequency on a weak, oppositely traveling probe wave. If a cavity were necessary to increase power incident on the gas, a ring cavity would probably be a good choice to enhance the saturating power. It would inherently eliminate feedback to the diode.

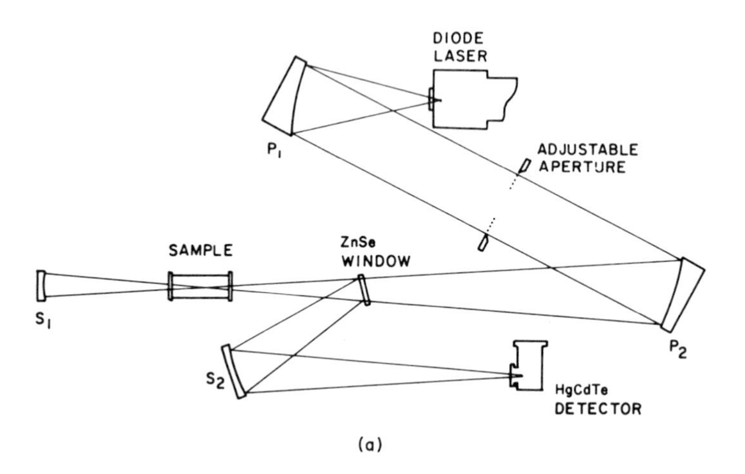

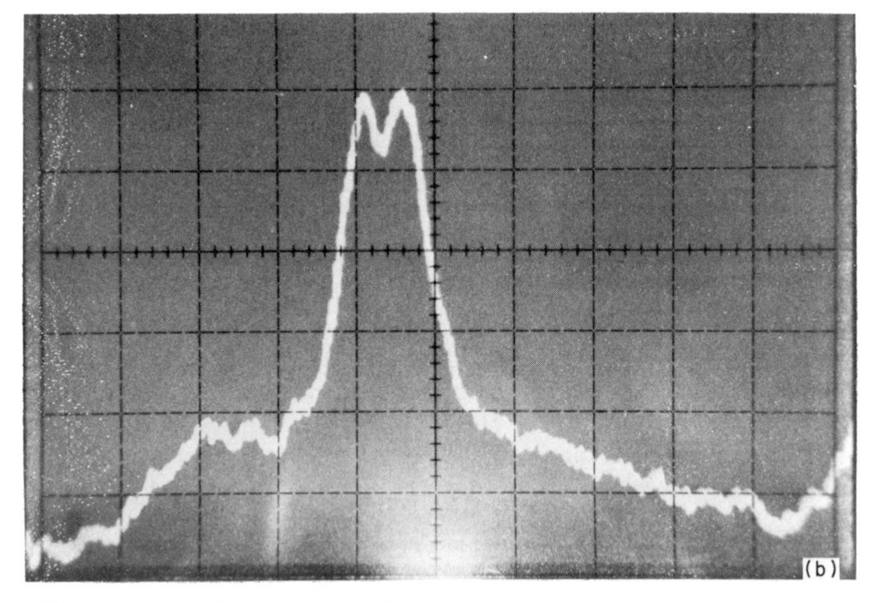

FIG. 4.36. (a) Optical arrangement for observing saturation with a tunable diode laser (TDL). Mirrors P_1 and P_2 are off-axis parabolas. S_1 and S_2 are spheres. Combination S_1, P_2, P_1 and the cleaved face of the TDL form a standing wave cavity. (b) Sub-Doppler saturation resonance as observed at the center of the $\nu_2[sP(4,3)]$ line of NH_3 in absorption at 887.878 cm^{-1}. The displayed signal is the dc amplified direct output of the detector, during a single sweep of the TDL current. Pressure is 1 Torr, path length 3 cm. Sweep scale is 2 msec/division. Zero transmission is at the bottom of the screen (Jennings, 1978b).

6. *Double Resonance Experiments*

The use of tunable diode laser spectroscopy to monitor transient processes is nicely illustrated by their application in laser double resonance experiments. In this kind of experiment, originally performed with fixed frequency lasers, a high-power pulsed laser is used to create a well-characterized nonequilibrium in the system, and absorption from a low power cw laser is used to probe the time dependence of populations in molecular energy levels as the system returns to equilibrium.

An example of a double resonance experiment performed with a diode laser is shown in Fig. 4.37 (Moulton *et al.*, 1977). The CO_2 laser pump beam was repetitively Q switched, giving pulses of about 150 nsec and 1 kW/cm^2 in the cell. Figure 4.38a,b,c,d shows four types of processes which were observed with various CO_2 pump lines and diode laser tunings: (a) The probe radiation is absorbed by molecules in level 2 (ν_3 = 1) populated by the pump laser. (b) The population of level 1 is decreased by the pump, leading to a decrease in absorption of probe radiation. Levels 2 and 3 are both in the v = 1, ν_3 level in this case. (c) There are no population changes by the pump, but a change in absorption of the probe is caused by the ac Stark splitting of level 2 by the pump. (d) A "hole burning" occurs in the molecular velocity distribution because of the pump, similar in nature to the Lamb dip effects discussed previously. The line shapes of observed double resonance probe signals corresponding to processes (a)–(d) are shown in Fig. 4.38e,f,g,h. By tuning the diode laser frequency with respect to the CO_2 laser frequency and noting its effect on the occurrence of processes (a) and (c), Moulton *et al.* were able to draw

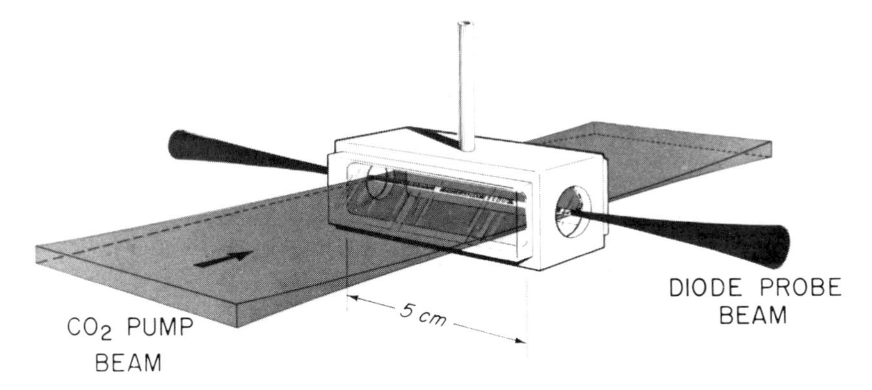

FIG. 4.37. Beam geometry used in a double-resonance experiments with tunable diode laser (Moulton *et al.*, 1977).

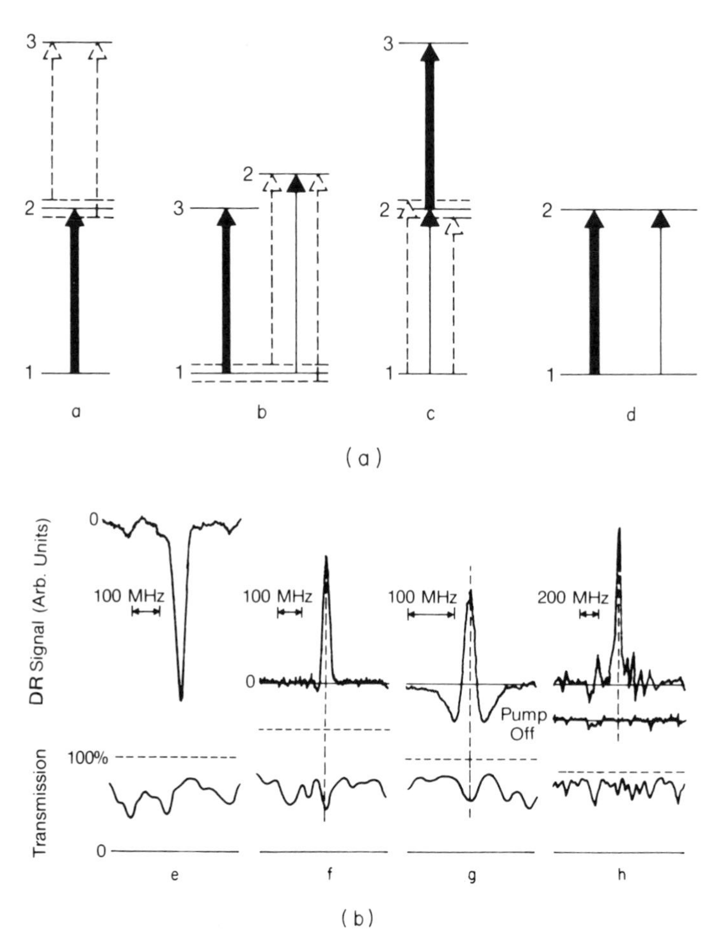

FIG. 4.38. (a)–(d) Schematic excitation diagrams for the four double resonance (DR) processes observed in SF_6. Thick arrows represent the pump radiation, thin arrows the probe radiation. The pump induces Rabi-split levels and associated transitions, which are indicated by dashed lines and arrows. (e)–(h) Line shapes of four representative transient DR signals, each with corresponding SF_6 average transmission spectra, taken simultaneously are shown below. Main peaks of DR signals in (e), (f), (g), and (h) occur at 944.91, 945.69, 943.59, and 951.19 cm^{-1}, respectively, and are found using CO_2 pump lines $P(18)$, $P(16)$, $P(22)$, and $P(12)$, respectively. Gas pressure was 0.2 Torr in (e), (f), and (h) and 0.3 Torr in (g). Dotted lines connect DR signals with associated absorption lines. In (h), DR signal is shown with pump laser off, to indicate system noise (Moulton *et al.*, 1977).

conclusions about multiphoton absorption processes at 10 μm, a subject of intense interest for laser chemistry and isotope separation.

4.4. Other Tunable Infrared Laser Sources

A. Color Center Lasers

One of the more recent developments in tunable infrared sources has been the advent of the color center (F-center) laser. These lasers employ dye-laser-like electronic transitions associated with vacancies in alkali halides (and recently alkaline earth metal compounds). The first report of an F-center laser was in 1965, but little progress was made until Mollenauer's work in 1974. Since then results have come rapidly, and a commercial product is now available. The lasers are typically operated continuously by optical pumping with the radiation from another laser. Pulse pumping can also be used for higher peak powers, and model locking has been demonstrated.

In this section we shall review briefly the physics of color centers, discuss the color center lasers reported to date, give some description of the technology of color center lasers, and describe some of the difficulties encountered in their practical use.

1. *Physics of Color Centers*

The centers that have exhibited laser operation are the F_2^+ and $F_A(II)$ centers. They are formed by first creating F centers, which are anion (halogen ion) vacancies. These positive vacancies can either combine with electrons (F center), hydrogen anions (U center), or halogen ions. If F centers combine with one or two impurity halogen ions, the resulting complexes are designated F_A or F_B centers. The $F_A(II)$ center has the property that there is a large configuration change in the excited state in which a symmetric orientation occurs with the appearance of two vacancies. A high density of ionized and neutral halogen vacancies can be created by, for example, electron bombardment. Charged and neutral F centers can combine at room temperature to form F_2^+ centers.

The laser active F centers [$F_A(II)$, $F_B(II)$, F_2^+, etc.] can make a radiationless nuclear configuration change when electronically excited. The new configuration lowers the electronic energy. The radiative transition associated with lasing preserves the configuration (Franck–Condon principle); however, subsequent nonradiative relaxation occurs to the equilibrium ground state configuration. Thus the energy level structure is equivalent to a generalized, four-level system. The pumping–lasing cycle is the following: (1) optical pumping from the ground state, (2) radiationless relaxation in the excited state, (3) laser emission, and (4) radiationless relaxation to ground state equilibrium. The emission is, of course, Stokes (red) shifted. The absorption and fluorescence bands of several of the color centers are shown in Fig. 4.39. Note the large wavelength shift due

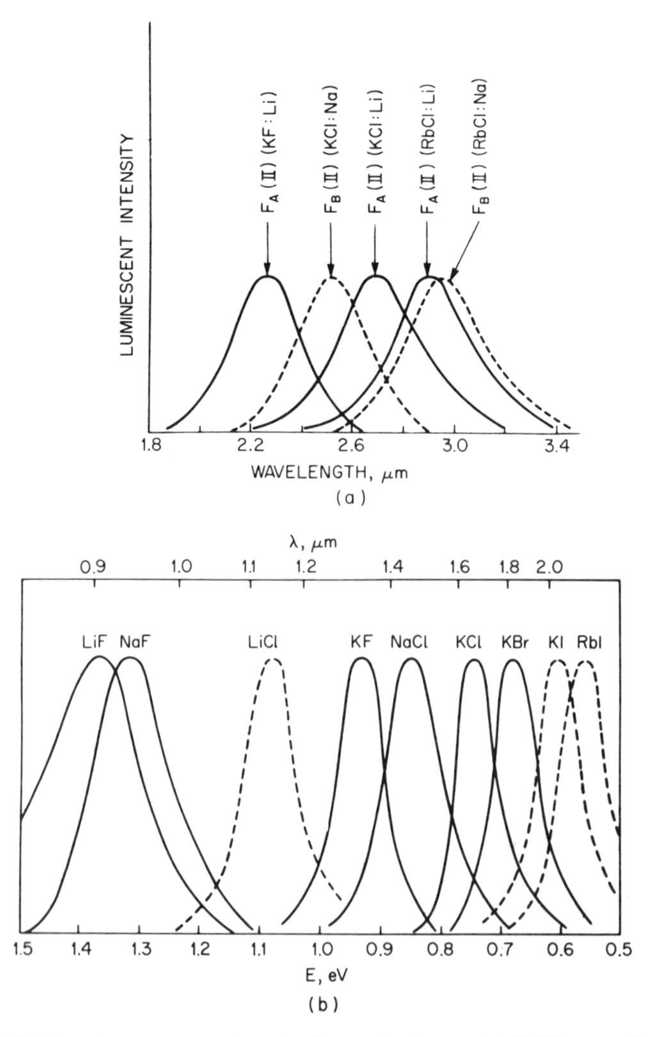

FIG. 4.39. Luminescence spectra of color center lasers. (a) F_A(II) center bands, (b) F_2^+ center bands, with the solid curves representing experimental data and the dashed curves representing predictions from an H_2^+ molecular ion model.

to the configuration change. Laser operation can be obtained over much of each fluorescence band.

Certain physical processes tend to destroy the laser active color centers. For example, $F_A(II)$ centers tend to dissociate into F centers plus impurities at room temperature. The stability is greatly enhanced by holding the crystals at about 200°K. The F_2^+ center also requires low temperatures (<200°K) for stability. It has been hypothesized that electron trapping may occur at higher temperatures or that the F_2^+ centers may wander in the crystal.

2. *Details of Color Center Laser Operation*

The standard configuration for color center laser operation is the three mirror cavity with either a prism, grating, or birefringent element for tuning (see Fig. 4.40 for a typical configuration). The laser is pumped longitudinally through one end of the cavity using a multilayer mirror which is highly transmitting at the pump wavelength and highly reflecting at the color center output wavelengths. Both Nd : YAG and Kr^+ lasers are typically used as pumps. Pump thresholds in the tens of milliwatts have been achieved, and slope efficiencies of 10% have been observed. Table 4.VII gives information on several color center lasers. It should also be noted that the alkaline earth fluorides and oxides have also been used as color center laser hosts, e.g., F^+ centers in CaO (Duran *et al.*, 1978) and $(F_2)_A$: Na centers in CaF_2 and SrF_2 (Arkhangelskaya *et al.*, 1979).

The output powers given in Table 4.VII are for cw operation. Pulse operation can also be obtained, and several authors have resorted to this method of operation to demonstrate laser action. This form of pulsed

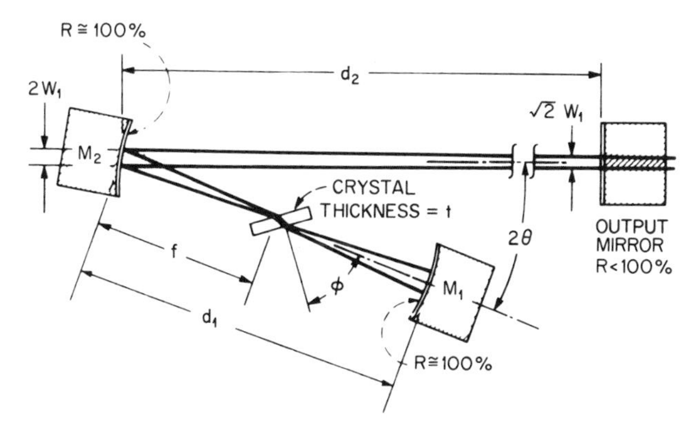

FIG. 4.40. A typical configuration for a color center laser.

TABLE 4.VII

SOME COLOR CENTER LASERS

Center	Host	Output wavelength (μm)	Pump	Output power (mW)	Reference
$F_A(II):Li$	KCl	2.6	Xenon lamp	—	Fritz and Menke (1965)
	KCl	2.5–2.9	Kr^+, Ar^+	240	Mollenauer and Olson (1974)
	RbCl	2.6–3.33	Kr^+	55	Mollenauer and Olson (1975)
$F_B(II):Na$	KCl	2.25–2.65	Ar^+, Kr^+	50	Litfin *et al.* (1977)
	RbCl	2.5–2.9	Kr^+	6	Litfin *et al.* (1977)
F_2^+	NaF	0.885–1.00	Kr^+	710	Mollenauer (1977)
	KCl	1.6–1.8	Nd : YAG	—	Mollenauer (1977)
	KF	1.23–1.46	Nd : YAG	2500	Mollenauer *et al.* (1978)
	LiF	0.82–1.07	Kr^+	>1000	Mollenauer *et al.* (1978)
	NaCl				Gellerman, W. R. (1979)
F_2^-	LiF	1.12–1.26	Nd : YAG	60	Gusev *et al.* (1977)
F_2	LiF	0.63–0.73	dye	—	Gusev *et al.* (1978)
$(F_2^+)_A:Na$	KCl	1.6–1.9	Nd : YAG	12	Schneider and Marrone (1979)
$(F_2^+)_A:Li$	KCl	2.0–2.5	Nd : YAG	5–10	Schneider and Marquardt (to be published)

operation is not particularly interesting in color centers, as the short (~100 nsec) radiative lifetime does not make them useful for pump energy storage and Q switching.

Of more significance is the achievement of mode-locked operation. Mode locking was obtained by synchronously pumping the F_2^+ color center in a fashion similar to that used for mode-locking dye lasers. A few watts of average power could be obtained with 100-psec pulses at a 300-MHz repetition rate.

The major drawbacks to the practical application of color center lasers are the difficulty of making the centers and the requirement of low-temperature operation to maintain the stability of the centers.

3. *Commercial Availability*

A commercial color center laser system is available from Burleigh Instruments. A photograph of the system is shown in Fig. 4.41. The laser uses the $F_A(II)$: Li center in KCl and RbCl as well as the $F_B(II)$: Na in KCl to provide tuning in the 2.2–3.1-μm range. Linewidths of 1 MHz can be obtained with continuous tuning of approximately 3 GHz. Prices are in the $19,000–$25,000 range.

FIG. 4.41. Photograph of commercial color center laser (Burleigh Instruments).

4. *Spectroscopy With Color Center Lasers*

The color center laser has yet to be widely applied in spectroscopy. An example of the broadband spectroscopy possible with this laser is shown in Fig. 4.42. The resolution of the optoacoustic absorption spectrum is of the order of one wavenumber.

B. Transition Metal Lasers

Broadly tunable, transition metal lasers were first reported in 1963 (Johnson *et al.*, 1963). Recently there has been a renewed interest in these lasers as sources for spectroscopic and other applications (Reed *et al.*, 1977; Moulton *et al.*, 1978). In this section we shall discuss the physics and some of the operating characteristics of the divalent transition metal lasers. To date nickel, cobalt, and vanadium have exhibited laser operation.

1. *Physics of Transition Metal Lasers*

Doubly ionized, transition metal ions have their outer electrons in 3d states and a single 4s state (Sugano, 1970). These electrons couple to form a ^3F ground state configuration in Ni and ^4F ground state configurations in the case of Co and V. Spin orbit coupling is relatively weak, but crystal

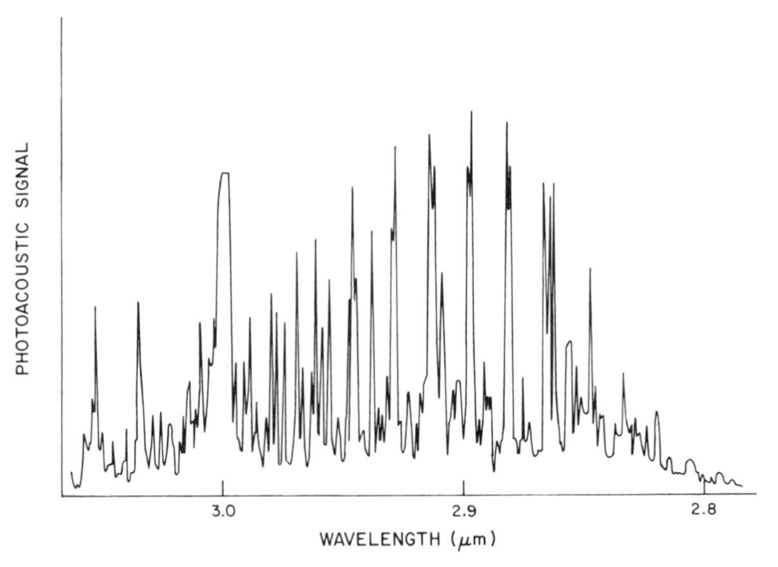

FIG. 4.42. Photoacoustic absorption spectrum of NH_3 in the 3-μm region obtained with a commercial color center laser system. (Figure courtesy of Burleigh Instruments Inc.)

field effects on the d electrons can be substantial (e.g., for Ni^{2+} in MgF_2 the $^3T_2$ state is split $\sim$7500 cm^{-1} from the $^3A_2$ state). Because the upper state minimum energy lies at a different local lattice configuration than the ground state, there can be strong phonon broadening (i.e., the Franck–Condon factors are small).

The transitions between these levels are magnetic-dipole in character and hence are weak. Optical absorption coupled with phonon production occurs on the high-frequency (anti-Stokes) side of the zero-phonon line, while optical emission, also coupled with phonon production, occurs on the low-frequency (Stokes) side. There are stronger transitions [e.g., $^3A_2(^3F)$–$^3T_1(^3P)$] lying in the visible and UV which can also be used for pumping, albeit with less efficiency than pumping on the lasing transition. The pump–lasing cycle consists of (1) optical pumping from the ground to excited electron state combined with phonon emission and (2) stimulated emission from the excited to the ground state with additional phonon emission.

Nonradiative decay plays a significant role at elevated temperatures (200–300°K). This decay occurs because of phonon *induced* nonradiative transitions between the excited and ground electronic states of the ion. To avoid the rapid lifetime decrease above 200–300°K, the lasers are operated at lower temperatures.

It is worth pointing out that trivalent Cr^{3+} is also an important lasing ion in the hosts ruby and alexandrite. In these materials the lowest lying upper level (^2E) has nearly the same dependence on local lattice configuration as the ground state, and there is little phonon broadening of the spectrum. There is also a nearby level ($^4T_2$) which is shifted relative to the ground state and hence its transition to the ground state is broad. At high temperatures this level can be substantially populated in alexandrite, and broadly tunable laser operation is possible in the fashion of the divalent transition metals.

2. *Laser Operation*

Lamp-pumped laser operation was obtained by Johnson *et al.* (1963, 1964, 1966; Johnson and Gugenheim, 1967). Their results are summarized in Table 4.VIII. Continuous operation was achieved with $Ni^{2+}:MgF_2$ and $NiMnF_2$ with temperature tuning of the laser observed in $Ni^{2+}:MgF_2$ from 1.623 to 1.797 μm. In this case the tungsten lamp clearly pumped the transition owing to its low color temperature. In pulsed operation with a xenon lamp, it is likely that, in addition, pumping occurs via the somewhat stronger short-wavelength $F \rightarrow P$ transition.

Recent work has concentrated on laser pumping (Reed *et al.* 1977;

TABLE 4.VIII

SUMMARY OF CHARACTERISTICS OF LAMP-PUMPED DIVALENT, TRANSITION METAL LASERS[a]

Active ion	Host	Temp. (°K)	Maser wavelength (μm)	Transition	Terminal state splitting (cm^{-1})	Lifetime (msec)	Pulse threshold[b] (J)	cw threshold[c] 85°K (W)
Ni^{2+}	MgF_2	77	1.623	${}^3T_2 \rightarrow {}^3A_2$	340	11.5	150	
		77–82	1.636		390		160	
		82–100	1.674–1.676		526–533		160–170	240
		100–192	1.731–1.756		723–805		170–570	65
		198–240	1.785–1.797		898–935		570–1650	
	MnF_2	77	1.915	${}^3T_2 \rightarrow {}^3A_2$	580[d]	11.1	840	
		20	1.865		560[e]	11.3	740	
		77	1.922		600[d]	11.1	210	
		85	1.929		620[d]		270	
		85	1.939		650[d]		240	
Ni^{2+}	MgO	77	1.3144	${}^3T_2 \rightarrow {}^3A_2$	398		230	
Co^{2+}	MgF_2	77	1.750[f]	${}^4T_2 \rightarrow {}^4T_1$	1087	1.3		
			1.8035[f]		1256			
			1.99		1780			
			2.05		1930			
	MnF_2	77	2.165	${}^4T_2 \rightarrow {}^4T_1$	1895	0.4	430	
	$KMgF_3$	77	1.821	${}^4T_2 \rightarrow {}^4T_1$	1420	3.1	530	
V^{2+}	MgF_2	77	1.1213	${}^4T_2 \rightarrow {}^4A_2$	1150	2.3	1070	

[a] Johnson *et al.*, 1966.
[b] The electrical energy into an FT-524 helical xenon lamp.
[c] 500 W T3Q iodine quartz lamp.
[d] Measured from 5800 cm^{-1}.
[e] Measured from 5926 cm^{-1}.
[f] Electronic (no-phonon) lines.

Moulton *et al.*, 1978; Moulton and Mooradian, 1979a,b). Ni : MgF_2 has been operated with ~40% power conversion efficiency when pumped with the 1.33-μm line of Nd : YAG. Thresholds of the order of 20 mW were observed, and output powers of up to 1.7 W obtained. Q-switched operation has given 140-W peak power at 100-Hz repetition rate. Other materials have also been successfully laser pumped (Ni : MgO and Co : MgF_2) with 9 W of output power produced in the case of Ni : MgO (Fig. 4.43 shows an example of output power versus pump power for this laser).

As discussed earlier, an important feature of these lasers is their broad tuning range. Using a three-mirror cavity (as described above for color center lasers), continuous tuning at fixed temperature (80°K) and with cw operation was seen from 1.61 to 1.74 μm in Ni : MgF_2 and from 1.63 to 2.08 μm in Co : MgF_2 (Fig. 4.44).

Other transition metal systems where renewed interest in their laser potential would be beneficial include Ni : MnF_2, Ni : ZnF_2, Co : ZnF_2, and V : MgF_2. It is to be anticipated that much higher Q-switched peak powers can be obtained, making these lasers suitable pumps for nonlinear processes. For example, with Co : MgF_2, first Stokes output from stimulated Raman scattering in hydrogen would encompass the range from 4.7 to 12.4 μm. Doubling would cover the range from 815 to 1040 nm.

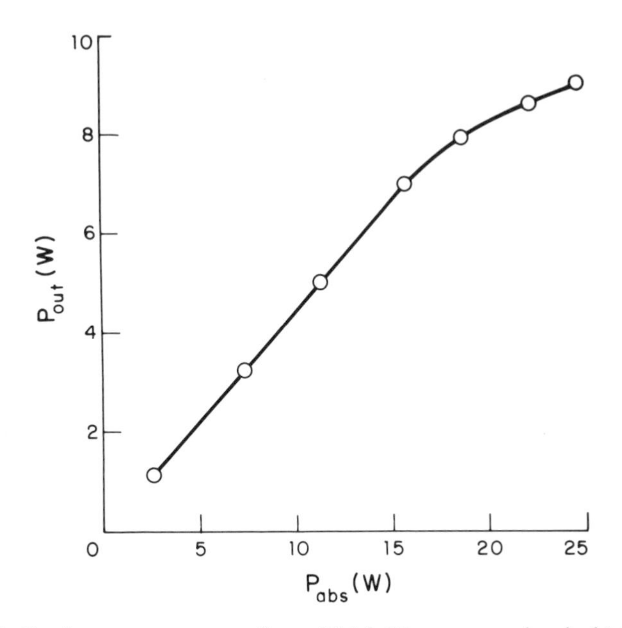

FIG. 4.43. Continuous power output from a Ni:MgO laser versus absorbed 1.06-μm pump power. η_q = 57%, T = 77%°K.

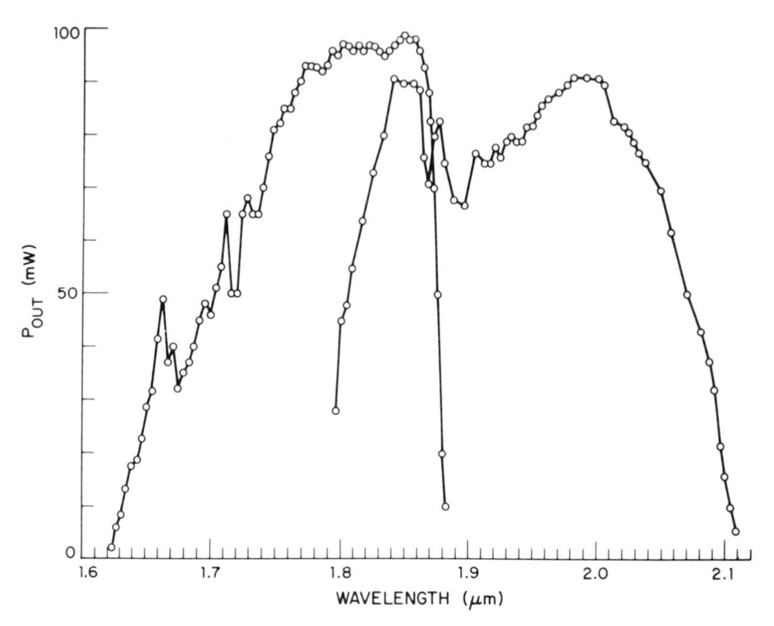

FIG. 4.44. Tuning curves for a Co:MgF_2 laser. P_{abs} = 1.5W, T = 80°K.

Transition metal lasers are being developed at several laboratories, notably M.I.T. Lincoln Laboratory. They are not commercially available now.

C. Nonlinear Optical Processes

Nonlinear optical processes can be used to generate new infrared frequencies using lasers as pump sources. To give an explanation of how these processes occur, we express the material polarization as a power series in applied electric field:

$$\begin{aligned} P(\omega_1) = {} & \chi^{(1)}(\omega_1)\, E(\omega_1) \\ & + \chi^{(2)}(\omega_2,\omega_3 = \omega_1 - \omega_2)\, E(\omega_2)\, E(\omega_3) \\ & + \chi^{(3)}(\omega_2,\omega_3,\omega_4 = \omega_1 - \omega_2 - \omega_3) E(\omega_2) E(\omega_3) E(\omega_4) \\ & + \cdots, \end{aligned}$$

where the ω's are allowed to take on negative as well as positive values. The first term is responsible for the normal linear dielectric response of the medium, while the second and third produce a variety of nonlinear phenomena. The $\chi^{(2)}$ term gives rise to mixing of optical fields with other optical fields as well as with microwave and radio frequency fields (modulation). Optical–optical mixing includes second harmonic generation, sum- and difference-frequency generation, and optical parametric oscillation. The $\chi^{(3)}$ term produces a number of effects; the most important for our purposes are four-wave mixing and stimulated Raman scattering. In this section we shall discuss various applications of nonlinear optics in tunable infrared generation.

1. *Sum and Difference Generators*

In noncentrosymmetric materials, $\chi^{(2)}$ is nonvanishing for electric dipole processes and hence may be sufficiently large that practical sum- and difference-frequency generation is possible. The stationarity restriction for the material medium implies that $\omega_1 = \omega_2 \pm \omega_3$, where ω_1 is the positive output and ω_2 and ω_3 are the positive input frequencies. In order to achieve coherent generation over substantial distances, phase matching ($\mathbf{k}_1 = \mathbf{k}_2 \pm \mathbf{k}_3$) should occur. It is necessary to use birefringent materials to overcome normal dispersive effects. For example, for visible to infrared conversion in the frequently used material, $LiNbO_3$, the input beams are ordinary while the output is extraordinary. The difference-frequency output power is given by

$$P_3 = [(8\pi\omega_3 dl)^2/n_1 n_2 n_3 A c^3]\, P_1 P_2,$$

where d is the nonlinear coefficient, l the crystal length, n the index of refraction, A the focal area, and c the velocity of light. For the case of $LiNbO_3$ and typical spot sizes and crystal lengths, the above expression reduces to

$$P_3 \approx 2 \times 10^{-4}\, P_1 P_2,$$

where the powers are in watts.

Dye lasers can be used as the tunable source in a difference-frequency generation system. In an early experiment, pulsed difference-frequency generation from the output of a ruby laser plus a dye laser mixed in $LiNbO_3$ achieved 6 kW of infrared power tunable between 3.1 and 4.5 μm, with a bandwidth of about ten wavenumbers (Dewey and Hocker, 1971). In this wavelength range further development of pulsed-laser mixing techniques has been carried out, including increasing average power and obtaining narrow linewidths (Meltzer and Goldberg, 1972; Bethea, 1973; Goldberg, 1975). Doubled Nd:YAG radiation has been used as the dye laser pump and $LiIO_3$ as an alternative nonlinear material.

Mixing in $LiNbO_3$ the output of a stable dye laser with that of a stable argon laser has yielded in the 2–4-μm region cw output powers of about 1 μW, covering a tuning range of over one hundred wavenumbers with a spectral precision and reproducibility of 15 MHz (Pine, 1974, 1976). Figure 4.45 shows a schematic of this system. With the exception of diode

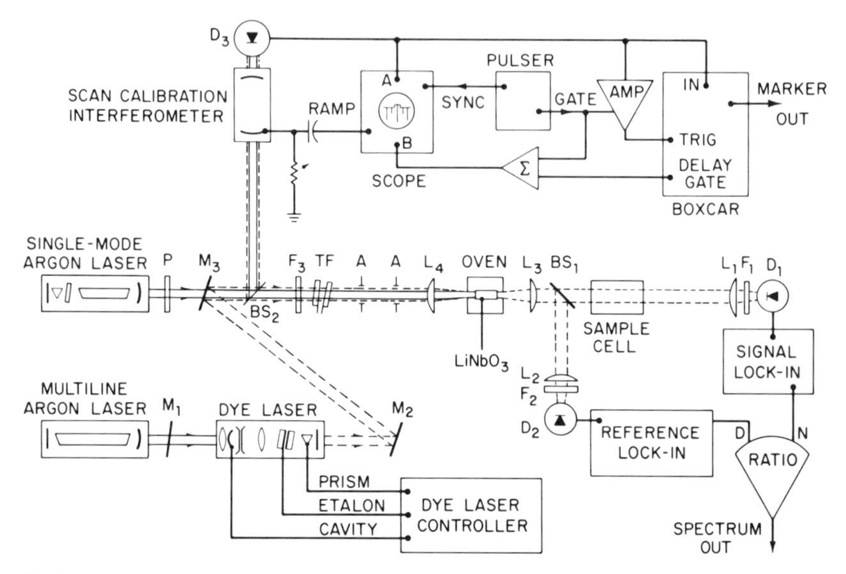

FIG. 4.45. Schematic of visible-to-infrared difference-frequency spectrometer system (Pine, 1976).

lasers, cw difference-frequency generation has been the most extensively used tunable coherent source for high-resolution infrared spectroscopy. The 2–4-μm region is not covered by available lead salt diode lasers. The difference-frequency spectrometer is now in use at several other laboratories, in addition to M.I.T. Lincoln Laboratory, where the initial development took place. The commercial availability of the components reduces the task of applying the system primarily to one of interfacing optics and electronics.

Using this system, Doppler-limited spectroscopy of NO, HF, HCl, HBr, N_2O, SO_2, H_2O, H_2CO, CH_4, and SF_6 has been carried out and analyzed for spectroscopic information.

Figure 4.46 gives the ν_3 $P(7)$ line of methane with the tetrahedral splittings fully resolved (Pine, 1976). The atmospheric pressure broadened lines are also shown and are similar to those observed with a spectrometer capable of 0.1-cm^{-1} resolution. It is interesting to note that the F_12 line is coincident with the 3.39-μm He–Ne laser line and that nuclear hyperfine structure has been further resolved by saturation spectroscopy.

A spectroscopic study of the Fermi resonance between the $K_A = 13$ subband of the $\nu_1 + \nu_3$ vibration and the $K_A = 15$ of the $2\nu_2 + \nu_3$ subband of $^{32}S^{16}O_2$ is shown in Fig. 4.47 (Pine *et al.*, 1977). A value of the interaction constant k_{eff}, of 8 cm^{-1} was obtained by fitting the data.

The location and character of the R-branch bandhead of the CO_2 ν_3

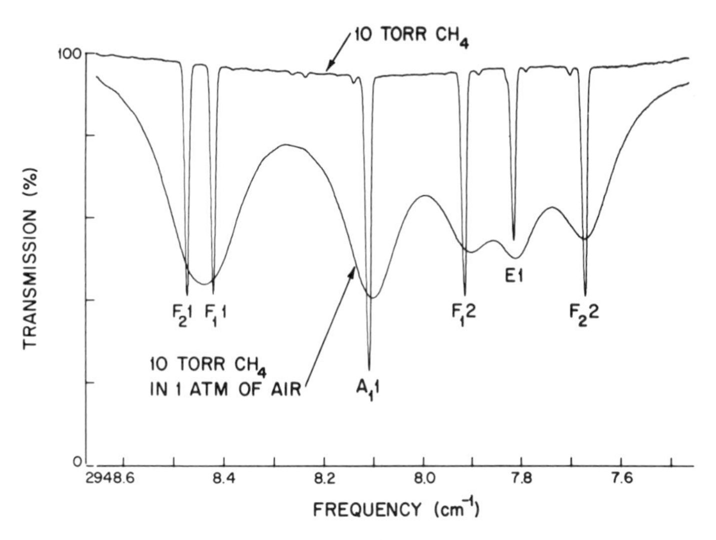

FIG. 4.46. Absorption spectra of the $\nu_3 P(7)$ line of CH_4. (Pine, 1976).

fundamental has long been of interest for practical applications of atmospheric spectroscopy. Figure 4.48 shows a Doppler-limited spectrum of the bandhead; J values to 140 were measured (Pine and Guelachvili, 1980). Band constants recently given by Rothman and Benedict (1978) predict

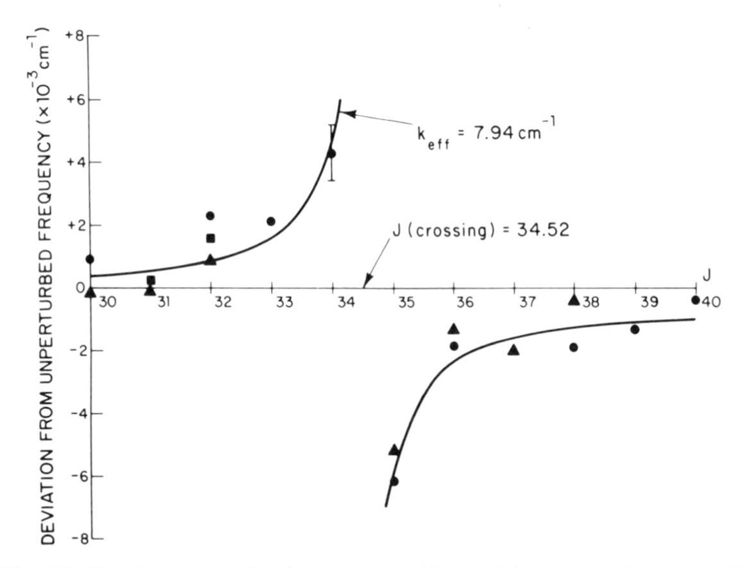

FIG. 4.47. Fermi resonance for the $K_A = 13$ subband of the $\nu_1 + \nu_3$ vibration of $^{32}S^{16}O_2$. The solid line is a theoretical fit. ●, $\Delta J = -1$, P branch; ■, $\Delta J = 0$, Q branch; ▲, $\Delta J = +1$, R branch.

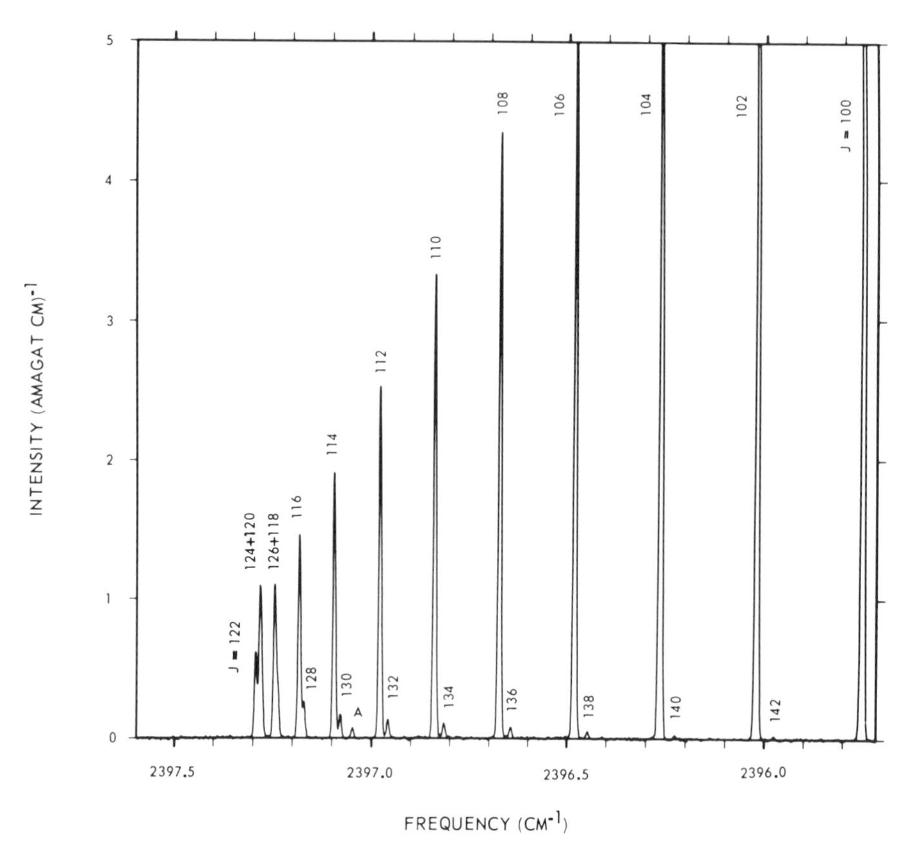

FIG. 4.48. R-branch bandhead for the ν_3 band of $^{12}C^{16}O_2$; the line labeled A is the $\nu_1 + \nu_3 \rightarrow \nu_1$ hot band line. $T = 930°K$.

$R(122)$ *at* 2397.3048 cm^{-1} and $R(140)$ at 2396.2446 cm^{-1}, as opposed to the tunable laser results of 2397.2932 cm^{-1} and 2396.2257 cm^{-1}, respectively.

The classical molecule for high-resolution infrared laser spectroscopy, SF_6, has also been studied with the cw tunable difference-frequency system (Pine and Robiette, 1980). In Fig. 4.49 a portion of the $3\nu_3$ spectrum of SF_6 is given showing two Q-branch bandheads. The total spectrum observed contained over one thousand lines in the range from 2819.5 to 2831.5 cm^{-1}. The anharmonic parameters determined from the spectra allow a more accurate modeling of the theory of the initial stages of multiphoton dissociation of this molecule.

Longer wavelengths can be generated by mixing in other crystals: proustite (Ag_3AsS_3) (Hanna *et al.*, 1971; Decker and Tittel, 1973) is usable to about 12 μm and CdSe (Bhar *et al.*, 1972; Hanna *et al.*, 1973) and Te

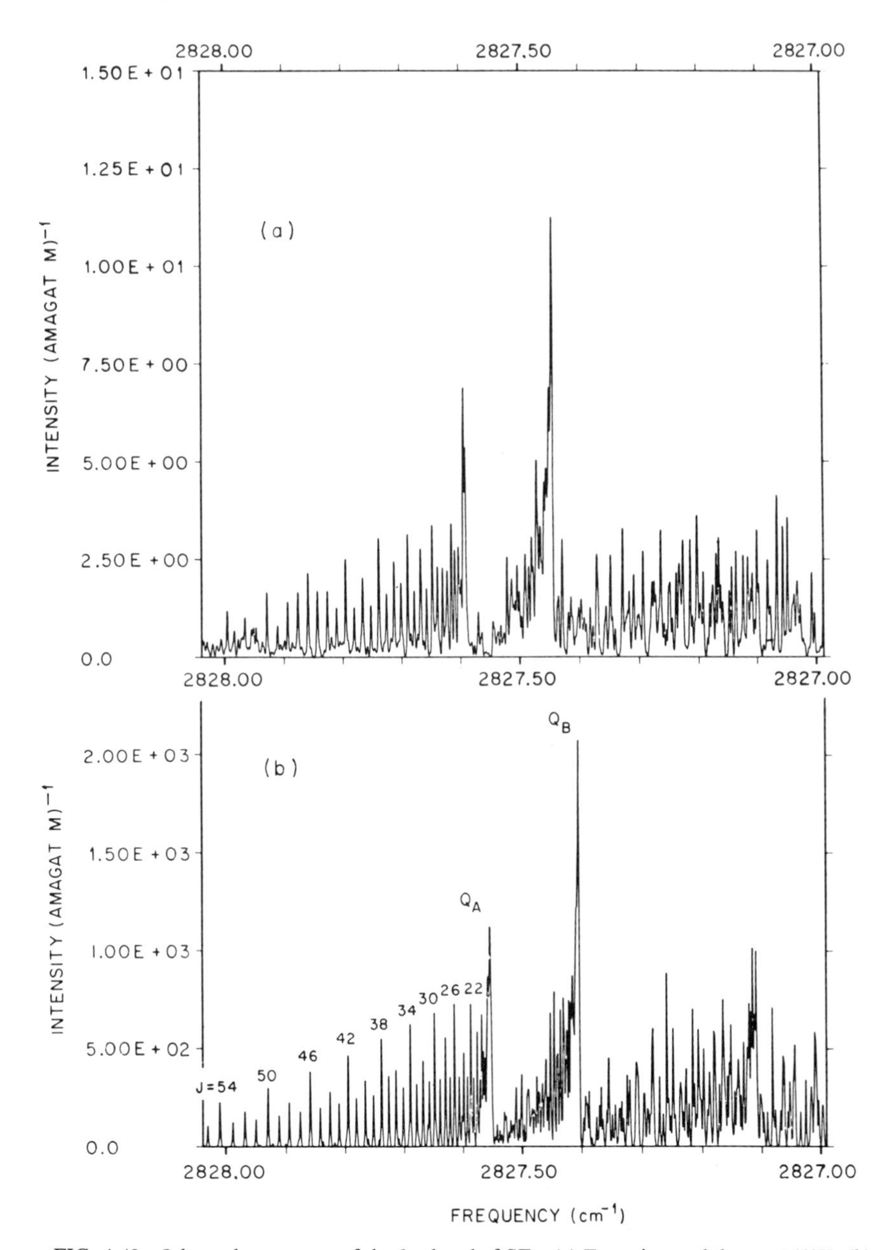

FIG. 4.49. Q-branch spectrum of the $3\nu_3$ band of SF_6. (a) Experimental data at 160°K. (b) A calculated simulation of the spectrum. Further refinements in the theoretical fitting are being made at Los Alamos Scientific Laboratory.

(Bridges *et al.*, 1975) to about 25 μm. Also used (Chemla *et al.*, 1971; Byer *et al.*, 1971; Boyd *et al.*, 1972a,b; Kildal and Mikkelsen, 1973, 1974; Kildal *et al.*, 1976; Seymour *et al.*, 1976) as nonlinear materials in the infrared are I–III–VI_2 and II–IV–V_2 compounds with the chalcopyrite structure (e.g., $AgGaS_2$, $AgGaSe_2$, $CdGeAs_2$, and $InGeP_2$). These phase-matchable materials can be applied to both sum- and difference-frequency generation. Mixing in $AgGaS_2$ of radiation from dye lasers has given tunable output in the 4–18-μm range.

Although primary tunable sources are not readily available in the infrared, high-power, step tunable, gas lasers (CO_2, N_2O, CO, HF/DF) can be used to generate nearly continuous coverage in many regions of the infrared by sum- and difference-frequency generation. If two gas lasers are used, the number of new frequencies is proportional to the product of the number of transitions of each laser and is substantially increased by including all the relatively abundant isotopic species. As an example (A. H. M. Ross, unpublished calculation), using known laser transitions in the four most abundant isotopes of CO_2 and CO, outputs with an average spacing of the order of 300 MHz are possible near 16 μm; and by increasing the gas pressure to broaden the gain bandwidths, continuous coverage should be possible. By mixing 97 mW of CO radiation with 1.25 W of CO_2 radiation in a crystal of $CdGeAs_2$, 4 μW of power has been generated at 13 μm (Kildal and Mikkelsen, 1974). For doubling of CO_2 radiation, 75 mW of cw power has been obtained with 17-W input power, and using a repetitively *Q*-switched CO_2 laser, up to 1.9 W of average doubled power has been obtained, and conversion efficiencies as high as 27% have been observed (Menyuk *et al.*, 1976; N. Menyuk personal communication). In addition, with 150-nsec CO_2 transversely excited atmospheric (TEA) laser pulses, 200 mJ of doubled radiation have been obtained (H. Kildal, personal communication) with an optical conversion efficiency of 10%. Recent results (N. Menyuk, personal communication) using a mini-TEA laser pump (Menyuk and Moulton, 1980) have given 500 mW of average doubled power at 29% conversion efficiency (see Fig. 4.50). This system has been used for remote sensing of atmospheric CO and NO. Dye laser outputs have been mixed in $AgGaS_2$ to obtain radiation between 4.6 and 18 μm (Hanna *et al.*, 1973; Seymour *et al.*, 1976). Noncollinear phase matching has also been used to mix infrared radiation in GaAs (Aggarwal *et al.*, 1973; Lee *et al.*, 1976). Large area beams and folded geometries are employed to provide long interaction distances.

The optical parametric oscillator (OPO) is closely related to difference-frequency generation. These devices are natural extensions of microwave parametric oscillators. The basic OPO consists of a nonlinear crystal between two wavelength-selective mirrors to form an optical cavity. A laser field at a frequency ω_p (the pump frequency) is applied to the crystal,

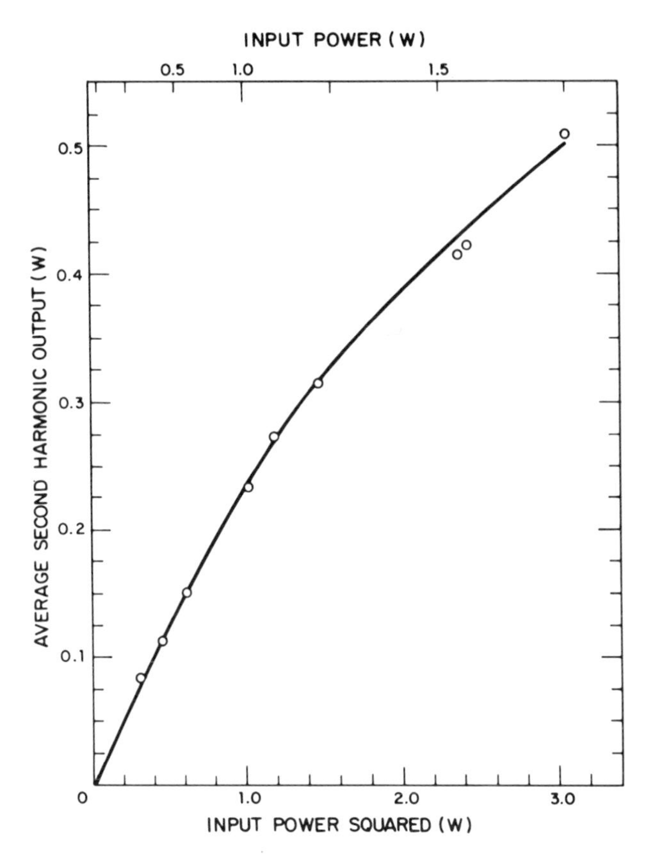

FIG. 4.50. Average second harmonic output from $CdGeAs_2$ versus average input power. The pump was a mini-TEA CO_2 laser with 100 nsec pulses and a PRF of 350 Hz.

usually through one of the end mirrors. Initially, the pump radiation mixes with the photon noise in the crystal, leading to a buildup of radiation at two frequencies: the signal frequency ω_s and the idler frequency $\omega_i = \omega_p - \omega_s$ phase matched for difference-frequency generation with the pump. If losses in the cavity are less than the gain of the buildup process, oscillation occurs. In order to change the phase-matched wavelengths (thereby tuning the oscillator), the indices of refraction of the crystal are varied with temperature, crystal rotation, or electric field. A singly resonant OPO is one in which either the signal or the idler is fed back in the optical cavity, while in a doubly resonant oscillator both the signal and idler are fed back.

Optical parametric oscillators operate cw as well as pulsed. Although cw thresholds can be as low as 2.8 mW (Harris, 1969; Smith, 1972; Byer, 1975), the cw OPO is generally difficult to operate, tends to be unstable, and requires careful geometric design. A singly resonant $LiNbO_3$ oscillator, pumped with a ruby laser, has yielded peak power up to 340 kW and has had up to 45% conversion efficiency. The average power was as high as 350 mW near 2.1 μm with a conversion efficiency of 70% when an internal oscillator with a $LiNbO_3$ crystal in a repetitively *Q*-switched Nd : YAG laser cavity was used (Ammann, 1970). $LiIO_3$ has also been used as the nonlinear material in parametric oscillators (Goldberg, 1970; Izrailenko *et al.*, 1970; Campillo and Tang, 1971); an internal Nd : YAG pumped oscillator has produced 13-kW peak power and 20-mW average power (Goldberg, 1972). A CdSe oscillator has been operated with a 2.36-μm Dy : CaF_2 laser pump (Davidov *et al.*, 1972), with a 1.83-μm Nd : YAG pump (Herbst and Byer, 1972), and with a 2.7–3.0-μm HF laser pump (Weiss and Goldberg, 1974; Wenzel and Arnold, 1976; see also Rockwood, 1976). In the HF pumped system tuning was obtained in the 14.1–16.4-μm range. No single OPO has been tuned over a very wide region because of difficulties with mirror coatings and materials. A $LiNbO_3$ oscillator pumped by intracavity doubled radiation from a *Q*-switched Nd : YAG laser was developed as a commercial device (Wallace, 1970). This system, with changes of mirrors on both the oscillator and pump laser, operates anywhere between 0.55 and 3.5 μm, with peak powers of up to several kilowatts and average powers of over 100 mW. Long-term linewidths of less than 0.001 cm^{-1} have been achieved (Hordvik and Sackett, 1974). The commercial OPO system was available until recently from the Chromatix Corporation of Mountain View, California. Because of the complexities in this technology, particularly in obtaining reliable narrow linewidth operation, OPOs are likely to be supplanted by other sources for most applications, e.g., as color center lasers, optical mixing, and frequency shifting by stimulated Raman scattering.

In cases where only moderate resolution is necessary but high-peak powers in short pulses are required, OPOs have proven to be practical. One example of this is a study of multiphoton dissociation of ethyl chloride (Dai *et al.*, 1979), where the authors used 3.5 mJ of OPO output energy in the 3.3 μm region (see Fig. 4.51). The OPO was a conventional Nd : YAG pumped $LiNbO_3$ system. Sharp resonances in photodissociation yield were found to coincide with overtone absorption bands.

Modulators are variants of sum- and difference-frequency generators in which one of the input frequencies is in the radio frequency or microwave range. Phase matching can be accomplished in the gigahertz modulation range using a waveguide. It is possible, for example, to fill in between the CO_2 transitions using a modulator. The technique has been used spectroscopically to measure pressure-broadened gain bandwidths for the CO_2 laser (Corcoran *et al.*, 1973). A tuning range of several gigahertz has been demonstrated at a frequency offset of 15 GHz with 30 mW of modulated

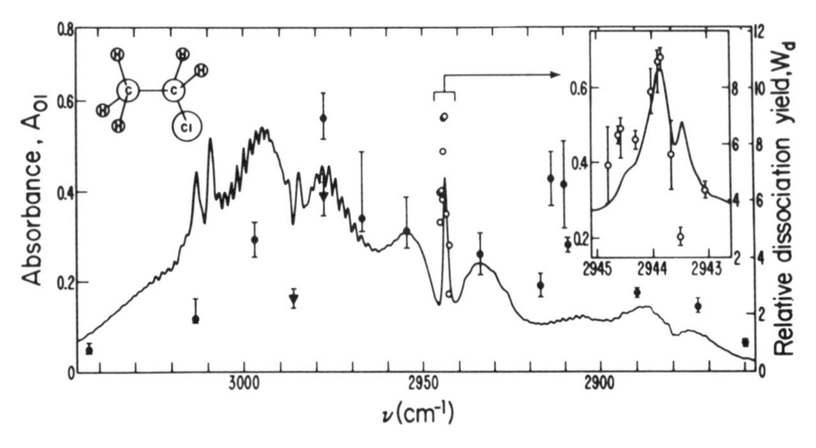

FIG. 4.51. Relative photodissociation yield (points) and linear absorption (solid curve) versus frequency for C_2H_5Cl. Points denoted by filled circles were measurements at absorption peaks while the filled triangles were taken at absorption minima. A sharp resonance, indicated by the open circles, is shown on an expanded scale in the inset.

power output for inputs of 50 W of CO_2 laser power and 100 W of microwave power (G. M. Carter and L. R. Tomasetta, personal communication). A pulsed, single sideband modulator has been constructed with 90% infrared-to-infrared conversion efficiency (Carter, 1980).

As pointed out earlier, processes involving χ^3 can also be used to generate infrared frequencies. Four-wave parametric mixing in alkali metal vapors using dye lasers has yielded tunable radiation in the 2–2.5-μm region with 0.2 cm^{-1} linewidth (Wynne *et al.*, 1974). In this case the average power was very low: 5×10^{-8} W at 2 μm and 5×10^{-11} W at 25 μm. This four-wave down conversion should not be considered to be a practical process. Radiation near 8.6 μm has been obtained by efficient four-wave mixing of 9.6 and 10.6 μm CO_2 radiation in Ge. With ~1 J in each CO_2 beam, about 10 mJ of 8.6-μm radiation has been achieved (Aggarwal *et al.*, 1976).

In another example of a four-wave process, 9.2-μm CO_2 laser radiation was tripled with 4% efficiency using a 20% concentration of CO dissolved in liquid oxygen (Brueck and Kildal, 1978). Carbon dioxide radiation at 9.2 μm has a two-photon resonance with the vibrational mode of CO. The use of CO dissolved in cryogenic liquids has several advantages over the use of gas phase CO. First, CO densities are high. Second, the two-photon linewidth is narrow compared to the overall gas-phase bandwidth (Brueck, 1977, 1978a). Finally, the two-photon resonance may be tuned for maximum tripling efficiency by changing CO concentrations (Kildal and Brueck, 1977).

The spin-flip Raman (SFR) laser (Patel and Shaw, 1971; Mooradian *et al.*, 1971; Eng *et al.*, 1974; Stattler *et al.*, 1974; Kruse, 1976) is a device that uses a fixed frequency laser (CO_2, CO, or HF gas laser) to pump a semiconductor crystal (InSb, $Hg_xCd_{1-x}Te$, or InAs) in a magnetic field. The pump-laser photons lose energy when they interact with an electron in the crystal and flip its spin. The downshifted Raman photon is separated in energy from the pump photon by the change in electron spin energy $g\beta H$, where g is the gyromagnetic ratio of the conduction electron, β the Bohr magnetron, and H the magnetic field strength. Consequently, the Raman photon frequency depends on magnetic field. At sufficiently high pump power, stimulated emission of Raman photons can exceed losses, and exponential gain and oscillation occurs. The first cw Raman laser was in InSb, where large gains were obtained by resonant pumping with a CO laser (Mooradian, *et al.*, 1971). The spin-flip laser necessitates cryogenic cooling of the semiconductor and substantial, tunable magnetic fields. These requirements, together with the need for an ultrastable pumping geometry to avoid frequency and amplitude variations, make the SFR laser an unlikely device for practical spectroscopic use.

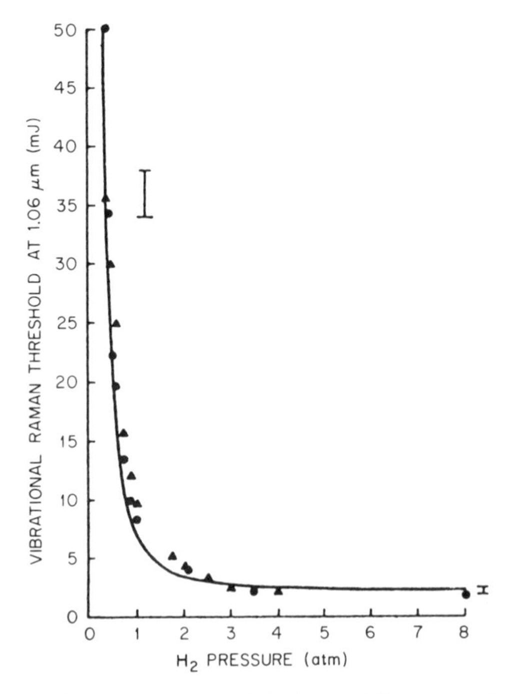

FIG. 4.52. Stimulated Raman scattering in hydrogen. The results give threshold energy for 1.06-μm pumping versus pressure (Byer, 1975).

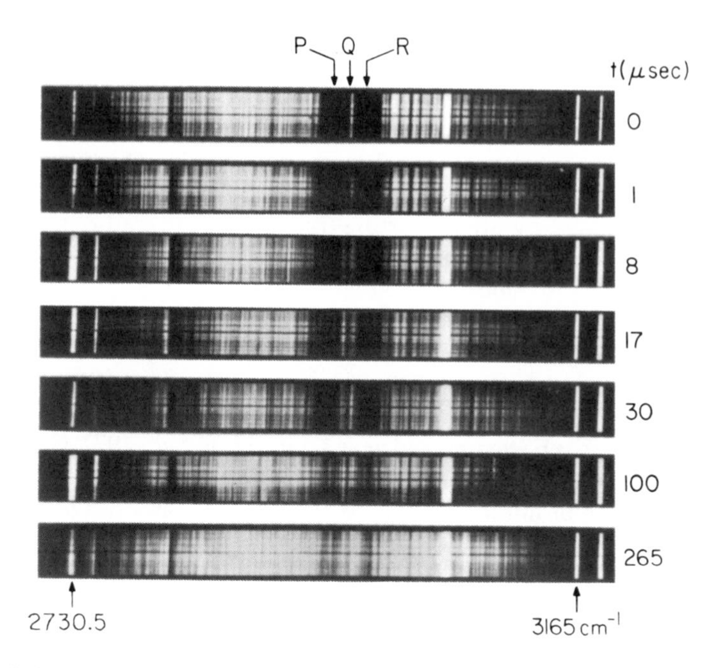

FIG. 4.53. Changes in up-converted spectra of 100 Torr CH_3NC as it thermally explodes (isomerizes) various times after application of CO_2 laser radiation (ν_{CO_2} = 960 cm^{-1}). Each spectrum represents 16 superimposed shots (Bethune *et al.*, 1979).

The use of stimulated Raman scattering (SRS) to downshift tunable dye laser radiation into the infrared has been carried out using high-pressure hydrogen (Schmidt and Appt, 1972, 1974) and atomic vapors (Wynne and Sorokin, 1975; Carlsten and Dunne, 1975; Cotter *et al.*, 1975, 1976). Hydrogen and deuterium are particularly useful on account of the large Raman gain arising primarily from the small pressure-broadening coefficient. In hydrogen, with 12 kW of dye laser pump power, third Stokes emission was observed at 2.1 μm (Schmidt and Appt, 1972, 1974). The threshold for 1.06-μm pumping versus pressure has been measured by Byer and co-workers (Byer, 1979), and the results are shown in Fig. 4.52. Stimulated Raman scattering in hydrogen of tunable neodymium glass laser radiation has given 0.8 J of output in the 8.33–9.1-μm range (Grasiuk *et al.*, 1976; Grasiuk and Zubarev, 1976). Four-wave mixing techniques can be used to further extend the utility of Raman scattering in hydrogen (Ducuing, 1976; Byer, 1976, 1979a). Byer and Trutna (1978) demonstrated 16 μm generation by scattering of CO_2 laser radiation from hydrogen rotational excitation produced by near infrared SRS (1.06-μm pump). As new, tunable, high-peak power sources, such as transition metal lasers, become available, further applications of SRS in H_2 and D_2 will be made. Output powers of 7 kW have been observed in the 5.67–8.65-μm range by atomic Raman scattering in cesium using 0.1–1.0 MW of dye laser pump power (Cotter *et al.*, 1976). A system for time resolved, infrared spectral photography has been developed (Bethune *et al.*, 1979) which uses broadband SRS in Rb to generate infrared radiation. Time evolving absorption spectra are obtained by up converting transmitted infrared radiation into the visible using four-wave mixing. Application to measurement of kinetic processes has been demonstrated (Fig. 4.53).

Stimulated Raman scattering is also of interest in glass fibers, where low losses (0.03 km^{-1}) and long interaction lengths are possible (Stolen *et al.*, 1972; Stolen and Ippen, 1973). Low-pump power, cw SRS has been observed in the visible (Jain *et al.*, 1977a,b; Johnson *et al.*, 1977), and pulsed operation has been reported near 1.1 μm (Lin *et al.*, 1977) (Fig. 4.54).

At this writing, the most practical and significant nonlinear techniques for the generation of infrared radiation appear to be three-wave sum- and difference-frequency generation and SRS in H_2 and D_2.

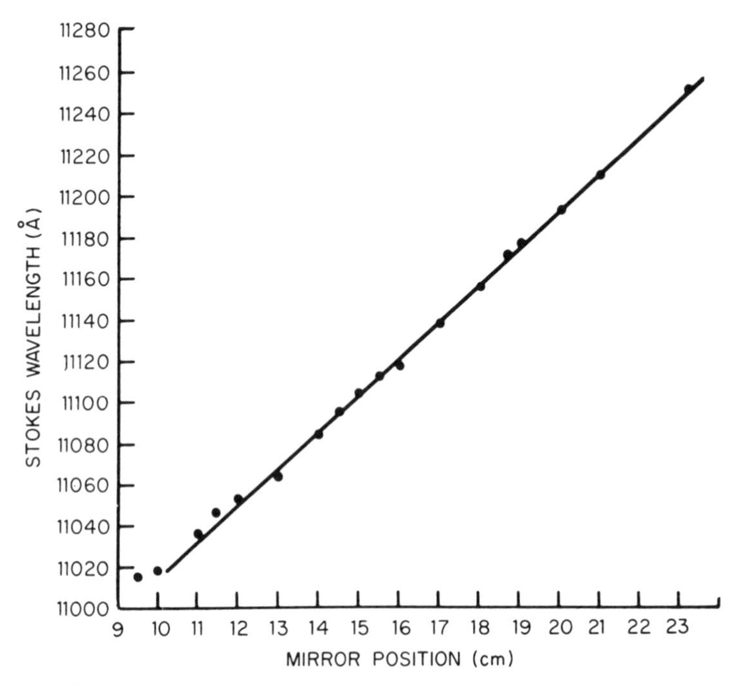

FIG. 4.54. Tuning curve for a time dispersion-tuned fiber optical Raman laser. Because of group velocity dispersion, the Stokes wavelength that is time synchronized with the mode-locked Nd:YAG pump depends on the cavity length (Lin *et al.*, 1977).

D. High Pressure Molecular Gas Lasers

Molecular gas lasers are most readily operated in the pressure range of tens of millitorr; however, tuning within a single vibration–rotation transition is limited to the 10–100-MHz range. Pressure broadening can be used to increase the tunability, generally at the expense of ease of operation. To produce overlapped transitions by pressure broadening requires 5–10-atm pressure in the case of CO_2. At these pressures, the longitudinal (~50 cm) discharges characteristic of low-pressure cw lasers are impossible to maintain in conventional tubes with localized arcing and "streamers" occurring, even in short distances. Pressure also increases the deactivation rate of the excited states and prevents the efficient transport of heat to the walls; both processes tend to raise the threshold for oscillation. A number of improvements in laser technology have allowed the development of high-pressure gas lasers to proceed. These include: (1) transversely excited atmospheric (TEA) configuration lasers with photoionization (Levine and Javan, 1973) and photopreionization (Alcock *et al.*, 1973); (2) optically pumped lasers (Chang and Wood, 1973, 1974; Kildal and Deutsch, 1976; Chang *et al.*, 1976); (3) electron-beam-pumped lasers (Bagratashvili *et al.*, 1973, 1976; Harris *et al.*, 1974; O'Neill and Whitney, 1975, 1976); and (4) capillary lasers (Abrams, 1974), including pulser-sustainer devices (Smith, 1974). The first three of these devices have operated in the 10-atm range, where continuous tuning between rotational transitions is possible, while the pulsed capillary system has operated to

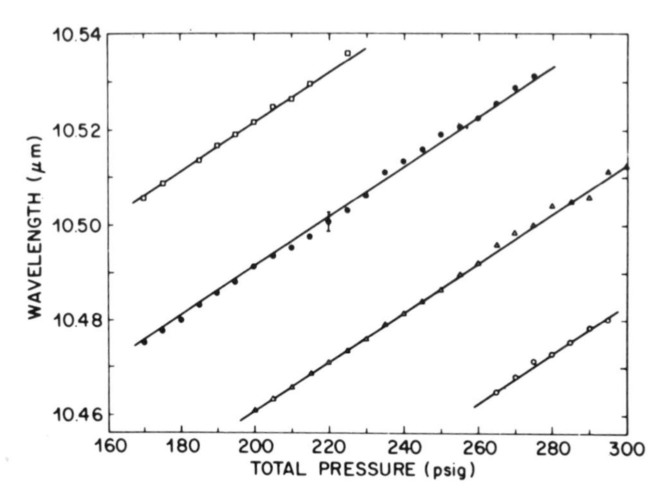

FIG. 4.55. Observed output wavelength from a high-pressure N_2O/Co_2 transfer laser as a function of laser operating pressure. Each line is a longtitudinal mode (Chang *et al.*, 1976).

several atmospheres pressure. A 15-atm electron-beam-pumped CO_2 laser emitting 0.1-J, 40-nsec pulses has been reported (O'Neil and Whitney, 1975), and using an étalon, $<0.2\ cm^{-1}$ linewidth radiation was obtained and tuned continuously over $2.3\ cm^{-1}$. Tunable high-pressure operation over a substantial fraction of the 9.2–12.3-μm region has also been obtained by this method using N_2O and several isotopes of CO_2 and CS_2 (O'Neil and Whitney, 1976). Continuous tuning over $5\ cm^{-1}$ has been obtained recently in an optically pumped, high-pressure, N_2O/CO_2 transfer laser (Chang *et al.,* 1976). The laser was tuned by changing pressure and a resolution of $0.014\ cm^{-1}$ obtained. Figure 4.55 shows output wavelength versus cell pressure for this system.

While these lasers have great potential for providing tunable infrared radiation, some effort remains in the engineering of them into practical devices. High-pressure gas lasers should have higher peak and average power than any other currently conceivable tunable infrared source. They are most appropriate for applications in remote sensing and photochemistry, including laser isotope separation.

References

Abrams, R. L. (1974). *Appl. Phys. Lett.* **24,** 304.

Aggarwal, R. L., Lax, B., and Favrot, G. (1973). *Appl. Phys. Lett.* **22,** 239.

Aggarwal, R. L., Lee, N., and Lax, B. (1976). *In* "Tunable Lasers and Applications" (A. Mooradian, T. Jaeger, and P. Stokseth, eds.), p. 96. Springer-Verlag, Berlin and New York.

Alcock, A. J., Leopold, K., and Richardson, M. C. (1973). *Appl. Phys. Lett.* **23,** 562.

Aldridge, J. P. *et al.* (1975). *J. Mol. Spectrosc.* **58,** 165.

Ammann, E. O., Yarborough J. M., Oshman, M. K., and Montgomery, P. C. (1970). *Appl. Phys. Lett.* **16,** 309.

Arkhangelskaya, V. A., Fedorov, A. A., and Feofilov, P. P. (1979). *Opt. Commun.* **28,** 87.

Bagratashvili, V. N., Knyazev, I. N., Kudryavtsev, Yu. A., and Tetokhov, V. S. (1973). *JETP Lett.* **18,** 62.

Bagratashvili, V. N., Knyazev, I. N., Letokhov, V. S., and Lobko, V. V. (1976). *Sov. J. Quantum Electron.* **5,** 541.

Besson, J. M., Paul, W., and Calawa, A. R. (1968). *Phys. Rev.* **173,** 699.

Bethea, C. G. (1973). *Appl. Opt.* **12,** 1104.

Bethune, D. S., Lankard, J. R., Loy, M. M., and Sorokin, P. P. (1979). *In* "Laser Spectroscopy (H. Walther and K. W. Rothe, eds.), Vol. IV, p. 416. Springer-Verlag, Berlin and New York.

Bhar, G. C., Hanna, D. C., Luther-Davies, B., and Smith, R. C. (1972). *Opt. Commun.* **6,** 323.

Bien, F. (1978). *J. Chem. Phys.* **69,** 2631.

Bischel, W. K., Kelly, P. J., and Rhodes, C. K. (1975). *Phys. Rev. Lett.* **34,** 300.

Boyd, G. D., Buehler, E., Storz, F. G., and Wernick, J. H. (1972a). *IEEE J. Quantum Electron.* **8,** 419.

Boyd, G. D., Kasper, H. M., McFee, J. H., and Storz, F. G. (1972b). *IEEE J. Quantum Electron.* **8,** 900.

Brewer, R. G. (1972). *Science* **178,** 247.

Bridges, T. J., Nguyen, V. T., Burkhardt, E. G., and Patel, C. K. N. (1975). *Appl. Phys. Lett.* **27,** 600.

Brockman, P., Bair, C. H., and Allario, F. (1978). *Appl. Opt.* **17,** 91.

Brueck, S. R. J. (1977). *Chem. Phys. Lett.* **53,** 516.

Brueck, S. R. J. (1978a). *Chem. Phys. Lett.* **54,** 273.

Brueck, S. R. J., and Kildal, H. (1978). *Appl. Phys. Lett.* **33,** 928.

Brunet, H., and Perez, M. (1969). *J. Mol. Spectrosc.* **29,** 472.

Butler, J. F. (1976). SPIE Meeting, "Laser Spectroscopy, Techniques and Applications", San Diego, California, (22–25 August).

Butler, J. F. (1979). Private communication; quoted in Wells *et al.* (1979).

Butler, J. F., Nill, K. W., Mantz, A. W., and Eng, R. S. (1978). *In* "New Applications of Lasers to Chemistry" ACS Symposium Series (G. M. Hieftje, ed.), No. 85, pp. 12–23. American Chemical Society.

Byer, R. L. (1975). *In* "Quantum Electronics" (H. Rabin and C. L. Tang, eds.), Vol. 18, p. 587. Academic Press, New York.

Byer, R. L. (1976). *In* "Tunable Lasers and Applications" (A. Mooradian, T. Jaeger, and P. Stokseth, eds.), p. 70. Springer-Verlag, Berlin and New York.

Byer, R. L. (1979a). *In* "Laser Spectroscopy" (H. Walther and K. W. Rothe, eds.), Vol. IV, p. 603. Springer-Verlag, Berlin and New York.

Byer, R. L. (1979b). Private communication.

Byer, R. L., and Trutna, W. R. (1978). *Opt. Lett.* **3,** 144.

Byer, R. L., Kildal, H., and Feigelson, R. S. (1971). *Appl. Phys. Lett.* **19,** 237.

Byer, R. L., Herbst, R. L., and Fleming, R. N. (1975). *In* "Laser Spectroscopy" (S. Haroche, J. C. Pebay-Peyroula, T. W. Hansch, and S. E. Harris, eds.), p. 207. Springer-Verlag, Berlin and New York.

Calawa, A. R., Dimmock, J. O., Harman, T. C., and Melngailis, I. (1969). *Phys. Rev. Lett.* **23,** 7.

Campillo, A. J., and Tang, C. L. (1971). *Appl. Phys. Lett.* **12,** 376.

Cappellani, F., and Restelli, G. (1979). *J. Mol. Spectrosc.* **77,** 36.

Cappellani, F., Restelli, G., and Melandrone, G. (1979). *Infrared Phys.* **19,** 195.

Carlsten, J. L., and Dunne, P. C. (1975). *Opt. Commun.* **14,** 8.

Carter, G. M. (1980). *Proc. SPIE Tech. Symp. East, April.*

Chaney, L. W., Rickel, D. G., Russwurm, G. M., and McClenny, W. A. (1979). *Appl. Opt.* **18,** 3004.

Chang, T. Y., and Wood, O. R. (1973). *Appl. Phys. Lett.* **23,** 524.

Chang, T. Y., and Wood, O. R. (1974). *Appl. Phys. Lett.* **24,** 182.

Chang, T. Y., McGee, J. D., and Wood, O. R. (1976). *Opt. Commun.* **18,** 279.

Chemla, D. S., Kupeck, P. J., Robertson, D. S., and Smith, R. C. (1971). *Opt. Commun.* **3,** 29.

Chraplyvy, A. R. (1978). *Appl. Opt.* **17,** 2674.

Chu, F. Y., and Oka, T. (1975). *J. Appl. Phys.* **46,** 1204.

Corcoran, V. J., Martin, J. M., and Smith, W. T. (1973). *Appl. Phys. Lett.* **22,** 517.

Cotter, D., Hanna, D. C., Karkkainen, P. A., and Wyatt, R. (1975). *Opt. Commun.* **15,** 143.

Cotter, D., Hanna, D. C., and Wyatt, R. (1976). *Opt. Commun.* **16,** 256.

DAI, H. L., KUNG, A. H., and MOORE, C. B. (1979). *Phys. Rev. Lett.* **43,** 761.
DAVIDOV, A. A., KULEVSKII, L. A., PROKHOROV, A. M., SAVELEV, A. D., and SMIRNOV, V. V. (1972). *JETP Lett.* **15,** 513.
DECKER, C. D., and TITTEL, F. K. .973). *Opt. Commun.* **8,** 244.
DEVI, V. M., DAS, P. P., and RAO, K. N. (1979). *Appl. Opt.* **18,** 2918.
DEWEY, C. F., Jr., (1977). *In* "Optoacoustic Spectroscopy and Detection" (Y. H. Pao, ed.), p. 47. Academic Press, New York.
DEWEY, C. F., Jr., and HOCKER, L. O. (1971). *Appl. Phys. Lett.* **18,** 58.
DUBS, M., and GUNTHARD, H. H. (1978). *Appl. Opt.* **17,** 3593.
DUCUING, J., FREY, R., and PRADERE, F. (1976). *In* "Tunable Lasers and Applications" (A. Mooradian, T. Jaeger, and P. Stokseth, eds.), p. 81. Springer-Verlag, Berlin and New York.
DURAN, J., EVESQUE, P., and BILLARDON, M. (1978). *Appl. Phys. Lett.* **33,** 1004.
ENG, R. S., and MANTZ, A. W. (1979a). *J. Mol. Spectrosc.* **74,** 388.
ENG, R. S., and MANTZ, A. W. (1979b). *J. Mol. Spectrosc.* **74,** 331.
ENG, R. S., CALAWA, A. R., HARMAN, T. C., KELLEY, P. L., and JAVAN, A. (1972). *Appl. Phys. Lett.* **21,** 303.
ENG, R. S., MOORADIAN, A., and FETTERMAN, H. R. (1974). *Appl. Phys. Lett.* **25,** 453.
ENG, R. S., NILL, K. W., and WHELAN, M. (1977). *Appl. Opt.* **16,** 3072.
ENG, R. S., PETAGNA, G., and NILL, K. W. (1978). *Appl. Opt.* **17,** 1723.
ENG, R. S., MANTZ, A. W., and TODD, T. R. (1979a). *Appl. Opt.* **18,** 1088.
ENG, R. S., MANTZ, A. W., and TODD, T. R. (1979b). *Appl. Opt.* **18,** 3438.
FLICKER, H., ALDRIDGE, J. P., NERESON, N. G., REISFELD, M. J., and WEBER, W. H. (1978). *Appl. Opt.* **17,** 851.
FRERKING, M. A., and MUEHLNER, D. J. (1977). *Appl. Opt.* **16,** 526.
FRITZ, B., and MENKE, E. (1965). *Solid State Commun.* **3,** 61.
GELLERMAN, W. R. (1979). Private communication, quoted in Koch *et al.* (1979).
GENTRY, W. R., and GIESE, C. F. (1978). *Rev. Sci. Instrum.* **49,** 595.
GOLDBERG, L. S. (1970). *Appl. Phys. Lett.* **17,** 489.
GOLDBERG, L. S. (1972). Digest Tech. Papers, *Int. Quantum Electron. Conf., 7th, Montreal, Canada* p. 55.
GOLDBERG, L. S. (1975). *Appl. Opt.* **14,** 653.
GOUGH, T. E., MILLER, R. E., and SCOLES, G. (1977). *Appl. Phys. Lett.* **30,** 338.
GOUGH, T. E., MILLER, R. E., and SCOLES, G. (1978). *J. Mol. Spectrosc.* **72,** 124.
GRASIUK, A. Z., and ZUBAREV, I. G. (1976). *In* "Tunable Lasers and Applications" (A. Mooradian, T. Jaeger, and P. Stokseth, eds.), p. 88. Springer-Verlag, Berlin and New York.
GRASIUK, A. Z., ZUBAREV, I. G., KOTOV, A. V., MIKHAILOV, S. I., and SMIRNOV, V. G. (1976). *Sov. J. Quantum Electron.* **6,** 568.
GROVES, S. H., NILL, K. W., and STRAUSS, A. J. (1974). *Appl. Phys. Lett.* **25,** 331.
GUSEV, Y. L., MARENNIKOV, S. I., and CHEBOTAEV, V. P. (1977). *Appl. Phys.* **74,** 121.
GUSEV, Y. L., KONOPLIN, S. N., and MARENNIKOV, S. I. (1978). *Sov. J. Quantum Electron.* **7,** 1157.
HANNA, D. C., SMITH, R. C., and STANLEY, C. R. (1971). *Opt. Commun.* **4,** 300.
HANNA, D. C., RAMPAL, V. V., and SMITH, R. C. (1973). *Opt. Commun.* **8,** 151.
HANSON, R. K. (1977). *Appl. Opt.* **16,** 1479.
HANSON, R. K., and FALCONE, P. K. (1978). *Appl. Opt.* **17,** 2447.
HANSON, R. K., KUNTZ, R., and KRUGER, C. H. (1977). *Appl. Opt.* **16,** 2045.
HARMON, T. C., and MCVITTIE J. P. (1974). *J. Electron. Mater.* **3,** 843.

Harris, S. E. (1969). *Proc. IEEE* **57,** 2096.
Harris, N. W., O'Neill, F., and Whitney, W. T. (1974). *Appl. Phys. Lett.* **25,** 148.
Herbst, R. L., and Byer, R. L. (1972). *Appl. Phys. Lett.* **21,** 189.
Hillman, J. J. *et al.* (1977). *Opt. Lett.* **1,** 81.
Hinkley, E. D. (1976). *Opt. Quantum Electron.* **8,** 155.
Hinkley, E. D., and Freed, C. (1969). *Phys. Rev. Lett.* **23,** 277.
Hinkley, E. D., and Kelley, P. L. (1971). *Science* **171,** 635.
Hinkley, E. D., Nill, K. W., and Blum, F. A. (1976a). *In* "Laser Spectroscopy of Atoms and Molecules" (H. Walther, ed.), pp. 125–196. Springer-Verlag, Berlin and New York.
Hinkley, E. D., Ku, R. T., Nill, K. W., and Butler, J. F. (1976b). *Appl. Opt.* **15,** 1653.
Hirota, E. (1979). *J. Mol. Spectrosc.* **74,** 209.
Hordvik, A., and Sackett, P. B. (1974). *Appl. Opt.* **13,** 1060.
Izrailenko, A. I., Kovrigin, A. I., and Nikles, P. V. (1970). *JETP Lett.* **12,** 331.
Jain, R. K., Lin, C., Stolen, R. H., Pleibel, W., and Kaiser, P. (1977a). *Appl. Phys. Lett.* **30,** 162.
Jain, R. K., Lin, C., Stolen, R. H., and Ashkin, A. (1977b). *Appl. Phys. Lett.* **31,** 89.
Jennings, D. E. (1978a). *Geophys. Res. Lett.* **5,** 241.
Jennings, D. E. (1978b). *Appl. Phys. Lett.* **33,** 493.
Jennings, D. E., and Hillman, J. J. (1977). *Rev. Sci. Instrum.* **48,** 1568.
Jennings, D. E., Hillman, J. J., and Weber, W. H. (1978). *Opt. Lett.* **2,** 157.
Johnson, L. F., and Gugenheim, H. J. (1967). *J. Appl. Phys.* **38,** 4837.
Johnson, L. F., Dietz, R. E., and Gugenheim, H. J. (1963). *Phys. Rev. Lett.* **11,** 318.
Johnson, L. F., Dietz, R. E., and Gugenheim, H. J. (1964). *Appl. Phys. Lett.* **5,** 21.
Johnson, L. F., Gugenheim, H. J., and Thomas, R. A. (1966). *Phys. Rev.* **149,** 179.
Johnson, D. C., Hill, K. O., Kawasaki, B. S., and Kato, D. (1977). *Electron. Lett.* **13,** 53.
Kauppinen, J. (1979). *Appl. Opt.* **18,** 1788.
Kelley, P. L. (1977). *In* "Optoacoustic Spectroscopy and Detection" (Y. H. Pao, ed.), p. 113. Academic Press, New York.
Kildal, H., and Brueck, S. R. J. (1977). *Phys. Rev. Lett.* **38,** 347.
Kildal, H., and Brueck, S. R. J. (1978). *Appl. Phys. Lett.* **32,** 173.
Kildal, H., and Deutsch, T. F. (1976). Conf. Digest, *Int. Quantum Electron. Conf., 9th, Amsterdam*, p. 124.
Kildal, H., and Mikkelsen, J. C. (1973). *Opt. Commun.* **9,** 315.
Kildal, H., and Mikkelsen, J. C. (1974). *Opt. Commun.* **10,** 306.
Kildal, H., Iseler, G. W., Menyuk, N., and Mikkelsen, J. C. (1976). Tech. Digest, *OSA Topical Meeting Opt. Phenomena Infrared Mater., Annapolis, Maryland.*
Kim, K. C., Griggs, E., and Person, W. B. (1978). *Appl. Opt.* **17,** 2511.
Kingston, R. H. (1978). "Detection of Optical and Infrared Radiation." Springer-Verlag, Berlin and New York.
Koch, K. P., and Lahmann, W. (1978). *Appl. Phys. Lett.* **32,** 289.
Koch, K. P., Litfin, G., and Nelling, H. (1979). *Opt. Lett.* **4,** 387.
Kolb, C. E. (1979). Private communication.
Kruse, P. W. (1976). *Appl. Phys. Lett.* **28,** 90.
Ku, R. T., and Spears, D. L. (1977). *Opt Lett.* **1,** 84.
Ku, R. T., Hinkley, E. D., and Sample, J. O. (1975). *Appl. Opt.* **14,** 854.
Lee, N., Lax, B., and Aggarwal, R. L. (1976). Conf. Digest, *Int. Quantum Electron. Conf., 9th, Amsterdam*, p. 50.
Letokhov, V. S. (1975). *Science* **190,** 344.

LETOKHOV, V. S., and CHEBOTAYEV, V. P. (1969). *Pis'ma Zh. Eksp. Teor. Fiz* **9,** 364.
LETOKHOV, V. S., and CHEBOTAYEV, V. P. (1977). "Nonlinear Laser Spectroscopy." Springer-Verlag, Berlin and New York.
LEVINE, J. S., and JAVAN, A. (1973). *Appl. Phys. Lett.* **22,** 55.
LEVY, D. H., WHARTON, L., and SMALLEY, R. E. (1977). *In* "Chemical and Biochemical Applications of Lasers" (C. B. Moore, ed.), p. 1. Academic Press, New York.
LIN, C., STOLEN, R. H., and COHEN, L. G. (1977). *Appl. Phys. Lett.* **31,** 97.
LITFIN, G., BEIGANG, R., and WELLING, H. (1977). *Appl. Phys. Lett.* **31,** 381.
MANTZ, A. W. (1979). Private communication.
MANTZ, A. W., and ENG, R. S. (1978). *Appl. Spectrosc.* **32,** 239.
MARGOLIS, J. S., MENZIES, R. T., and HINKLEY, E. D. (1978). *Appl. Opt.* **17,** 1680.
MATTICK, A. T., SANCHEZ, A., KURNIT, N. A., and JAVAN, A. (1973). *Appl. Phys. Lett.* **23,** 675.
MAX, E., and ROSENGREN, L. G. (1974). *Opt. Commun.* **6,** 422.
MCCLELLAND, J. (1979). *Opt. News* **5,** No 1, 18.
MCDOWELL, R. S., GALBRAITH, H. W., KROHN, B. J., CANTRELL, C. D. and HINKLEY, E. D. (1976). *Opt. Commun.* **17,** 178.
MCDOWELL, R. S. *et al.* (1977). *J. Mol. Spectrosc.* **68,** 288.
MCDOWELL, R. S. *et al.* (1978). *J. Chem. Phys.* **69,** 1513.
MELTZER, D. W., and GOLDBERG, L. S. (1972). *Opt. Commun.* **5,** 209.
MENYUK, N., and MOULTON, P. F. (1980). *Rev. Sci. Instrum.* **51,** 216.
MENYUK, N., ISELER, G. W., and MOORADIAN, A. (1976). *Appl. Phys. Lett.* **29,** 422.
MENZIES, R. T. (1971). *Appl. Opt.* **10,** 1532.
MENZIES, R. T. (1976). *In* "Laser Monitoring of the Atmosphere" (E. D. Hinkley, ed.), pp. 297–353. Springer-Verlag, Berlin and New York.
MOLLENAUER, L. F. (1977). *Opt. Lett.* **1,** 164.
MOLLENAUER, L. F. (1979). *In* "Methods of Experimental Physics: Quantum Electronics (Volume 15, Part A)", p. 1. Academic Press, New York.
MOLLENAUER, L. F., and OLSON, D. A., (1974). *Appl. Phys. Lett.* **24,** 386.
MOLLENAUER, L. F., and OLSON, D. A. (1975). *J. Appl. Phys.* **46,** 3109.
MOLLENAUER, L. F., BLOOM, D. M., and DEL GOUDIO, A. M. (1978). *Opt. Lett.* **3,** 48.
MONTGOMERY, C. P., and MAJKOWSKI, R. F. (1978). *Appl. Opt.* **17,** 173.
MOORADIAN, A. (1974). *In* "Laser Spectroscopy" (R. G. Brewer and A. Mooradian, eds.), p. 223. Plenum, New York.
MOORADIAN, A., BRUECK, S. R. J., and BLUM, F. A. (1971). *Appl. Phys. Lett.* **17,** 481.
MOULTON, P. F., and MOORADIAN, A. (1979a). "Laser Spectroscopy" (H. Walther and K. W. Rothe, eds.), Vol. IV, p. 584. Springer-Verlag, Berlin and New York.
MOULTON, P. F., and MOORADIAN, A. (1979b). *Appl. Phys. Lett.* **35,** 838.
MOULTON, P. F., LARSEN, D. M., WALPOLE, J. N., and MOORADIAN, A. (1977). *Opt. Lett.* **1,** 51.
MOULTON, P. F., MOORADIAN, A., and REED, T. B. (1978). *Opt. Lett.* **3,** 164.
NERESON, N. G. (1978). *J. Mol. Spectrosc.* **69,** 489.
NILL, K. (1977). *Laser Focus* **13,** No. 2, p. 32.
NILL, K. W., BLUM, F. A., CALAWA, A. R., and HARMAN, T. C. (1973). *J. Nonmet.* **1,** 211.
O'NEIL, F., and WHITNEY, W. T. (1975). *Appl. Phys. Lett.* **26,** 454.
O'NEIL, F., and WHITNEY, W. T. (1976). *Appl. Phys. Lett.* **28,** 539.
PAO, Y. H. (ed.) (1977). "Optoacoustic Spectroscopy and Detection." Academic Press, New York.
PATEL, C. K. N., and SHAW, E. D. (1971). *Phys. Rev. B* **3,** 1279.
PERSON, W. B., and KIM, K. C. (1978). *J. Chem. Phys.* **69,** 2117.

PINE, A. S. (1974). *J. Opt. Soc. Am.* **64,** 1683.
PINE, A. S. (1976). *J. Opt. Soc. Am* **66,** 97.
PINE, A. S., and GUELACHVILI, G. (1980). *J. Mol. Spectrosc.* **79,** 84.
PINE, A. S., and NILL, K. W. (1979). *J. Mol. Spectrosc.* **74,** 43.
PINE, A. S., and ROBIETTE, A. G. (1980). *J. Mol. Spec.* **80,** 388.
PINE, A. S., GLASSBRENNER, C. J., and KAFALAS, J. A. (1973). *IEEE J. Quantum Electron.* **QE-9,** 800.
PINE, A. S., DRESSELHAUS, G., PALM, B., DAVIES, R. W., and CLOUGH, S. A., (1977). *J. Mol. Spectrosc.* **67,** 386.
RADZIEMSKI, L. J., BEGLEY, R. F., FLICKER, H., NERESON, N. G., and REISFELD, M. J. (1978). *Opt. Lett.* **3,** 241.
RALSTON, R. W., WALPOLE, J. N., CALAWA, A. R., HARMON, T. C., and MCVITTIE, J. P. (1974). *J. Appl. Phys.* **45,** 1323.
REDDY, S. P., IVANCIC, W., DEVI, V. M., BALDACCI, A., RAO, K. N., MANTZ, A. W., and ENG, R. S. (1979a). *Appl. Opt.* **18,** 1350.
REDDY, S. P., DEVI, V. M., BALDACCI, A., IVANCIC, W., and RAO, K. N. (1979b). *J. Mol. Spectrosc.* **74,** 217.
REED, T. B., FAHEY, R. E., and MOULTON, P. F. (1977). *J. Crystal Growth* **42,** 569.
REID, J., and MCKELLAR, A. R. W. (1978). *Phys. Rev.* **18,** 224.
REID, J., SHEWCHUN, J., GARSIDE, B. K., and BALLIK, E. A. (1978a). *Appl. Opt.* **17,** 300.
REID, J., GARSIDE, B. K., SHEWCHUN, J., EL-SHERBINY, M., and BALLIK, E. A. (1978b). *Appl. Opt.* **17,** 1806.
RESTELLI, G., CAPPELLANI, F. and MELANDRONE, G. (1978). *Pure Appl. Geophys.* **117,** 531.
ROCKWOOD, S. (1976). *In* "Tunable Lasers and Applications" (A. Mooradian, T. Jaeger, and P. Stokseth, eds.), Springer-Verlag, Berlin and New York.
ROTHMAN, L. S., and BENEDICT, W. S., (1978). *Appl. Opt.* **17,** 2605.
SATTLER, J. P., and RITTER, K. J. (1978). *J. Mol. Spectrosc.* **69,** 486.
SATTLER, J. P., and SIMONIS, G. J. (1977). *IEEE J. Quantum Electron.* **QE-13,** 461.
SATTLER, J. P., WORCHESKY, T. L., and RIESSLER, W. A. (1978). *Infrared Phys.* **18,** 521.
SCHLOSSBERG, H. R. (1976). *J. Appl. Phys.* **47,** 2044.
SCHLOSSBERG, H. R., and KELLEY, P. L. (1972). *Phys. Today,* **25,** No. 7.
SCHMIDT, W., and APPT, W. (1972). *Z. Naturforsch.* **27a,** 1373.
SCHMIDT, W., and APPT, W. (1974). Post Deadline Paper R6, *Int. Quantum Electron. Conf., 8th, San Francisco, California.*
SCHNEIDER, I., and MARRONE, M. J. (1979). *Opt. Lett.* **4,** 390.
SCHNEIDER, I., and MARQUARDT, C. L. (to be published).
SEYMOUR, R. J., ZERNIKE, F., and SAM, D. L. (1976). Conf. Digest, *Int. Quantum Electron. Conf., 9th, Amsterdam* p. 49.
SHIMODA, K., and SHIMIZU, T. (1972). "Nonlinear Spectroscopy of Molecules." Pergamon, Oxford.
SMITH, P. W. (1974). *In* "Laser Spectroscopy" (R. G. Brewer and A. Mooradian, eds.), p. 247. Plenum, New York.
SMITH, R. G. (1972). *In* "Laser Handbook" (F. T. Arrechi and E. O. Schulz-DuBois, eds.), p. 837. North-Holland Publ., Amsterdam.
SPEARS, D. L., MELNGAILIS, and HARMAN, T. C. (1975). *IEEE J. Quantum Electron.* **QE-11,** 79.
STANTON, A. C., and KOLB, C. E. (1980). *J. Chem. Phys.* **72,** 6637.
STATTLER, J. P., WEBER, B. A., and NEMARICH, J. R. (1974). *Appl. Phys. Lett.* **25,** 491.
STOLEN, R. H., and IPPEN, E. P. (1973). *Appl. Phys. Lett.* **22,** 276.
STOLEN, R. H., IPPEN, E. P., and TYNES, A. R., (1972). *Appl. Phys. Lett.* **20,** 62.

SUGANO, S., TANABE, Y., and KAMIMURA, H. (1970). "Multiplets of Transition Metal Ions in Crystals." Academic Press, New York.

SULZMANN, K. G. P., LOWDER, J. E. L., and PENNER, S. S. (1973). *Combust. Flame* **20,** 177.

TODD, T. R., and OLSON, W. B. (1979). *J. Mol. Spectrosc.* **74,** 190.

TOWNES, C. H., and SCHAWLOW, A. (1955). "Microwave Spectroscopy." McGraw-Hill, New York.

TRAVIS, D. N., MCGURK, J. C., MCKEOWN, D., and DENNING, R. G. (1977). *Chem. Phys. Lett.* **45,** 287.

VASILENKO, L. S., CHEBOTAEV, V. P., and SCHISSHAEV, A. V. (1970). *Pis'ma Zh. Eksp. Teor. Fiz.* **12,** 161.

WALLACE, R. W. (1970). *Appl. Phys. Lett.* **17,** 497.

WALPOLE, J. N., CALAWA, A. R., RALSTON, R. W., HARMAN, T. C., and MCVITTIE, J. P. (1973). *Appl. Phys. Lett.* **23,** 620.

WALPOLE, J. N., CALAWA, A. R., HARMAN, T. C., and GROVES, S. H. (1976). *Appl Phys. Lett.* **28,** 552.

WEISS, J. A., and GOLDBERG, L. S. (1974). *Appl. Phys. Lett.* **24,** 289.

WELLS, J. S., PETERSON, F. R., and MAKI, A. G. (1979). *Appl. Opt.* **18,** 3567.

WENZEL, R. G., and ARNOLD, G. P. (1976); *Appl. Opt.* **15,** 1322. See also Rockwood, S. D. (1976). *In* "Tunable Lasers and Applications" (A. Mooradian, T. Jaeger, and P. Stokseth, eds.), p. 140. Springer-Verlag, Berlin and New York.

WHELAN, M., ENG, R. S., and NILL, K. W. (1977). *Appl. Opt.* **16,** 2350.

WORCHESKY, T. L., RITTER, K. J., SATTLER, J. P., and RIESSLER, W. A. (1978). *Opt. Lett.* **2,** 70.

WYNNE, J. J., and SOROKIN, P. P. (1975). *J. Phys.* **B8,** L37.

WYNNE, J. J., SOROKIN, P. P., and LANKARD, J. R. (1974). "Laser Spectroscopy" (R. G. Brewer and A. Mooradian, eds.), p. 103. Plenum, New York.

YAMADA, C., and HIRODA, E. (1979). *J. Mol. Spectrosc.* **74,** 203.

Chapter 5

Absolute Photon Counting in the Ultraviolet

LEON HEROUX

AERONOMY DIVISION
AIR FORCE GEOPHYSICS LABORATORY
HANSCOM AFB
BEDFORD, MASSACHUSETTS

5.1. Introduction

The measurement of absolute values of ultraviolet (UV) radiation in laboratory and space experiments requires the use of a detector whose spectral response is both stable with time while operating in vacuum and

ISBN 0-12-710402-X

reproducible after repeated exposure to air. In addition, to measure low intensities, the detector must have a response linear over a wide range of intensities and have low background noise. These requirements can be met by photomultipliers and gas-flow counters operated as photon counters.

Photon counting techniques are applied frequently to the measurement of dispersed radiation behind the exit slit of a vacuum spectrometer, where the measured counting rates are within the linear range of the counting system. When the entire spectrometer and its counting system are radiometrically calibrated, absolute spectral measurements of radiation can be made from steady-state sources of flux such as those produced with dc gas-discharge and x-ray sources and in beam-foil and laser spectroscopy experiments. Photon counting techniques are now used routinely in spectrometers flown in rockets and satellites for absolute measurements of solar and stellar UV radiation and also in photometers to measure UV airglow radiation. Windowless photomultiplier detectors can also be used as charged particle counters in electron energy analyzers and mass spectrometers. For these applications, the operation of the detection system as a charged particle counter is similar to that of a photon counter. Photon counting techniques cannot always be used for radiation measurements from pulsed sources that produce high levels of flux for very short intervals of time, because the count rates produced during the pulse may exceed the linear range of the counting system.

The operation of several types of photomultipliers and gas-flow counters as photon counters in the UV region between about 30 and 3500 Å will be discussed here. The techniques commonly used to obtain radiometric calibrations of photon-counting detection systems in this wavelength region will also be discussed. In addition, several light sources that can be used to obtain radiometric calibrations will be described. Those discussed are limited to sources that are either commercially available or simple to construct. Synchrotron radiation can be used as an accurate standard source of radiation throughout the ultraviolet. However, this source requires a special facility and, therefore, is not readily available. The use of synchrotron radiation for obtaining radiometric calibrations has been described in detail by Madden *et al.* (1967) and Ederer *et al.* (1975).

5.2. Photon Counting with Photomultipliers

Because of the limited spectral response of individual detectors, different types of detectors are used for flux measurements in the wavelength

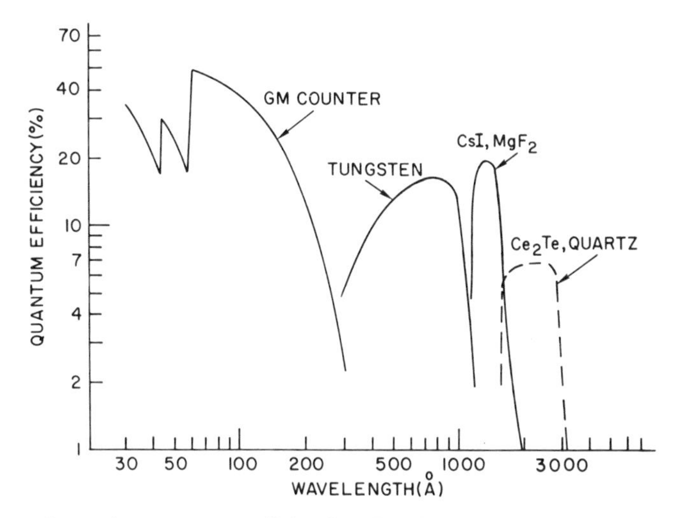

FIG. 5.1. Approximate quantum efficiencies of a thin-window gas-flow Geiger counter, a windowless detector with a tungsten photocathode, a sealed detector with a CsI photocathode and MgF_2 window, and a sealed detector with a Ce_2Te cathode and fused quartz window.

range extending from about 30 to 3500 Å. Photomultiplier detectors can be grouped into two categories depending upon their spectral response. One category consists of sealed photomultipliers with windows for use at wavelengths between about 3500 and 1100 Å, where the lower wavelength is established by the lower limit for transmission of the window. The other category consists of windowless photomultipliers for use at wavelengths shorter than ~1200 Å. Approximate quantum efficiencies of two sealed photomultipliers and a windowless photomultiplier having a tungsten photocathode are given in Fig. 5.1, where the quantum efficiency is defined as the number of counts recorded for a single photon incident on the detector. To estimate the quantum efficiencies, it was assumed that each photoelectron emitted from the cathode is counted. For comparison, the quantum efficiency of a gas-flow counter with a thin window is also included in Fig. 5.1. Again, it was assumed that each ionization event in the gas-flow counter is recorded.

A. Operation of the Photomultipliers

Sealed photomultipliers that are used as photon counters at wavelengths longer than about 1100 Å are usually electrostatic electron multipliers. Windowless channel electron multipliers (CEM) and magnetic electron multipliers (MEM) are widely used as photon counters for wavelengths shorter than about 1200 Å. Electrostatic multipliers have also been used occasionally as windowless detectors (Heroux and Hinteregger, 1962), and CEMs have been used as sealed detectors with windows (Timothy and Lapson, 1974). Figure 5.2 shows schematic illustrations of an electrostatic photomultiplier, a CEM, and a MEM. All of these photomultipliers operate in the same manner. A photoelectron ejected from the photocathode is accelerated so that it strikes the input stage of an electron multiplier with sufficient energy to eject, on the average, more than one secondary electron. This multiplication process is repeated in successive dynode stages, and the charge pulse that originates from a single cathode photoelectron is collected by the anode of the multiplier. The anode charge pulse is then coupled to a charge sensitive or voltage sensitive pulse amplifier and counter. For the electrostatic photomultiplier illus-

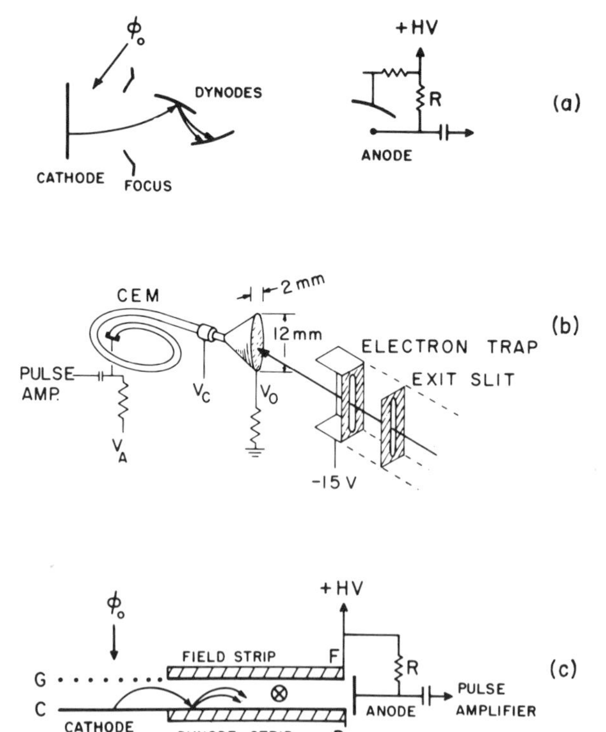

FIG. 5.2. Schematic illustrations of three photomultipliers commonly used as UV detectors: (a) electrostatic photomultiplier, (b) channel electron multiplier, and (c) magnetic electron multiplier. ϕ_0 represents the incident radiation. [(a) and (c) reproduced from Heroux (1968, p. 2352).]

trated in Fig. 5.2a, electrostatic electron focusing is used between the photocathode and the first dynode stage of the electron multiplier and also between the successive discrete dynode stages of the multiplier. For the CEM illustrated in Fig. 5.2b, the photoelectrons emitted from the inner surface of the elliptical cathode cone are directed toward the input end of the tubular electron multiplier by an electric field established along the axis of the cone by a voltage V_C. The electron multiplier consists of a semiconducting glass channel having a resistance of about $10^9\ \Omega$. An electric field is established along the inner surface of the channel by applying a high voltage V_A to the anode of the multiplier. The electrons are therefore accelerated within the channel, producing secondary electrons as they strike the inner surface of the channel. Detailed discussions of the operation of CEM's are given by Evans (1965) and Timothy and Lapson (1974). For the magnetic electron multiplier illustrated in Fig. 5.2c, the multiplier consists of two parallel glass plates having semiconducting inner surfaces. An electric field established in the region between the field and dynode strips and a magnetic field perpendicular to this electric field focus the cathode photoelectrons so that they strike the input of the continuous strip electron multiplier. These crossed fields also focus the secondary electrons between successive multiplication stages. The electric field along the dynode strip also accelerates the electrons so that secondary multiplication occurs.

Because of a spread in the transit times of the electrons which form the single anode pulse, the anode charge pulse will have an intrinsic width that is determined by the focusing properties of the electron multiplier. The pulse widths are generally small and can range from less than 2 nsec to about 20 nsec. The total charge collected by the anode is $Q = eG$, where e is the electron charge and G, the multiplier gain, is the number of electrons collected by the anode for the emission of a single cathode photoelectron.

The anode of the photomultipliers should be operated at positive high voltage, and the cathode near or below ground potential to reduce pickup of environmental charged particles at the photocathodes of the windowless photomultipliers and to eliminate intense electric fields at the windows of the sealed photomultipliers. Stray electron pickup for the CEM's can be reduced further, as shown in Fig. 5.2b, by placing an electron trap biased at about -15 V immediately before the CEM cone. To reject positive ions, the opening of the cathode cone is maintained at a potential of about $+20$ V by inserting a biasing resistor between the cone of the CEM and ground as shown in Fig. 5.2b. For the windowless magnetic photomultiplier, the field grid in front of the cathode should be operated at a positive potential with respect to ground to repel positive ions. The cathode is held negative to repel low-energy electrons. The input and output differential voltages ($V_G - V_C$) and ($V_F - V_D$), indicated in Fig. 5.2c, must be determined for individual magnetic electron multipliers, since these differential voltages significantly influence the operation of the multiplier (Timothy *et al.*, 1967).

B. The Quantum Efficiency of the Detection System

The rate of photoelectron emission $N_c(\text{sec}^{-1})$ (for a given wavelength λ) from the cathode of a photomultiplier is related to the incident photon rate ϕ (sec^{-1}) as

$$N_C(\lambda) = T(\lambda)Y(\lambda)\phi(\lambda), \tag{5.1}$$

where $T(\lambda)$ is the transmittance of the window and $Y(\lambda)$ the photoelectric yield of the cathode, defined as the number of photoelectrons emitted per photon incident on the cathode. The measured counting rate N_0 of the entire photon detection system consisting of the photomultiplier and counting circuitry is related to $\phi(\lambda)$ as

$$N_0(\lambda) = \eta T(\lambda)Y(\lambda)\phi(\lambda), \tag{5.2}$$

where the fraction η, defined as the photoelectron counting efficiency of the system, is the number of counts recorded per photoelectron emitted from the cathode. For an ideal photomultiplier in which each emitted photoelectron enters the multiplier and is counted, η would be unity. In practice, however, η is less than unity because of both imperfect focusing of the photoelectrons between the photocathode and the input stage of the multiplier and electron loss at the first dynode of the multiplier. The fraction η can approach unity for electrostatic photomultipliers (Young and Schild, 1971) having good focusing properties and a high secondary electron gain of about four at the first stage of the multiplier. Measurements of the quantum efficiencies of CEM's (Timothy and Lapson, 1974; Mack *et al.*, 1976) indicate that the fraction η also approaches unity for the CEM's. Because of the inferior focusing properties of the magnetic electron multiplier and the low value of gain at the input stage of the multiplier, η is usually significantly less than unity for the MEM's (Heroux, 1968). The efficiency of a photon counting system is generally characterized by its absolute quantum efficiency E (or spectral response), defined as the counts recorded per photon incident on the detector. The quantum efficiency (counts/photon) is, therefore, given by

$$E(\lambda) = \eta T(\lambda)Y(\lambda). \tag{5.3}$$

For an ideal windowless detection system for which T and η are unity, the quantum efficiency is identical to the photoelectric yield of the cathode.

C. Sealed Photomultipliers

Sealed electrostatic photomultipliers are commercially available in several sizes with different photocathode and window materials and with either side window or end-on window configurations from Electro-Mechanical Research, Inc., Princeton, New Jersey; EMI Gencom Inc., Plainview, New York; Hamamatsu Corp., Middlesex, New Jersey; and from other sources. The short-wavelength cutoff in the spectral response of the detectors is determined by the transmittance of the window material. The long-wavelength cutoff is determined by the photoelectric yield of the photocathode material. The spectral response and the configuration of the different photomultipliers can be obtained from the manufacturer's brochures. In principle, a single detector having a magnesium fluoride (MgF_2) window and a cesium telluride (Cs_2Te) photocathode can be used for measurements over the entire wavelength region 1150–3500 Å. However, when measurements are to be made over regions of wavelength narrower than this extended range, it is preferable to select a detector with a limited spectral response that brackets the wavelength region being measured, since the detector will then discriminate against scattered light of wavelengths outside the region of significant spectral response.

The approximate spectral responses of two sealed photomultipliers frequently used in laboratory and space experiments for UV radiation measurements are given in Fig. 5.1. The photomultiplier with the MgF_2 window and cesium iodide (CsI) photocathode is well suited for radiation measurements between 1150 and 1850 Å. Because the spectral response of the CsI cathode is very low at wavelengths longer than about 1900 Å, this cathode is effective in rejecting long-wavelength scattered light. Thermal emission of electrons from a CsI cathode is also lower than that from a cathode such as Ce_2Te because of the higher work function of CsI, and therefore this source of background noise is also low. The reduction of signals from both scattered light and thermal emission is particularly important for measurements of low flux levels since these noise sources limit the accuracy of the measurements.

The use of a LiF window rather than a MgF_2 window would extend the short-wavelength cutoff to 1050 Å. However, the stability of MgF_2 in humid air and intense illumination is superior to that of LiF. The use of either a CaF_2 window rather than a MgF_2 window or a CaF_2 filter inserted before the MgF_2 window will further limit the spectral response of the detector to the region 1250–1850 Å. This may be important, for example, if one wishes to exclude scattered light background arising from the H Lyα emission line near 1216 Å, which often appears as an intense emission line in UV sources of radiation. The photomultiplier with the fused quartz window and Ce_2Te photocathode is well suited for radiation measurements in the overlapping region 1700–3500 Å, as is evident from Fig. 5.1. Again, different windows or additional filters can be used either to extend or limit the short-wavelength spectral response of the detector.

D. Windowless Photomultipliers

Although there are numerous thin films available that transmit radiation below 1200 Å (Samson, 1967), none has sufficient strength for use as a photomultiplier window because the thickness of these films is less than about 2000 Å. Windowless open-structure photomultipliers, therefore, must be used for the detection of UV radiation below about 1200 Å. These detectors are usually constructed with either high work-function metals or semiconductors for both the photocathode and the dynode elements of the multiplier. Evaporated MgF_2 on a conducting or semiconducting substrate can also be used as a stable photocathode in a windowless photomultiplier.

The photoelectric yields of many smooth-surface metallic and semiconductor photocathodes are similar in that the yields increase rapidly with decreasing wavelength below ~1300 Å, attain a broad maximum value of approximately 15% near 700 Å, and then decrease slowly with a further decrease in wavelength (Cairns and Samson, 1966; Samson, 1967). The spectral response of the tungsten-cathode windowless photomultiplier illustrated in Fig. 5.1 (in which a collection efficiency η of unity is assumed) is fairly representative of the photoelectric yields of metal and semiconducting cathodes. Many of these metals, such as tungsten and nickel, and semiconducting materials, such as those used in CEM's, have yields that are stable and reproducible when cycled repeatedly between operation in vacuum and exposure to air. Because the work functions of both the photocathode and the dynode elements of the electron multiplier are high, background noise from thermionic emission of electrons from the detector surfaces is low at room temperature, and typically less than 0.1 counts/sec. Because the yield of the cathode decreases rapidly at wavelengths longer than about 1300 Å, background noise from long-wavelength scattered light is also low.

It is evident from Fig. 5.1 that the photoelectric yield of a high work-function metal or semiconducting cathode decreases gradually with decreasing wavelength and approaches a yield of about 5% near 300 Å. Measurements of photoelectric yields exceeding 50% have been reported for the alkali halides and the alkali-earth halides for wavelengths below about 1000 Å (Lukirskii *et al.,* 1960a,b; Duckett and Metzger, 1965; Heroux *et al.,* 1966). Except for MgF_2, these materials are not satisfactory for use as cathode materials in windowless photomultipliers, because they are not stable when exposed to air. Magnesium fluoride, when evaporated on a metal or semiconductor substrate, can be used as a high quantum efficiency, low-noise photocathode that is stable after an initial period of exposure to air and burn-in. Lapson and Timothy (1973) and Canfield *et al.* (1973) have obtained similar values for the yields of evaporated films of MgF_2 when illuminated at normal incidence in the wavelength region of about 500–1000 Å. Figure 5.3 from Lapson and Timothy (1973) compares the detection efficiency of an uncoated and a MgF_2-coated CEM. If it is assumed that η for the CEM is unity, the detection efficiency of the CEM is identical to the photoelectric yield of the photocathode. The yield of an uncoated semiconducting cathode of a CEM is also similar to that of other

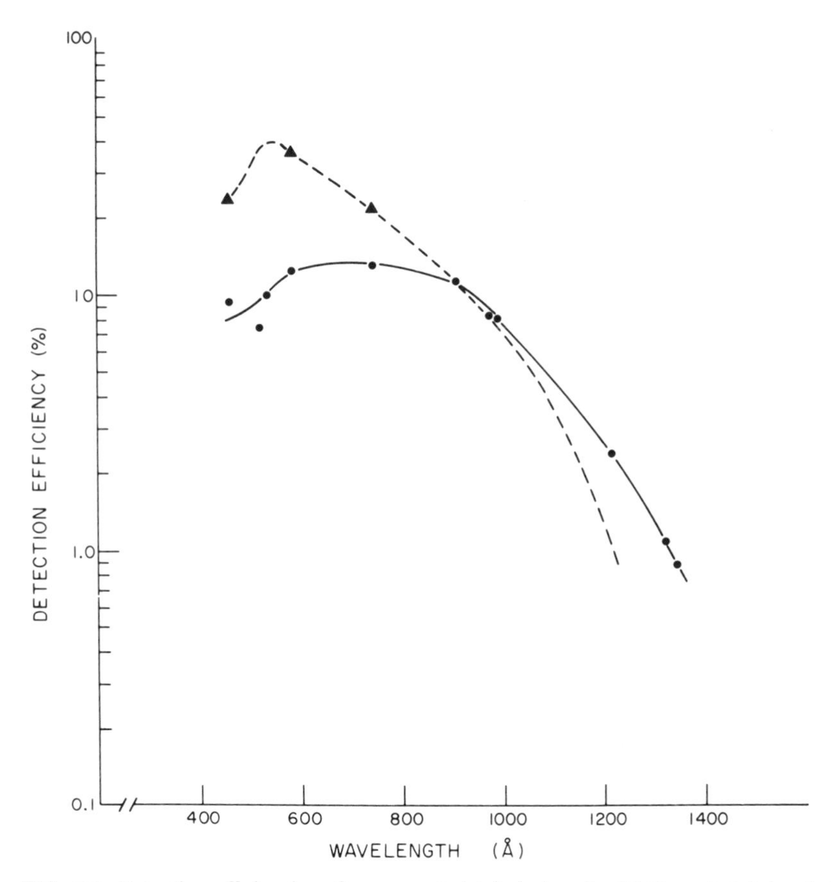

FIG. 5.3. Detection efficiencies of an uncoated (circles) and a MgF_2-coated (triangles) channel electron multiplier. The thickness of the MgF_2 coating was 3000 Å. [Courtesy of Lapson and Timothy (1973, p. 391).]

high work-function metal cathodes, so that a comparison of the yields of coated and uncoated CEM's also applies to coated and uncoated metal cathodes. As is evident from Fig. 5.3, the photoelectric yield of MgF_2 near the wavelength 900 Å is about 10%; which is similar to the yield of a typical metallic photocathode such as tungsten. At wavelengths below 900 Å, the yield of MgF_2 becomes substantially greater than that of the uncoated cathode and exceeds the yield of the uncoated cathode by a factor of about 2.5 in the wavelength region 400–600 Å.

Lapson and Timothy (1976) have extended the measurements of the response of MgF_2 down to a wavelength of 44 Å. Their measurements, shown in Fig. 5.4, were made by comparing the detection efficiency of two identical channel electron multipliers illuminated at an angle of incidence of 45°. One of the CEM's had a normal semiconducting glass photocathode and the other a 1400-Å thick coating of MgF_2 evaporated on the cathode. If one assumes that the overall photoelectron detection efficiency of both CEM's is the same, the ratio of the detection efficiencies of the CEM's gives the ratio of the photoelectric yields of the two photocathodes. These data indicate that the enhancement in the yield of MgF_2 increases linearly with photon energy for the wavelength range 950–

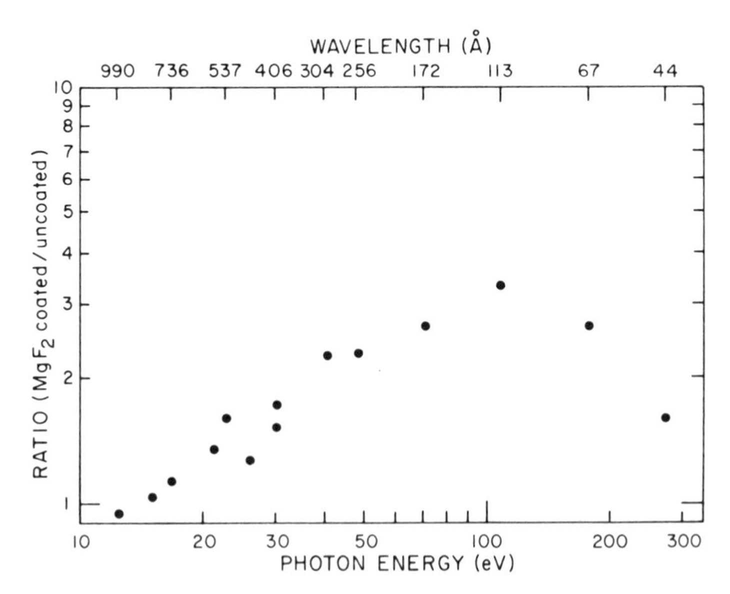

FIG. 5.4. Variation of the ratio of the detection efficiencies of a MgF_2-coated and an uncoated channel electron multiplier as a function of wave-length and photon energy. [Courtesy of Lapson and Timothy (1976, p. 1219).]

113 Å. At 113 Å, the enhancement in detection efficiency of the MgF_2-coated cathode relative to the uncoated cathode is a factor of 3.3. This enhancement at shorter wavelengths is particularly important, since the short-wavelength intensities measured in laboratory and space experiments are often low. Although the detection efficiency of the MgF_2-coated CEM decreases at wavelengths below 113 Å, its detection efficiency is still superior to that of the uncoated CEM down to 44 Å.

The photoelectric yields of metallic and alkali-halide cathodes also increase with increasing angle of incidence χ measured from the cathode normal. Rumsh *et al.* (1960) observed that for wavelengths below 13 Å, the yield of tungsten, nickel, and beryllium varied as sec χ for $\chi \leq 80°$. Heroux *et al.* (1965) observed a similar angular variation in yield for a semiconducting cathode illuminated with 304 Å radiation, as χ was varied from 0 to 60°, and a systematic decrease in the angular variation of yield for longer wavelengths. Lapson and Timothy (1976) have measured the angular dependence of the detection efficiency of MgF_2-coated CEM's at several wavelengths in the region 44–990 Å. The maximum yield of MgF_2 at wavelengths longer than about 600 Å is obtained near 45°. This increase in yield near 45° compared to the yield at normal incidence is about 20%. A similar increase in the yield of MgF_2 was also observed by Canfield *et al.* (1973) for an angle of illumination near 60°. Both Lapson and Timothy and Canfield *et al.* have observed a decrease in yield for MgF_2 when the angle of incidence was increased to values near 75°.

Although the increase in yield with angle of incidence is small at the longer wavelengths, the increase is generally much greater at shorter wavelengths, particularly at wavelengths below 300 Å. As discussed by Lapson and Timothy (1976), the maximum yield at wavelengths shorter than 600 Å is obtained when the angle of incidence is slightly less than the critical angle for total external reflection. For the wavelength region 300–600 Å, the critical angle for MgF_2 is about 65°. At 113 Å, the maximum yield will be obtained for an angle of incidence near 75°, while for 44 Å, the angle of incidence should approach 85°. The magnitude of the enhancement of yield at wavelengths below about 300 Å can be significant, as can be seen in the data of Lapson and Timothy (1976) reproduced in Fig. 5.5. These data show an enhancement in detection efficiency of a factor of 3.4 for a MgF_2-coated CEM at 113 Å when the angle of illumination is increased from 15 to 70°. Channel electron multipliers that are used to measure solar UV fluxes at wavelengths below about 150 Å are often coated with MgF_2 and illuminated at angles near 75° to increase their spectral response in this region of the solar spectrum where the radiation levels are low.

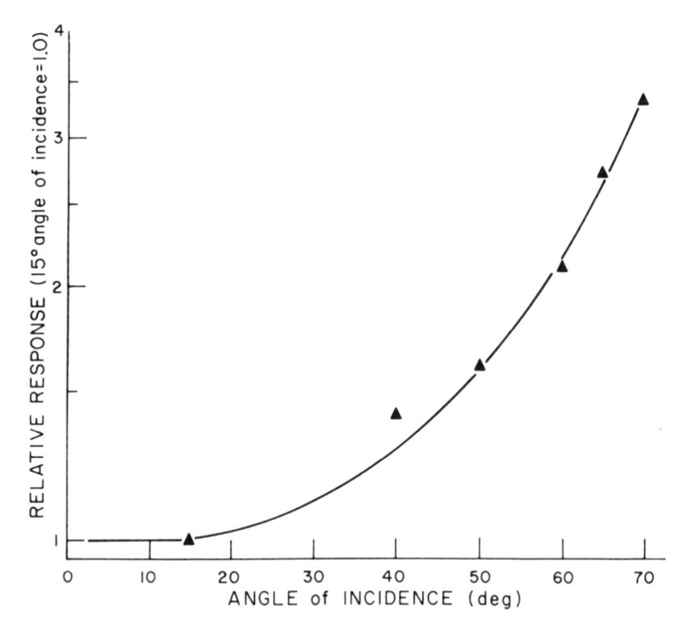

FIG. 5.5 Variation of the detection efficiency of a MgF_2-coated channel electron multiplier at 113 Å as a function of the angle of incidence from the cathode normal [Courtesy of Lapson and Timothy (1976, p. 1220).]

E. The Anode Pulse Height Distribution

Because of a variation in the photoelectron multiplication for individual photoelectrons emitted from the cathode, a photomultiplier will exhibit a distribution in the magnitude of the anode charge pulses. The distribution in charge is given by $P_K(n)$, which is the probability of observing n electrons at the K dynode stage for a single electron entering the multiplier. Lombard and Martin (1961) have calculated $P_K(n)$ for an ideal multiplier having K stages with an identical gain m per stage. They assumed a Poisson charge distribution at each stage given by

$$P_1(n) = m^n e^{-m}/n!. \tag{5.4}$$

The calculated anode distribution shows a peak for single electrons entering the multiplier. They found that the shape of the anode pulse height distribution is determined predominantly by the gain of the first few input stages, the distribution being broad for $m = 1.5$ and narrowing appreciably as the gain increases to $m = 5.0$. This is illustrated in Fig. 5.6, which gives the probability $P(n)$ of observing n secondary electrons from the second stage of a multiplier for the three values of gain per stage $m = 1.5$,

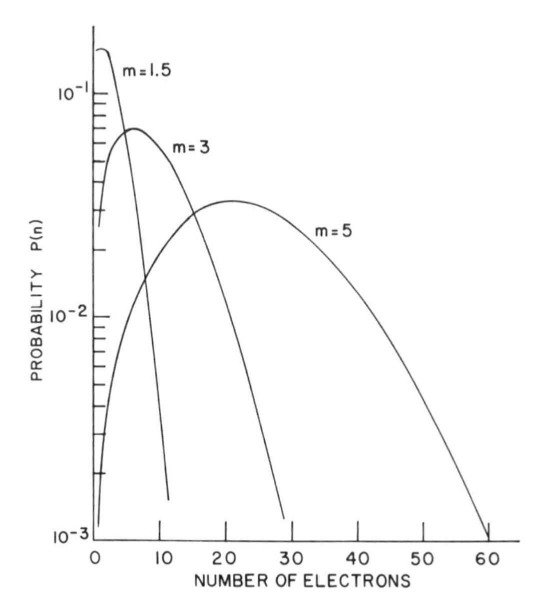

FIG. 5.6. The probability of observing n secondary electrons emitted from the second stage of an ideal multiplier for three values of the gain per stage of $m = 1.5$, 3.0, and 5.0. A Poisson charge distribution was assumed for each stage and the curves normalized so that $\Sigma P(n) = 1$.

3.0, and 5.0. The curves in Fig. 5.6 were calculated by using the Poisson distribution given in Eq. (5.4), and the data were normalized so that the area under the curve, $\Sigma P(n)$, is unity. The probability of losing an incident electron to the multiplication process is also determined primarily by the gain of the first few stages of the multiplier. Lombard and Martin show that the percentage of electron loss is about 40% for $m = 1.5$, 6% for $m = 3.0$, and less than 1% for $m = 5.0$.

The observed shapes of the anode voltage pulse height distributions for different photomultipliers, however, have been reported to vary from a peaked distribution that very nearly approaches the theoretical distribution of Lombard and Martin to a decreasing exponential distribution with no apparent peak. These differences in the distributions have been shown by Prescott (1966) to originate from a nonuniform gain across the dynode surface. His calculation for $P_K(n)$ also assumed a Poisson electron distribution for the secondary electron emission process. For a uniform value of m across the dynode surface, Prescott also obtained the Poisson distribution of Lombard and Martin, whereas for a high degree of nonuniformity in m, he obtained an exponential distribution. Again Prescott's calculations indicated that the value of m at the first stage of the electron multiplier is the predominant influence on the shape of the distribution and on the number of incident electrons lost to the multiplication process.

The anode pulse height distribution of a photomultiplier is obtained by coupling the anode pulses to a preamplifier and pulse height analyzer. The pulse height distribution for a windowless electrostatic photomultiplier fabricated by Electro-Mechanical Research (EMR), Inc. is illustrated in Fig. 5.7. The multiplier was a 20-stage venetian-blind structure having a first stage gain of 4.5 and was fabricated for use in an UV retarding potential analyzer (Heroux *et al.*, 1968). Similar multipliers having even better focusing properties than the multiplier for which the data of Fig. 5.7 were taken are now used in many commercially available EMR sealed photomultipliers. The ordinate in Fig. 5.7 gives the relative number of anode pulses with voltage heights between V and $V + \Delta V$. The distribution in voltage pulses is directly related to the distribution in anode charge pulses. The voltage axis has been normalized to the average pulse height $e\overline{G}$ of the anode distribution, where e is the electron charge and $\overline{G}$ the average value of multiplier gain. As can be seen from Fig. 5.7, the shape of the pulse height distribution approximates the computed distribution for an ideal multiplier in which Poisson statistics are assumed to be valid. The narrow distribution would be expected because of the high gain at the first stage of the multiplier and the good overall focusing properties of the multiplier.

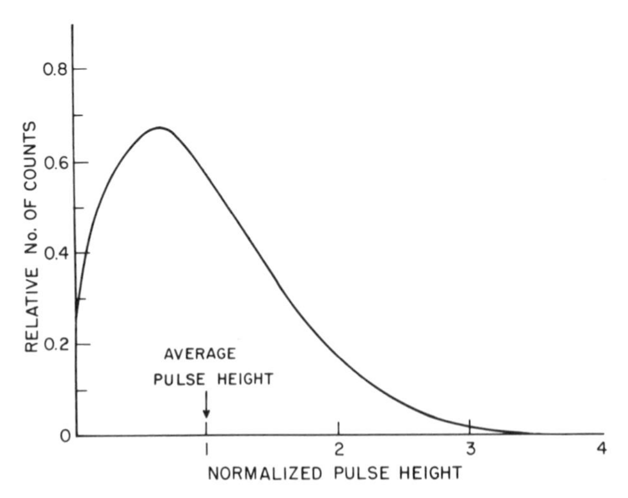

FIG. 5.7. Anode pulse height distribution for a 20-stage electrostatic photomultiplier manufactured by Electro-Mechanical Research, Inc. [Reproduced from Heroux (1968, p. 2353).]

In contrast to the relatively narrow distribution possible in an electrostatic multiplier, the anode pulse height distribution for a magnetic electron multiplier is considerably broader, as shown in Fig. 5.8. These data were obtained by Heroux and Hinteregger (1960) with an early model Bendix M306 detector, which is essentially identical to the Model 306 MEM now manufactured by Galileo Electro-Optics Corp., Sturbridge, Massachusetts. The broad distribution results from a low value of gain at the input stages of the multiplier and gain variations at the first and subsequent multiplication stages. The variations in gain are associated with the inferior focusing properties of the MEM in comparison to those of electrostatic multipliers. The very rapid increase in counts for pulse heights less than ~0.4 mV originates from regeneration pulses. These pulses are caused by the ionization of environmental gases near the output of the multiplier, where the electron density of the charge pulse is high. These ions are accelerated toward the input of the multiplier, where they strike the dynode strip and eject additional electrons, which in turn produce additional anode pulses. Electron excitation of environmental gases and subsequent photon decay can also contribute to photoelectron emission from the input of the multiplier and thus produce additional anode pulses. The onset of regeneration depends on the gain of the photomultiplier and on the environmental gas pressure. The regeneration pulses have a distribution similar to that of the noise pulse distribution produced by pulse amplifiers. However, these two distributions can be identified unambigu-

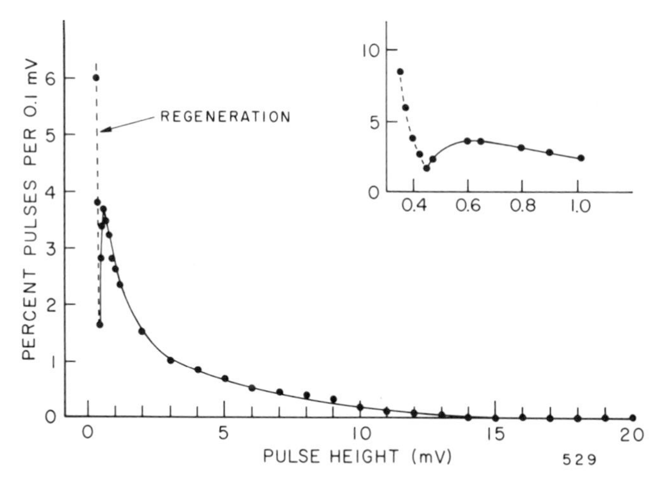

FIG. 5.8. Anode pulse height distribution for a magnetic electron multiplier. The data were obtained with a Bendix M306 detector. This detector is now available as a Galileo M306 magnetic electron multiplier. [Reproduced from Heroux and Hinteregger (1960, p. 285).]

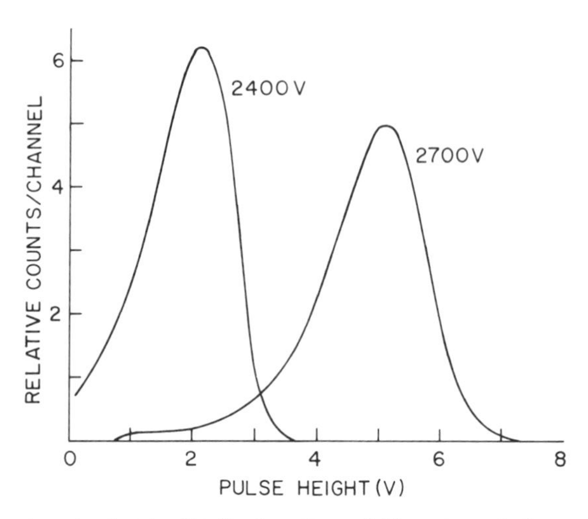

FIG. 5.9. Anode pulse height distributions for a Galileo channel electron multiplier for two values of high voltage.

ously since the amplifier noise will be present without cathode illumination. Regeneration is particularly apparent in the open-structure magnetic photomultipliers of the type illustrated in Fig. 5.2c, for which the data of Fig. 5.8 were obtained, because of the unobstructed path between the output and input stages of the multiplier. Regeneration is a less serious problem in electrostatic photomultipliers because the output and input stages are isolated by intermediate dynode stages. For the channel electron multiplier, regeneration is essentially eliminated even at relatively high gains by curving the channel to isolate the input and output stages.

For the channel electron multiplier, the number of electrons within the charge pulse is limited by space charge (Andresen and Page, 1971). Because of this saturation, the shape of the anode pulse height distribution for a sufficiently high gain is considerably narrower than that of an ideal electrostatic photomultiplier. The anode pulse height distribution for a particular Galileo CEM, which has been used in several AFGL rocket experiments, is given in Fig. 5.9. Distributions are shown for two different values of high-voltage (HV) or overall gain and approximate a Gaussian distribution rather than the Poisson distribution characteristic of the electrostatic photomultiplier.

F. The Counting Plateau

To operate the photomultiplier as a photoelectron counter, the anode pulses are amplified, shaped, and coupled to a counter that records all pulses exceeding the discrimination level of the counter. The magnitude of the pulse that is fed into the counter depends on the gain of the photomultiplier and the pulse amplifier. When regeneration processes are important, the gain of the electron multiplier should be as low as possible, and therefore the gain of the pulse amplifier should be as high as possible, with the restriction that the amplifier noise not exceed the discrimination level of the counter. For a particular value of amplifier gain and discrimination level, the optimum conditions for operating the counting system, which consists of the photomultiplier, amplifier, and electronic counter, are determined by examining the dependence of the counting rate on the gain of the photomultiplier when the photocathode is illuminated with a constant level of flux. This procedure is illustrated in Fig. 5.10, which gives typical data for an electrostatic photomultiplier. These data were obtained with the 20-stage windowless multiplier, which was used to obtain the anode pulse height distribution shown in Fig. 5.7.

The data of Fig. 5.10 were obtained with a charge-sensitive preamplifier coupled to a linear amplifier and counter. As the gain of the multiplier is increased by increasing its high voltage, the anode pulse height distribution shifts toward higher voltage. This shift can be seen in the pulse height distributions of the CEM given for two values of multiplier high voltage in Fig. 5.9. The onset of counting occurs when the largest pulses in the distribution exceed the discrimination level of the counter. The counting rate continues to increase with increasing gain as a larger fraction of the anode pulses exceeds the discrimination level. When the majority of the anode pulses is counted, a further increase in gain generally causes only a

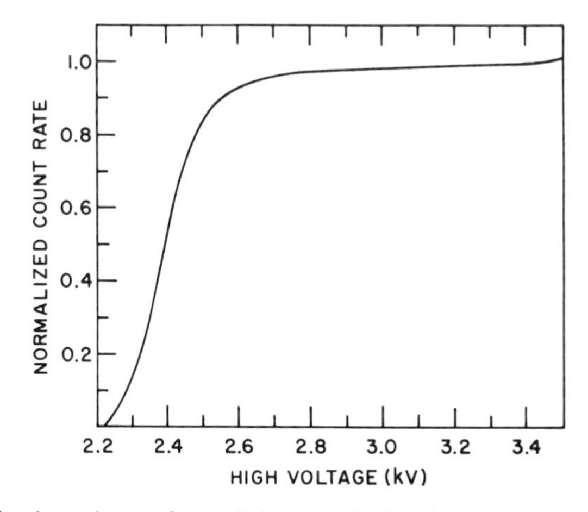

FIG. 5.10. The dependence of counting rate on high voltage for a 20-stage electrostatic photomultiplier manufactured by Electro-Mechanical Research, Inc.

slight increase in the counting rate. This region of gain, or HV, where the counting rate becomes nearly insensitive to high voltage, is the plateau region of the counting system. If one operates in the plateau region, the counting rate for a constant level of flux becomes relatively insensitive to moderate gain changes in the multiplier and in the pulse amplifier. It is this domain of operation in which the counting systems are stable over long periods of time and in which windowless detectors are also reproducible in quantum efficiency after repeated exposure to air. It is apparent from Fig. 5.10 that the plateau region of this electrostatic photomultiplier is broad and that regeneration is not a problem, at least for the range of high voltages plotted. Although the data were obtained with a windowless detector, a similar counting plateau is observed for many sealed electrostatic photomultipliers.

Because of their narrow anode pulse height distributions, high gains, and freedom from regeneration, channel electron multipliers are particularly well suited for photon counting. Because of their small size, they are widely used in space instrumentation as UV photon counters for radiation measurements at wavelengths shorter than ~1200 Å. As can be seen from Fig. 5.11, the counting rate of a CEM increases rapidly with increasing voltage after the onset of counting, and the plateau extends over a wide voltage range with very little slope. These data were obtained with a Galileo CEM of the type illustrated in Fig. 5.2b and a voltage sensitive preamplifier that produced a shaped output pulse of constant amplitude when the anode voltage pulse developed across the anode resistor exceeded 25 mV. Similar data are obtained with a charge sensitive

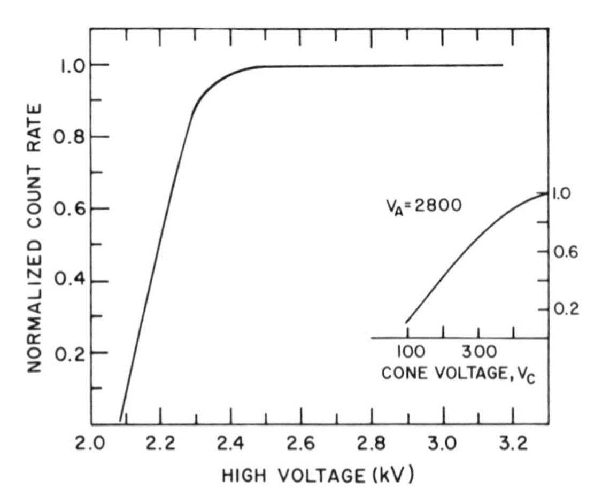

FIG. 5.11. The dependence of counting rate on high voltage for a Galileo channel electron multiplier.

preamplifier and counter for which the discrimination level corresponds to an anode pulse of approximately 2×10^6 electrons. It should be pointed out that the voltage across the channel of the multiplier is 500 V less than the plotted value of the anode voltage given in Fig. 5.11, because of the bias voltage applied to the cone. A relatively high cone voltage of 500 V was necessary to ensure that an appreciable fraction of the photoelectrons ejected from the surface of the cone entered the multiplier channel, as illustrated in the inset of Fig. 5.11. The cone bias will vary with different CEM's, depending on the geometry of the cone and the position of illumination within the cone.

The space-charge saturated domain of the CEM, where the anode charge distribution becomes narrow, occurs at gains of $\sim 10^7$. Typically the CEM's are operated at gains in the range of 10^6–10^8. The secondary electrons ejected in the multiplication process within the CEM are supplied by the current in the semiconducting channel. The replacement of charge following a pulse is determined by the resistance of the channel, which for the Galileo 4000 series CEM's is $\sim 10^9\ \Omega$, and also by the capacitance of the channel. For high count rates, as the signal current increases, the wall current in the CEM becomes depleted, and the CEM gain decreases. This decrease in gain limits the use of CEM's to count rates less than $\sim 10^6$ counts/sec. If the quantum efficiency of the CEM at a particular wavelength is 10%, this limiting count rate corresponds to a photon rate of about 10^7 photons/sec. Timothy and Bybee (1978) give a detailed discussion of the count-rate capability of channel electron multipliers and the characteristics of the Galileo 4800 series CEM's developed to extend the counting range of channel electron multipliers. This series of detectors is constructed with semiconducting channels having conductivities of a factor of about 10 greater than that used in the 4000 series CEM's. Because of the higher conductivity of these CEM's, the degradation of gain at high count rates is less severe, and the detector can be used at significantly higher count rates. The characteristics of the 4800 series CEM's and other CEM's are given in data sheets available from Galileo Electro-Optics Corp., Sturbridge, Massachusetts.

The counting plateau for a magnetic electron multiplier obtained by Heroux and Hinteregger (1960) is shown in Fig. 5.12. These data were obtained with the same MEM for which the anode pulse height distribution of Fig. 5.8 was obtained. As mentioned above (Section 5.2,A), the counting characteristics of this detector depend on the input and output differential voltages applied to the detector. From Fig. 5.12 it is apparent that the onset of counting for the MEM is less rapid than that for the electrostatic multiplier and the CEM. The slope of the counting plateau for the MEM is also appreciably greater than that for the electrostatic

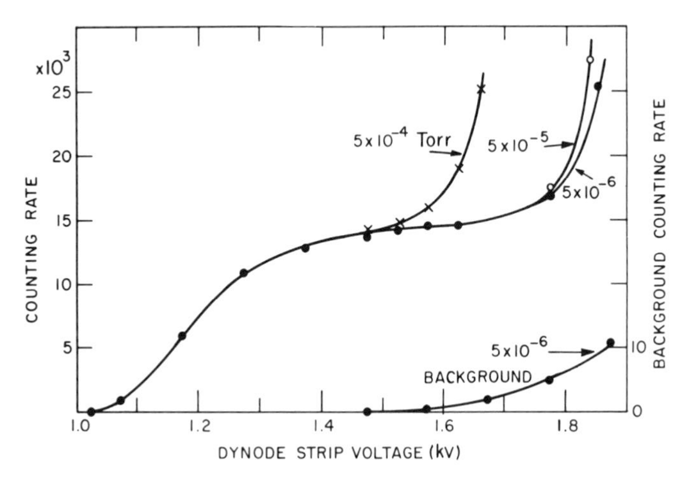

FIG. 5.12. The dependence of counting rate on high voltage for a Bendix M306 magnetic electron multiplier at several pressures. [Reproduced from Heroux and Hinteregger (1960, p. 284).]

multiplier and the CEM. These differences would be expected because of the very broad anode pulse height distribution for the magnetic electron multiplier.

The effect of regeneration pulses on the counting plateau is apparent in Fig. 5.12 as a rapid increase in counting rate that sets in at higher voltages after the development of the counting plateau. The effect of the environmental pressure on regeneration can significantly reduce the width of the counting plateau, as can be seen from the data obtained at a pressure of 5×10^{-4} Torr. To reduce the effects of regeneration, it is particularly important to use a high-gain amplifier so that the plateau region is reached at a low multiplier gain before the regeneration pulses become important.

Although the counting characteristics of the MEM are inferior to those of CEM's and electrostatic photomultipliers, the MEM's have the advantage of being rugged detectors with large cathode areas. When the detector has a well-defined plateau, such as the plateau shown in Fig. 5.12 for pressures less than about 5×10^{-5} Torr, the detector is stable and reproducible and can be used as a photon counter for many laboratory applications.

G. The Electronic Circuitry

The width of the anode pulse produced by a single cathode photoelectron is usually less than 20 nsec. The total anode pulse charge is $Q = eG$, where the electron charge e is 1.6×10^{-19} C, and G is the multiplier gain.

When the anode is terminated with a resistance R as illustrated for the multipliers shown in Fig. 5.2, the anode circuit will have a time constant RC, where C is the total distributed capacitance of the anode circuit that includes the input capacitance of an amplifier attached to it. Because the time constant will usually be large compared to the width of the charge pulse, the anode voltage V will increase to the value Q/C during the arrival time of the charge pulse and then decay with the characteristic time constant RC of the anode circuit. When the anode is operated at HV, the anode pulse is coupled through a capacitor to the input of a pulse amplifier followed by a pulse height discriminator and counter. The discriminator level establishes the minimum voltage pulse to be counted. Typically, this level would be high enough to reject amplifier noise. When operating in the plateau region, the majority of the anode pulses exceed the discrimination level. Often a pulse shaper is also used so that any pulse that exceeds the discrimination level produces a standardized shaped pulse that is fed to the counter.

When a voltage-sensitive pulse amplifier is used for pulse counting, the magnitude of the amplifier output pulse, for a particular value of amplifier and multiplier gain, will depend on the total capacitance of the anode circuitry, since the voltage pulse at the amplifier input varies inversely with capacitance. For this reason, it is important to have the input of the amplifier placed as closely as possible to the multiplier anode to minimize the capacitance of the anode circuitry.

A charge-sensitive amplifier, rather than a voltage-sensitive amplifier, is now used almost routinely for pulse counting. For this type of amplifier, the capacitance of the anode circuitry does not influence the output voltage resulting from the anode charge pulse, and the output voltage is a function only of the charge fed into the input of the amplifier. These amplifiers are characterized by their charge sensitivity V/Q (in volts/coulomb), defined as the output voltage V resulting from an input pulse charge Q. Typically the charge sensitivity of commercially available charge-sensitive preamplifiers is 10^{12} V/C. For a photomultiplier gain of 10^6, which corresponds to an anode charge of 1.6×10^{-13} C, this would produce an output pulse of 0.16 V. This pulse would then be fed into a linear amplifier followed by a discriminator and counter. Several manufacturers of photomultipliers also supply small pulse amplifier discriminator (PAD) packages designed specifically to convert photomultiplier anode pulses into output pulses of fixed width and amplitude for use with counters. These PAD's usually provide a 5 V output pulse for an input pulse of $\sim 10^6$ electrons. The width of the output pulse of the PAD establishes the pulse pair resolution of the amplifier and, therefore, the linear counting range of the detection system. The pulse widths of PAD's manufactured by Electro-Mechanical Research and Galileo Electro-Optics, for example, are ~50 nsec. Pulse amplifier discriminators having a similar charge sensitivity but an output pulse width of 220 nsec are also available from Amptek Inc., Bedford, Massachusetts.

A pulse amplifier designed by Lampton and Primbsch (1971) and shown in Fig. 5.13 is a particularly simple and compact charge-sensitive amplifier that can be used for photon counting with photomultipliers and also with Geiger-Müller (GM) and proportional counters. The operation of the amplifier is described by Lampton and Primbsch. The design is based on an RCA integrated circuit CA 3035 that provides three amplifiers consisting of a charge-sensitive input amplifier, a variable gain amplifier whose gain is determined by R_1, and a fixed-gain output stage coupled to an emitter follower to provide an output pulse of up to 7 V without clipping. The emitter follower has a 2N3904 transistor or equivalent. The rise-and-fall times of the output pulse are 0.3 and 1.0 μsec, respectively. The charge sensitivity of the amplifier can be varied from 10^{11} to 10^{14} V/C by varying R_1, a range that is adequate for most photon counting systems.

The determination of the linear range of the counting system is an important characteristic that should be determined experimentally. Within the linear range, the measured count rate for operation in the plateau region will increase linearly with the incident photon rate. Any counting system will become nonlinear at some high level of counting. Gain reduction of the photomultiplier at high count rates is a possible source of nonlinearity. However, when the counting system is operated well above the onset of the counting plateau, this gain reduction in the multiplier will not influence the linearity significantly unless it is sufficient

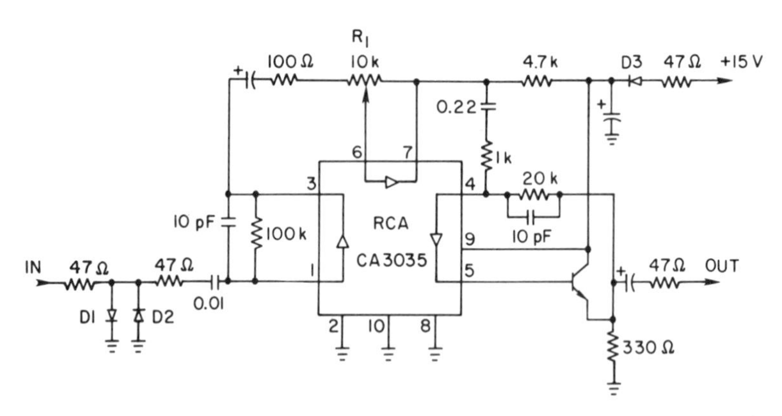

FIG. 5.13. Schematic diagram of pulse amplifier. Diodes are 1N914A, polarized capacitors are 3.3 μF, 15-V tantalum, and the transistor is a 2N3904. [Courtesy of Lampton and Primbsch (1971, p. 732).]

to reduce the magnitude of a fraction of the anode pulses to a level below the discrimination level of the counting system.

As mentioned previously, charge-sensitive amplifiers and discriminators typically have thresholds of 10^6 electrons or less. The Galileo 4000 series CEM's have gains of about 10^8 at low counting rates when operated near 2800 V, and this gain drops to about 4×10^6 at counting rates of about 5×10^6 counts/sec. The Galileo high-current 4800 series CEM's have gains of about 10^7 at counting rates of 10^7 counts/sec. Both series of multipliers, therefore, have sufficient gain at relatively high counting rates to permit linear operation within the plateau region when a counting system having a threshold of $\sim 10^6$ electrons is used. Rather than being limited by gain reduction in the photomultiplier, the linearity of a counting system is usually limited by the pulse pair resolution of the amplifier and discriminator. The pulse pair resolution is established by the width of the standardized output pulse produced by a PAD or the output pulse of a linear amplifier. When the pulse pair resolution is 250 nsec, for example, the PAD will record periodic pulses at rates up to 4×10^6 sec^{-1}. However, because the rate of arrival of photons on the photomultiplier, and the corresponding anode pulses, has a random Poisson distribution rather than a periodic one, nonlinearity for the counting system will occur at a significantly lower rate than that for periodic pulses. This effect occurs because the most probable time interval between successive pulses for the random photon distribution is zero. As an approximation, nonlinearity of a photon counting system will set in at a count rate that is about one-tenth of the rate for periodic pulses established by the pulse pair resolution of the amplifier and discriminator. The above amplifier and discriminator system that records periodic pulses up to 4×10^6 sec^{-1}, therefore, would become nonlinear for photon counting at rates in excess of $\sim 4 \times 10^5$ sec^{-1}.

A convenient method for determining the linear range of the entire counting system consists of measuring the transmittance of a nickel mesh grid over a wide range of photon fluxes. The filter is inserted into and removed from a photon beam that illuminates the photomultiplier. The transmittance of the filter T is obtained from the ratio N/N_0, where N and N_0 are the measured count rates with the filter in and out of the photon beam, respectively. The measured ratio N/N_0 will be constant and equal to T as the flux of the incident photon beam is varied until the counting rate N_0 exceeds the linear range of the counting system. When N_0 exceeds the linear range, counts are lost and the ratio N/N_0 will increase. The use of this technique to determine the linear range of counting is illustrated in the data of Heroux (1968) shown in Fig. 5.14. These data were obtained with a nickel mesh grid having a transmittance of ~ 0.45 inserted between

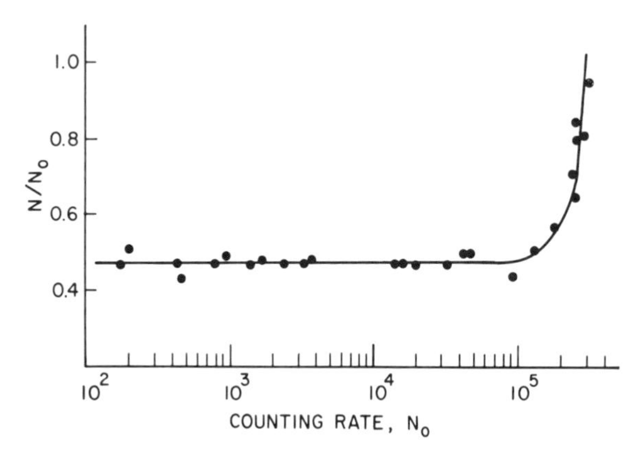

FIG. 5.14. The determination of the linear range of a counting system by measuring the transmittance of a nickel mesh filter. Nonlinearity at high counting rates near 2×10^5 Hz is evident as the measured transmittance N/N_0 departs from the value 0.47. [Reproduced from Heroux (1968, p. 2358).]

a photomultiplier and the exit slit of a laboratory spectrometer. Because the transmittance of the grid is independent of wavelength, the flux could be varied over a wide range by selecting different emission lines. The measured ratio N/N_0 plotted as a function of N_0 can be seen to deviate from the value 0.45 when N_0 approaches the counting rate 10^5 counts/sec. The counting rate of the system, therefore, can be assumed to vary linearly with photon flux if the measured counting rate does not exceed values of about 10^5 counts/sec. This range of linear counting is typical when the counting system has a pulse pair resolution of ~ 1 μsec.

The filter measurement described above and illustrated in Fig. 5.14 determines the linearity of the entire counting system consisting of the detector and pulse counting circuitry. Since the linearity of the system is usually limited by the counting circuitry rather than by the photomultiplier, one can check the linearity of the circuitry alone by using a random pulse generator. This technique is illustrated in Fig. 5.15, where the output count rate of a voltage-sensitive amplifier, discriminator, and counter is plotted as a function of the pulse frequency applied to the amplifier input. The input pulses were from a pulse generator that provided either periodic or random pulses, 100 nsec wide. The counting system, which has been used frequently with CEM's in rocket-borne spectrometers, has a pulse pair resolution of ~ 300 nsec and will record periodic pulses without loss to a rate of ~ 3 MHz. The data plotted in Fig. 5.15 give the ratios of the output count rate of the counting system to the input count rate

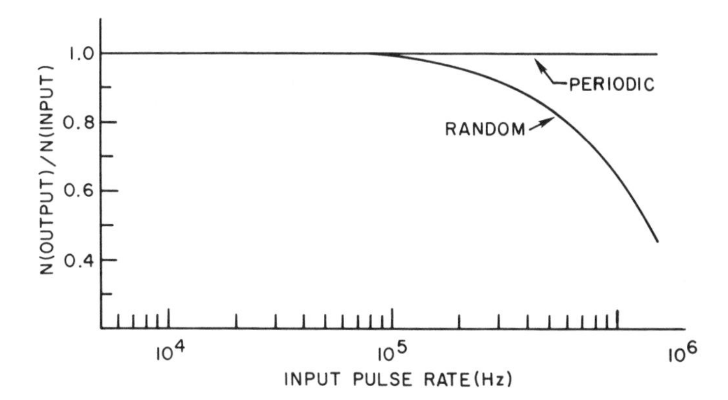

FIG. 5.15. The determination of the linearity of the electronic pulse counting circuitry when periodic or random pulses are applied to the amplifier input. Nonlinearity for random pulses is evident near 10^5 Hz as the ratio N(output)/N(input) departs from unity. The data were obtained with 100 nsec wide pulses. The pulse pair resolution of the counting circuitry was 300 nsec.

when either periodic or random pulses (of the frequency plotted on the abscissa) are applied to the amplifier input. As expected, the counting system is linear when periodic pulses in the range of pulse frequencies of the measurement are applied to the amplifier input. For this range of pulse frequencies, a constant ratio of unity is observed. Nonlinearity becomes apparent for random pulses near 10^5 Hz as the ratio begins to decrease below unity with increasing pulse frequency. At a random pulse frequency 3×10^5 Hz, approximately 10% of the pulses are lost, which is the loss expected when an anode pulse rate of 3×10^5 counts/sec is measured with an amplifier having a pulse pair resolution of 300 nsec.

5.3. Photon Counting with Gas-Flow Counters

As is evident from Fig. 5.1, the quantum efficiency of windowless photomultipliers having high work-function cathodes is low for wavelengths shorter than ~300 Å. Coating the cathode with MgF_2 and illuminating the cathodes at grazing angles can increase the yields significantly, but this is not always convenient, and care must be taken to prevent contamination of the coated cathodes. On the other hand, thin-window, gas-flow GM counters and proportional counters, because of their high efficiencies, are particularly well suited for absolute measurements of low-level steady-state fluxes at wavelengths shorter than ~300 Å.

A. Geiger Counters

Ederer and Tomboulian (1964) have established that GM counters can be used as absolute photon counters at wavelengths below ~300 Å. When rare gases are used as counter gases at wavelengths shorter than their ionization thresholds, the absorption of one photon within the counter produces at least one electron either by photoionization or Auger processes. If the gas pressure within the counter is high enough to insure that all photons are absorbed within the active volume of the counter, then each photon transmitted by the counter window will produce one pulse at the anode of the GM counter. The quantum efficiency of the counter, therefore, will be given by the transmittance of the window.

The gas pressure required to achieve total absorption of the photon beam in a counter can be calculated when the photoionization cross section of the gas is known. Argon is a convenient counter gas for use at wavelengths below ~300 Å. The wavelength dependence of the absorption cross section of Ar is given in Fig. 5.16, where the data are from a compilation by Hudson and Kieffer (1971). This compilation also gives data for helium, which also can be used as a counter gas between about 100 and 300 Å. To quench the Geiger pulse, 4% isobutene is generally added to the rare gas. The pressure region for proper operation of the counter can be determined by calculating the ratio $N(p)/N_s$ for a range of pressures. This ratio represents the fraction of the photon beam absorbed in a counter at pressure p having an absorption length L between the counter window and the back wall of the counter. $N(p)$ is the plateau counting rate at pressure p, and N_s the saturated counting rate for sufficiently high pressure when the photon beam is completely absorbed by the

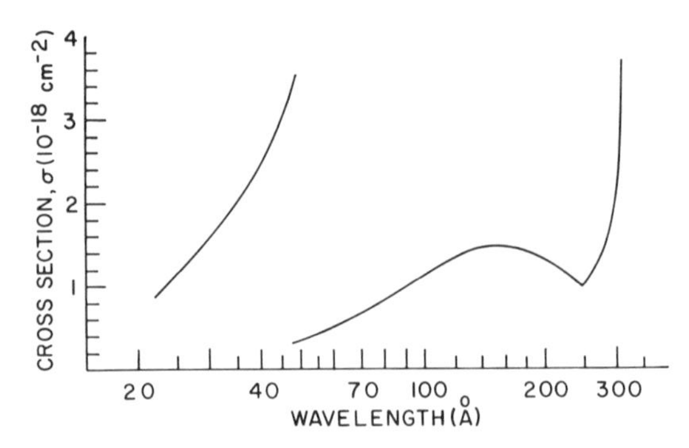

FIG. 5.16. The absorption cross sections of argon between 20 and 310 Å. [Data from compilation of Hudson and Kieffer (1971, pp. 223, 224).]

counter. The ratio is obtained from the equation

$$N(p)/N_s = 1 - \exp(-\sigma NL), \tag{5.5}$$

where σ (cm^2) is the photoabsorption cross section, N (cm^{-3}) the number density of the gas, and L (cm) the absorption path length in the counter. For a counter operated at a temperature T (°K) and pressure p (Torr), N is related to the Loschmidt number $N_0(2.69 \times 10^{19}$ cm$^{-3})$ defined for $T_0 = 273$°K and $p_0 = 760$ Torr, and Eq. (5.5) can be expressed as

$$N(p)/N_s = 1 - \exp[-\sigma N_0(T_0/T)(p/p_0)L\,]. \tag{5.6}$$

This ratio is plotted as a function of pressure inFig. 5.17 for a cylindrical side window counter having an inner diameter of 1.2 cm and argon as the counter gas. The calculation is given for wavelengths of 67 Å, where the cross section is low, and of 304 Å, where the cross section is high. For complete absorption of the photon beam, the counter should be operated at pressures beyond the pressure where $N(p)/N_s$ is unity. Usually the counter pressure can be considerably above the minimum pressure for total absorption. However, if the gas pressure is too high, an appreciable fraction of the photon absorption will occur close to the counter window, and the ejected electrons may not enter the active region of the counter and produce anode pulses. To insure proper operation of the counter at a particular wavelength, the plateau counting rate should be examined experimentally for a wide range of pressures, and the plateau rate should remain constant over a range of pressures for constant illumination. A

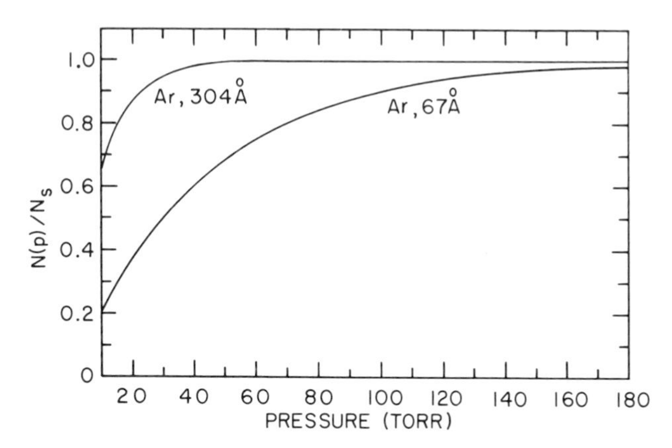

FIG. 5.17. Calculated ratio of the counting rate $N(p)$ at pressure p to the saturated counting rate N_s for an argon filled GM counter when exposed to 304 and 67 Å radiation. The absorption length of the counter is 1.2 cm.

decrease in counting rate with decreasing pressure would indicate incomplete absorption of the photon beam by the counter gas, while a decrease in the rate with increasing pressure would indicate loss of electrons near the counter window.

A GM counter that has been used at the Air Force Geophysics Laboratory for obtaining photometric calibrations of rocket spectrometers at wavelengths between about 30 and 300 Å (Heroux *et al.,* 1972) is illustrated in Fig. 5.18. This counter is similar to a compact GM counter designed for use as a detector in a rocket spectrometer to measure solar fluxes between 30 and 130 Å (Manson, 1967). These counters have been used behind the slits of spectrometers so that the windows are elongated, with the length parallel to the anode wire and to the spectrometer slit. The

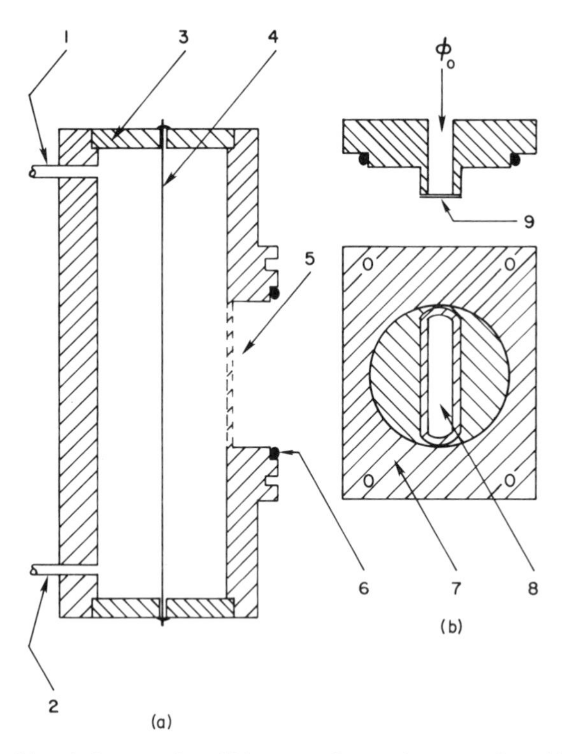

FIG. 5.18. A thin-window gas-flow GM counter for use between about 30 and 300 Å. (a) Cross section of the interior. The inner diameter of the stainless steel counter body is 1.2 cm and the length 4.4 cm. Gas inlet and outlet tubes (1,2), ceramic insulators (3), 50 μ stainless steel anode wire (4), cutout for window support (5), and O ring (6). (b) Window. Stainless steel window support (7), window area, 0.24 cm wide and 1.3 cm long (8), and thin film and nickel mesh grid (9), shown on top view of window (upper right). ϕ_0 represents the incident radiation. [Courtesy of J. E. Manson (unpublished).]

length of the window is slightly longer than the slit length and considerably wider than the slit width to insure unobstructed passage of a diverging photon beam and to reduce the need for accurate window positioning behind the spectrometer slit. The window of the counter, shown in Fig. 5.18(b), has a length of 1.3 cm and a width of 0.24 cm. Both the counter and window support were constructed of 303 stainless steel. The interior of the counter, shown in cross section in Fig. 5.18(a), is cylindrical, with a diameter of 1.2 cm and a length of 4.4 cm. Ceramic insulators on each end of the counter support a 50 μ stainless steel anode wire along the axis of the cylinder.

The gas inlet and outlet connections are attached to a gas-flow control system, which controls the gas flow and pressure within the counter. The outer wall of the counter is machined to accept a window support to position the window close to the inner cylindrical wall of the counter. This feature was introduced in the GM counter developed by Ederer and Tomboulian (1964) and is necessary to minimize the dead space between the counter window and the active volume of the counter, where electrons can be lost. This reentrant window design should be incorporated in any GM counter used as an absolute photon counter between about 30 and 300 Å. For the counter described here, the distance between the window and inner wall was ~0.05 cm. A slot having approximately the dimensions of the window is cut out of the remaining thin counter wall in the area that accepts the window support. The window support is fastened to the counter and sealed with an O ring.

When the counter is operated in its plateau region at pressures high enough to absorb all photons entering the active volume of the counter, its quantum efficiency is determined by the transmittance of the window. There are a variety of plastic films, which have a sufficiently high transmittance for thicknesses near 1000 Å, that can be used as counter windows (Samson, 1967). These windows are often supported on Buckby-Mears nickel mesh that is fastened to the window support with conducting epoxy cement to insure good electrical contact. The mesh provides support for the thin window material and also extends the ground plane of the counter to the window area. A commonly used mesh has 400 lines/in. and a transmittance of approximately 60%.

Manson (1973) describes the fabrication and gives the characteristics of a VYNS window, which is a copolymer of vinyl chloride and vinyl acetate. This film is particularly well suited for gas-flow counters for use between ~30 and 300 Å, because of its strength and ease of fabrication. It is prepared by dipping a microscope slide into a solution of the resin VYNS, floating the film on a water surface, and transferring the film to the mesh-covered window support. The J. E. Manson Company, Concord,

Massachusetts manufactures gas-flow counters with VYNS windows as well as pulse amplifiers and gas-flow control systems to operate the counters.

Figure 5.19 gives the transmittance of a window made from VYNS on 400 line/in. mesh. The solid curve gives empirically calculated values of the transmittance. Several experimental values of transmittance are also given in the figure. The discontinuities in the structure are due to the chlorine L absorption edge near 61 Å and the carbon K edge at 43.77 Å. There is also structure in the region of the Cl L edge that is not shown in the transmittance curve. Usually only a few wavelengths are available for obtaining experimental measurements of transmittance. However, calculated values of transmittance can be obtained with relatively high accuracy when the transmittance is measured at a single wavelength in each region between the absorption edges and at one wavelength longer than the L edge. The transmittance of the window is given by

$$T = S \exp(-\rho x \mu_{\mathrm{m}}), \tag{5.7}$$

where S is the transmittance of the nickel mesh, ρ (g/cm^3) and x (cm) are the density and thickness of the film, respectively, and μ_{m}(cm^2/g) is the mass absorption coefficient of the film. The product ρx (g/cm^2) is often referred to as the area density of the film and given in units of micrograms

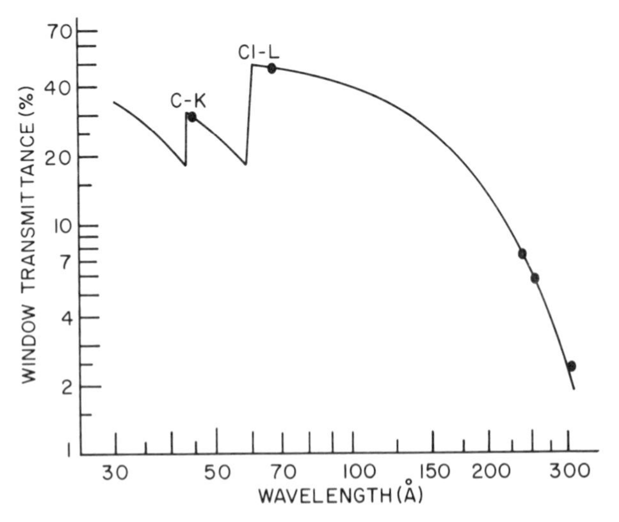

FIG. 5.19. Transmittance of a VYNS window having an area density of 25 μg/cm^2 supported on 400 line/in. mesh. The transmittance of the mesh alone was 57%. The solid curves are empirically calculated values of the window transmittance. Several experimental values of transmittance are given for comparison with the calculated values.

per square centimeter. Manson (1973) gives an empirical formula that can be used to calculate the mass absorption coefficient of VYNS for wavelengths longer than the chlorine L edge as

$$\mu_m = 1.4315\ \lambda^{1.994}\ \mathrm{cm^2/g}. \qquad (5.8)$$

When the mesh transmission S is known and the window transmittance is measured at any one wavelength, the area density ρx can be determined from Eqs. (5.7) and (5.8) by calculating μ_m for that wavelength. The window transmittance as a function of wavelength beyond the L edge can then be calculated from Eqs. (5.7) and (5.8). The solid curve in Fig. 5.19 for wavelengths between 61 and 304 Å was calculated in this manner. For this calculation, S was 56%, and the area density obtained from a single transmittance measurement at 67 Å was 25 μg/cm^2. Experimental values of the window transmittance at 67 Å and several other wavelengths are indicated in the figure to illustrate the accuracy of the calculated values of the window transmittance. Calculated values for the window transmittance in the regions between the C K and Cl L edges and below the C K edge can also be obtained by using the "Universal" absorption functions tabulated by Henke *et al.* (1957). These tables allow one to calculate the mass absorption coefficients for wavelengths between the K and L edges and also below the K edge when the mass absorption coefficients are known at one wavelength in each region between the edges. The calculated window transmittance between 44 and 61 Å given in Fig. 5.19 were obtained from the tables of Henke *et al.* For these calculations, the window transmittance at 45 Å was measured to determine μ_m at this wavelength. The calculated transmittance of the window below the C K edge is an estimate based on previous transmittance measurements of a VYNS window at 31.6 Å made by Manson (1972).

The anode pulse of a GM counter is of the order of 10^7 electrons. When operated in the Geiger region, the pulse height distribution is narrow since the charge pulses are nearly identical in magnitude. The magnitude of the charge pulse is also independent of wavelength. After the pulse is produced, the counter is completely insensitive to an additional ionizing event until the discharge is quenched and the positive ion sheath travels back to the cathode. After this recovery time, the counter is capable of producing a second pulse of full amplitude. The time interval during which the counter is incapable of producing a second pulse is referred to as the counter dead time. The dead time depends upon the dimensions of the counter, the gas pressure, and the type of gas used. The dead time is usually greater than ~50 μsec, and this relatively long dead time limits the counting rate of a Geiger counter. Because the dead time is appreciable, the measured counting rates should be corrected for dead time losses by using the equation

$$N = N_0/(1 - N_0\tau), \tag{5.9}$$

where N is the corrected counting rate, N_0 the measured rate, and τ the dead time of the counter. For example, a counting rate of 2000 sec^{-1} measured with a GM counter having a dead time of 50 μsec should be increased by 10% to give the true counting rate of the measurement. The dead time can be measured at the output of a linear amplifier by observing the time interval required for the successive pulses to build up to the discrimination level of the counting system after an initial pulse of normal amplitude. The pulse amplifier of Lampton and Primbsch (1971) shown in Fig. 5.13 can be used with a GM counter. Because the magnitude of the Geiger pulse is large, the amplifier must be operated at low gain to prevent clipping of the output pulse.

Thin-window counters require a gas-flow system to establish the pressure and rate of gas flow in the counter. The counter is often located within a vacuum chamber so that care must be taken to avoid window damage during the system pump down and pressurization of the counter. Similar precautions must be taken also when the vacuum chamber and counter are brought up to atmospheric pressure. A convenient gas handling system is illustrated in Fig. 5.20 and is similar in operation to that described by Ederer and Tomboulian (1964) and to a gas-flow system available from the J. E. Manson Company. The components of the flow

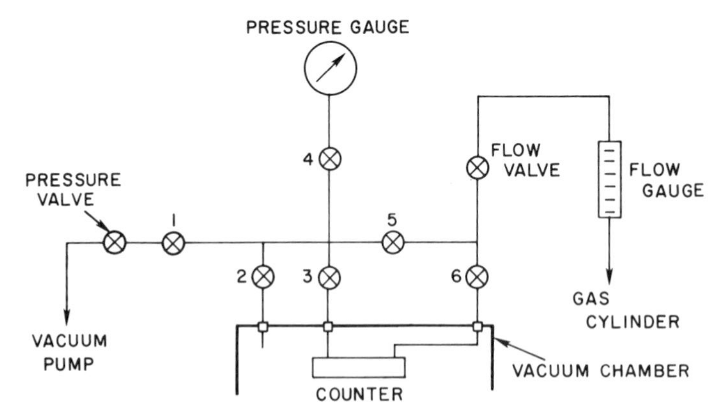

FIG. 5.20. Gas-flow system to control pressure and rate of gas flow in a thin-window gas-flow counter. Shut-off valves are numbered 1–6. The pressure and flow control valves are adjustable needle valves. The counter is mounted within a vacuum chamber and evacuated simultaneously with the vacuum chamber by opening valves 2 and 3 with the remaining valves closed, except for valve 4, which monitors the counter and vacuum chamber pressure. The gas cylinder is fitted with a standard pressure reduction regulator.

system external to the vacuum chamber are panel mounted for ease of operation. Copper or stainless steel tubing is used for the control system. Flexible tubing can be used to connect the control panel to the counting gas cylinder, the auxiliary vacuum pump, and the three inlets to the vacuum chamber that contains the counter. The counter is evacuated simultaneously with the vacuum chamber by closing all valves except valves 2 and 3. This procedure minimizes the pressure differential across the window during evacuation. After evacuation of the chamber and counter, valves 2 and 3 are closed, valve 6 remains closed, and the other valves are opened so that the flow system can be flushed and evacuated with the auxiliary vacuum pump. To flow gas through the counter, valve 5 is closed and valves 3 and 6 opened. The gas flow rate is controlled with a flow-control needle valve and measured with the flow gauge. Typical values of the flow rate are 0.04 ft^3/hr. The gas pressure is controlled with a pressure control needle valve and measured with an aneroid pressure gauge. An equilibrium pressure is usually established after several minutes. A procedure similar to that used for evacuation of the counter is also used to bring the counter and chamber up to atmospheric pressure. The flow control needle valve is closed and the counter and control system are evacuated with the auxiliary vacuum pump. After counter evacuation, valve 1 is also closed to isolate the flow system, and valve 2 is opened so that the vacuum chamber and counter will rise in pressure nearly simultaneously when air is admitted at a low rate to the vacuum chamber.

B. Proportional Counters

The GM counter described in Section 5.3,A can also be operated as a proportional counter at wavelengths below about 300 Å when a suitable counter gas is used. When complete photon absorption occurs in the active volume of the counter, the quantum efficiency of the proportional counter is identical to that of the GM counter and determined by the transmittance of the window. In contrast to the GM counter, in which the anode pulse is of fixed magnitude and independent of the photon energy, the magnitude of the anode pulse of a proportional counter is proportional to the energy of the photon absorbed so that the counter has the capability of providing limited energy resolution. Because the width of the anode pulse is typically less than 1 μsec, the proportional counter can be used over a much wider dynamic range of intensities than is possible with a GM counter. When operated in its region of proportionality, the magnitude of the anode charge pulse is about 10^4. Because of this relatively low gain, a low-noise charge sensitive amplifier is used with a proportional counter.

A mixture of argon and 50% methane is a convenient proportional counter gas for use in the wavelength region 30–300 Å. A proportional counter operates in a voltage region where gas multiplication increases with voltage in a controlled manner. The gas multiplication is defined as the ratio of total ion pairs produced in the counter to the initial ion pairs produced by photon absorption. The anode charge pulse resulting from the multiplication will depend upon the geometry of the counter, the type of gas, the gas pressure, and the anode voltage applied to the counter. The anode charge pulse is also proportional to the number of initial ion pairs formed when the photon is absorbed. This characteristic allows the counter to provide energy resolution.

The absorption of a photon of energy E by the proportional counter gas produces a mean number of initial electron–ion pairs of E/W, where W is the mean energy required to form an ion pair. For a methane proportional counter, W has the value 27 eV (Henke and Tester, 1974). As an example, the mean number of initial electrons is ~1.5 when a 300 Å photon is absorbed within the counter, and ~11 when a 45 Å photon is absorbed. For the same values of gas multiplication and amplifier gain, the mean voltage of the anode pulse for 45 Å radiation, therefore, would be larger by a factor of about 7 than that measured for 300Å radiation. Because of statistical fluctuations, both in the number of primary electrons formed and in the gas gain, a distribution in the magnitudes of the anode pulses will occur. The width of the pulse height distribution also varies significantly with wavelength, and the width of the distribution determines the energy resolution that can be achieved with a proportional counter. Henke and Tester (1974) have shown that the full width at half maximum (FWHM) of the pulse height distribution is given by the empirical equation

$$(\Delta V/\overline{V})_{\mathrm{FWHM}} = \sqrt{\lambda}/9 = 12/\sqrt{E}, \tag{5.10}$$

where $\overline{V}$ is the mean pulse height, λ the wavelength in angstroms, and E the photon energy in electron volts. The quantity $(\Delta V/\overline{V})$ as a function of both wavelength and photon energy is plotted in Fig. 5.21. It is apparent from this figure that the energy resolution of a proportional counter for wavelengths longer than ~150 Å is poor, because of the broad pulse height distributions, given by $\Delta V/\overline{V}$. With decreasing wavelength below ~100 Å, the pulse height distribution narrows considerably, and therefore the energy resolution capability of the proportional counter is often used at these shorter wavelengths. The dependence of the width and mean voltage of the pulse height distribution on the wavelength of the incident radiation is illustrated in Fig. 5.22, in which plots are given of the anode pulse height distributions obtained with a proportional counter illuminated with radiation of wavelengths 45 and 113 Å.

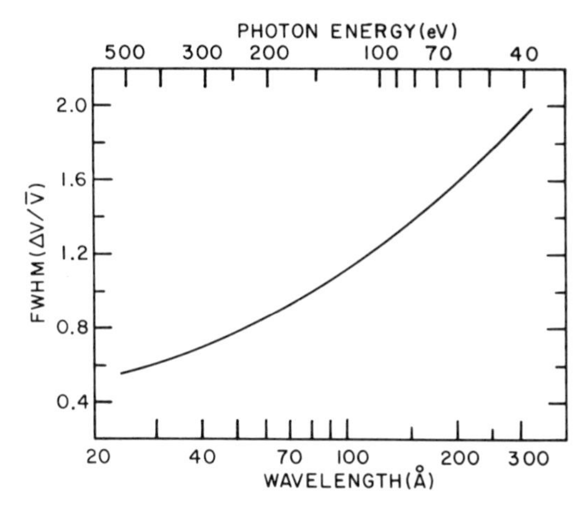

FIG. 5.21. The dependence on photon energy and wavelength of empirically calculated values of the full width at half maximum (FWHM) of the pulse height distribution for a proportional counter.

Although the proportional counter does not have the capability for energy resolution above ~150 Å, it can be used to measure photon rates that are considerably higher than those possible with a GM counter, because of its narrow pulse width and low dead time, which is usually less than 1 μsec. The proper operation of the proportional counter for photon counting is similar in principle to the operation of photomultipliers, discussed in Section 5.2,F. The discrimination level of the counting system is set to reject amplifier noise. The gain of the proportional counter is in-

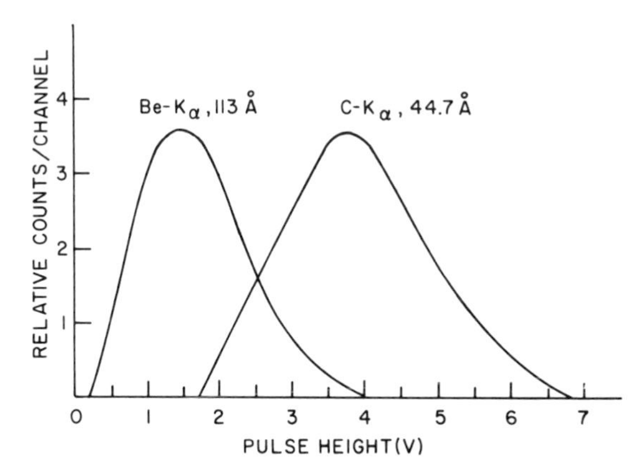

FIG. 5.22. The anode pulse height distributions of a proportional counter for the wavelengths 45 and 113 Å.

creased by raising its anode voltage to shift the anode pulse height distribution above the discrimination level. When the majority of pulses exceeds the discrimination level, a counting plateau will be obtained. Charge-sensitive amplifiers are used with proportional counters because of their low noise and high gain. The techniques used to determine the linearity of the counting system are similar to those used with photomultipliers (Section 5.2,G). However, the linearity is usually limited by the dead time of the proportional counter rather than by the pulse pair resolution of the electronic circuitry. Complete absorption of radiation in the region between about 30 and 300 Å will occur in a proportional counter when operated with a gas mixture of argon and 50% methane at a pressure of 200 Torr. To insure that radiation of a particular wavelength is absorbed within the active volume of the counter, one should examine the plateau counting rate over a range of pressures with constant illumination, as discussed in Section 5.3,A.

5.4. Spectroradiometric Calibration

The spectral efficiency $K(\lambda)$ of a detection system, such as a spectrometer instrumented for photon counting, must be known when the system is used to determine the absolute spectral intensity of radiation emitted from a source. The spectral efficiency can be defined as

$$K(\lambda) = N_0(\lambda)/S(\lambda), \tag{5.11}$$

where $N_0(\lambda)$ is the output count rate and $S(\lambda)$ the rate of photons entering the aperture of the system. The parameter $K(\lambda)$ is obtained by measuring the output count rate of the system when the aperture of the system is illuminated with radiation of known intensity and wavelength. When defined in this manner, $K(\lambda)$ in units of counts/photon is proportional to the product of the optical transmission of the system and the overall photon counting efficiency of the counting system. One would expect that the determination of the single parameter $K(\lambda)$ would be more accurate than the alternative method of determining the optical transmission and photon detection efficiency separately. Several methods of obtaining the spectral efficiency of spectrometers instrumented for photon counting will be discussed here. When $K(\lambda)$ has been found, the detection system can then be used to determine the absolute rate of photon emission from sources of radiation. To measure the spectral intensity emitted by the source, the geometrical parameters of the spectrometer, such as its aperture area and field of view, must also be considered since these parameters determine the rate at which photons enter the aperture and the source area seen by the spectrometer (Fastie and Kerr, 1975).

The radiometric calibration of the system that determines $K(\lambda)$ is made by using standard sources of spectral irradiance, standard sources of radiance, and transfer standard detectors to provide absolute values of the incident photon rate $S(\lambda)$. Sources of spectral irradiance give the radiation incident on a surface, while sources of spectral radiance give the radiation emitted from the surface of the source. Quartz halogen standard sources of spectral irradiance are widely used to obtain calibrations at wavelengths longer than 2500 Å. Spectral radiance sources include tungsten ribbon filament lamps for use at wavelengths above 2500 Å, deuterium lamps for use between 1300 and 3000 Å (Pitz, 1969), wall-stabilized hydrogen arcs for use between 1300 and 3000 Å (Ott *et al.*, 1973), and argon arcs for use between 1140 and 3300 Å (Bridges and Ott, 1977). Synchrotron radiation is also particularly well suited as a standard source for calibration for the entire x-ray and ultraviolet region above 0.1 Å (Ederer *et al.*, 1975). All of these standard sources emit continuum radiation with known spectral distribution. The aperture of the system being calibrated is illuminated directly with the standard source without predispersion. When transfer standard detectors are used for calibration, the aperture of the system being calibrated is illuminated with monochromatic emission lines or narrow emission bands produced with an auxiliary monochromator. The absolute flux of the dispersed radiation entering the aperture of the spectrometer is measured with a transfer standard detector. Transfer standard detectors calibrated for the wavelength region between 200 and 3000 Å are available from the National Bureau of Standards. For the wavelength region below 200 Å, gas-flow counters can be used as standard detectors. Dispersed radiation and transfer standard detectors are often used at wavelengths below ~1600 Å, since neither synchrotron sources of radiation nor wall-stabilized arcs that produce radiation at wavelengths shorter than 1600 Å are generally available. Radiometric calibrations at wavelengths below ~2300 Å must be carried out in vacuum chambers. Calibrations at longer wavelengths can be done at atmospheric pressure, since negligible absorption of radiation occurs for the distances required between the standard sources and the apertures of the systems being calibrated.

A. Calibration with Dispersed Radiation Sources

When radiometric calibrations of a spectrometer are made by using dispersed radiation and transfer standard detectors, the spectrometer being calibrated is coupled to an auxiliary monochromator such that the entrance slit of the spectrometer replaces the exit slit of the auxiliary monochromator, as illustrated in Fig. 5.23. This permits the entrance aperture of the spectrometer to be illuminated with dispersed radiation. The angular divergence of the incident radiation is limited with optical stops placed before the aperture of the spectrometer to insure that all radiation entering the entrance aperture is completely intercepted by the grating. The photon rate $S(\lambda)$ entering the aperture is obtained by inserting a standard detector into the photon beam between the entrance slit and grating. The standard detector also must completely intercept the incident radiation. After extraction of the standard detector, the counting rate $N_0(\lambda)$ of the system is measured, and the radiometric efficiency of the system obtained from Eq. (5.11). When spectrally resolved sharp atomic emission lines are used for the incident radiation, the linewidth measured with the spectrometer usually will be considerably wider than the linewidth of the incident spectral line. The full width at half maximum (FWHM) intensity of the spectral line measured with the spectrometer determines the spectral resolution of the instrument and is often referred to as the instrumental linewidth $\Delta\lambda_i$. When the exit slit is moved in small steps through a spectral line and the distance moved for each step is less than the instrumental linewidth, a triangular line profile will be obtained

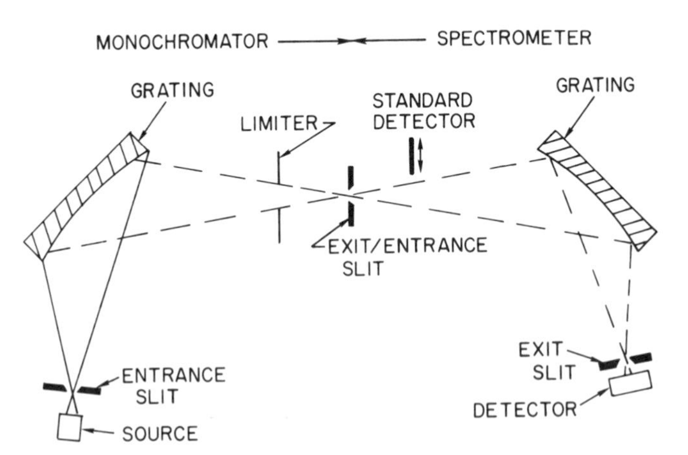

FIG. 5.23. Optical arrangement used to calibrate a spectrometer with a dispersed source of radiation and a standard detector. Monochromatic radiation is produced with an auxiliary monochromator. The entrance slit of the spectrometer to be calibrated replaces the exit slit of the auxiliary monochromator. The limiter restricts the angular divergence of the photon beam to insure that all radiation entering the spectrometer aperture is intercepted by the grating. The photon rate entering the spectrometer aperture is measured with the standard detector that can be inserted and removed from the incident beam.

for matched entrance and exit slits, and a flat topped profile will be obtained when the exit slit width is wider than that of the entrance slit. The number of steps corresponding to the FWHM intensity of the line profile will depend upon the dispersion of the spectrometer. For a grazing incidence spectrometer, the spectral resolution $\Delta\lambda_i$ of the instrument will be constant with λ. However, the dispersion will vary with wavelength so that the number of steps corresponding to the FWHM will also vary with wavelength. For a monochromatic emission line, $N_0(\lambda)$ is given by the peak reading of the line profile. $N_0(\lambda)$ can also be obtained by summing the counts as the slit is stepped through the line and dividing the sum by the steps corresponding to the FWHM intensity of the instrumental linewidth. This integrated value of N_0 is identical to that measured at the line peak. The integrated values of counts under the line will provide better statistical accuracy than a peak reading, when the recorded count rates are low. If the incident radiation $S(\lambda)$ is not monochromatic, but instead is made up of unresolved spectral lines or emission bands such that the spectral width $\Delta\lambda$ of the radiation entering the aperture of the spectrometer being calibrated is wider than the instrumental line shape, the integrated value of $N_0(\lambda)$ must be used to determine $K(\lambda)$. This would be necessary, for example, when the entrance aperture is illuminated with dispersed radiation from an x-ray emission band. The incident flux measured with the standard detector will record photons in the wavelength band $S(\Delta\lambda)$. The integrated output counts will then measure the total counts $N_0(\Delta\lambda)$ corresponding to the incident wavelength increment $\Delta\lambda$. This measurement will then give the radiometric efficiency at a wavelength λ near the center of the band $\Delta\lambda$.

The calibration technique described above, which is based on illumination of the entrance aperture with a dispersed source of radiation, requires stable transfer standard detectors to determine the absolute photon rate of the incident flux $S(\lambda)$. The gas-flow GM and proportional counters discussed in Sections 5.3,A and 5.3,B are excellent absolute detectors for the wavelength region between about 30 and 300 Å. The detector shown in Fig. 5.18 has been used for calibrating rocket-borne spectrometers in this wavelength region. Because of its small size, it can be inserted between the entrance slit and grating of the spectrometer to determine $S(\lambda)$. Transfer standard detectors calibrated at the National Bureau of Standards (NBS) are now generally used for photometric calibrations in the wavelength region between about 200 and 2500 Å. The NBS transfer standard used for the wavelength region 200–1200 Å is a windowless photodiode having an Al_2O_3 cathode (Canfield *et al.*, 1973; Saloman and Ederer, 1975). The quoted probable error in the calibration of this detector varies from about 6 to 10%, depending upon the wavelength. To avoid contamination of the photodiode, which can result from its continual use in a vacuum system during the long times generally required for calibration, it is advisable to use the NBS standard detector to calibrate a secondary photodiode detector having an aged high-work cathode such as tungsten. This secondary detector can then be used to calibrate the detection system and can be recalibrated periodically with the NBS standard detector. The NBS detector should also be recalibrated periodically at NBS.

Any photodiode detector should be operated so that the emitted cathode photoelectrons are measured rather than the electrons collected by the anode. This is necessary because a significant fraction of the emitted electrons may not be collected by the anode. This mode of operation for a photodiode is illustrated in Fig. 5.24. The photodiode should be operated at a voltage well above the voltage corresponding to current saturation, where the emitted current becomes constant with voltage. The current saturation region for a particular photodiode, which is also plotted in Fig. 5.24, occurs at voltages greater than ~20 V. The photoelectron emission current from a photodiode is generally small. The current I is given by the relationship

$$I = S(\lambda)Y(\lambda)e, \tag{5.12}$$

where $S(\lambda)$ is the incident photon rate, Y the photoelectric yield of the cathode, and e, the electron charge, is 1.6×10^{-19} C. As an example, when S is 10^6 photons/sec and Y 0.1 electrons/photon, the cathode current

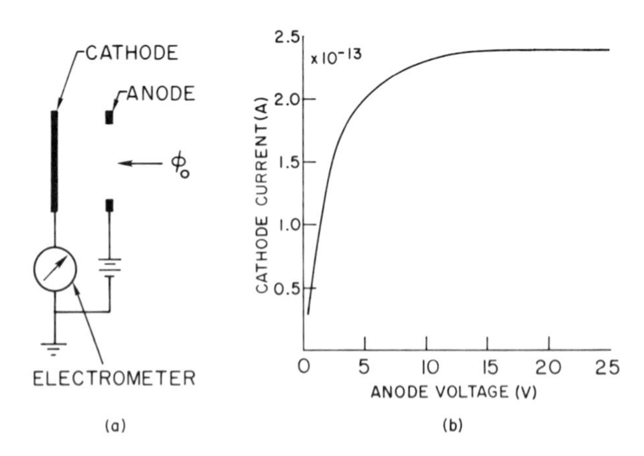

FIG. 5.24. Operation of a photodiode detector. (a) The electrometer is connected to the photocathode to measure all emitted cathode photoelectrons. (b) The photocathode current of a particular photodiode as a function of anode voltage for constant illumination. Current saturation is apparent near 20 V, where the current becomes constant with increasing anode voltage. ϕ_0 represents the incident radiation.

will be 1.6×10^{-14} A. This current can be measured with commercially available electrometers. The calibration data supplied with a transfer standard detector gives the quantum efficiency of the diode as a function of wavelength. To calibrate a detection system, the quantity $S(\lambda)$ is, therefore, obtained from the measured current and the quantum efficiency of the standard detector, as is evident from Eq. (5.12).

Current measurements with electrometers are difficult and often impossible when the incident photon flux is considerably less than 10^6 photons/sec. For these low-level flux measurements, photon counting detectors can be used. In this case, a photomultiplier detector, e.g., the Galileo M306 magnetic electron multiplier, operated as a photon counter, is first calibrated with the NBS standard detector. The calibration of the standard detector is said to be "transferred" to the multiplier which now becomes a secondary standard. The photomultiplier is calibrated at several wavelengths for which the photon flux is sufficient to obtain reliable current readings with the standard detector. It is necessary to insure that the photon counting system is operated within its linear region of counting during this calibration. If the counting rate exceeds the linear range of counting, nickel mesh grids having a known transmission can be mounted in front of the photomultiplier to reduce the response of the detector. The use of a photomultiplier as a secondary standard will then allow valid photometric calibration to be made for values of $S(\lambda)$ less than 10^3 photons/sec. Mechanical fixtures usually can be designed to permit the insertion of either a photodiode or a photomultiplier into the incident beam for obtaining photometric calibrations. This allows calibrations to be made over a wide range of incident flux and wavelength.

Transfer standard photodiode detectors for use at wavelengths between about 1100 and 3000 Å are also described by Canfield *et al.* (1973). These detectors are sealed and have MgF_2 windows and either Cs_2Te or Rb_2Te photocathodes. When calibrated at the National Bureau of Standards, the photodiode Model 543 P-09-00 manufactured by Electro-Mechanical Research Inc. is a stable transfer standard for the wavelength region 1150–2500 Å. This detector has a typical quantum efficiency of about 6% in the region 1150–2600 Å. Its quantum efficiency decreases rapidly outside this wavelength region because of a decrease in the window transmittance below 1150 Å and a decrease in cathode quantum efficiency above ~2600 Å. The quoted probable errors in the NBS calibration is ~6%, This photodiode should be operated in a manner similar to that of the windowless photodiode. The emitted cathode photoelectrons should be measured, and the diode should be operated in its current saturation region. The photodiode should also be recalibrated periodically at NBS.

Fastie and Kerr (1975) describe a versatile calibration system that has been used to obtain radiometric calibrations of spectrometers in the wavelength region 1200–1700 Å. The system also uses dispersed radiation to illuminate the aperture of the spectrometer being calibrated and transfer standard detectors to determine the flux incident on the aperture of the spectrometer. However, with this system, a narrow calibrated beam can be directed to different points of the entrance slit of the spectrometer being calibrated. The spectrometer is also mounted on a gimbaled platform so that the calibrated beam can be directed to different areas of the grating. In contrast to the previously described calibration procedure that gives the radiometric efficiency of the spectrometer averaged over the aperture, the calibration procedure described by Fastie and Kerr gives the efficiency of the spectrometer as a function of position of illumination on the entrance slit and on the grating. A calibration of this type would be necessary, for example, to measure spatially resolved radiation emitted from the solar disk rather than the integrated flux emitted from the full solar disk.

B. Calibration with Nondispersed Radiation Sources

Radiometric calibrations at wavelengths longer than ~1500 Å that are made with standard sources of spectral radiance or irradiance do not require dispersion of the radiation. When these sources are used, the entrance aperture of the spectrometer being calibrated is illuminated directly with the standard source. There are several standard sources that have been used for radiometric calibrations. A survey of the sources available for use at wavelengths above 1500 Å is given by Cann (1969). Since that survey, many of the sources and techniques described by him have been improved considerably. The use of the synchrotron as a standard source of radiation is now well established (Madden *et al.,* 1967; Ederer *et al.,* 1975). The wall-stabilized arc standard source (Ott *et al.,* 1973) is also an accurate standard source of spectral radiance. Other discharge sources, such as the deuterium lamp (Pitz, 1969), have also been used as standard sources of spectral radiance. The synchrotron source and the wall-stabilized arc source, however, are not portable and are available at a few facilities only. The most widely used standard sources are the quartz halogen lamp standard of spectral irradiance and the tungsten ribbon filament lamp standard of spectral radiance. Both sources are commercially available with calibrations based on NBS standards. These are convenient standard sources because of their accuracy, stability, and simplicity of operation. These sources are ordinarily used for calibrations for wavelengths longer than ~2500 Å. However, Buckley (1971) has shown that the tungsten ribbon filament lamp, when fitted with a sapphire window, can be used as a stable and reproducible calibration source down to 1500 Å.

Quartz halogen lamp standards of spectral irradiance and power supplies to operate them are commercially available from Eppley Laboratories, Inc., Newport, Rhode Island. Because the brightness of these lamps is not uniform over the emitting area, these sources are not focused on the aperture of the spectrometer being calibrated. Figure 5.25 illustrates the geometry that is used to calibrate spectrometers with a spectral irradiance standard source. These continuum radiation sources are supplied with calibration curves that give the spectral irradiance B measured at a fixed distance D_1 from the lamp in units of photons/cm^2 sec Å. The flux in the same units measured at a distance D_2 from the lamp is obtained by using the inverse square law for the ratio of the distance D_2 to D_1. To obtain a radiometric calibration, the aperture of the spectrometer being calibrated is illuminated directly with the standard lamp as shown in Fig. 5.25. It is important that the radiation from the standard lamp does not overfill the grating of the spectrometer being calibrated. This condition is met by positioning the spectrometer at a distance D_2 from the lamp such that the emitting area of the lamp falls within the angular aperture Ω of the spectrometer. Under this condition, the projected area viewed by the spectrometer at the lamp distance will be greater than the emitting area of the lamp itself. The counting rate of the spectrometer, $N_0(\lambda)$, at wavelength λ is given by

$$N_0(\lambda) = (D_1/D_2)^2 B(\lambda) A \, \Delta\lambda_i \, K(\lambda), \tag{5.13}$$

where A and $\Delta\lambda_i$ are the entrance slit area and instrumental linewidth of the spectrometer, respectively, and $K(\lambda)$ is the radiometric efficiency of the spectrometer in units of counts/photon. The quantity $\Delta\lambda_i$ for the spectrometer can be either calculated or measured experimentally. If wave-

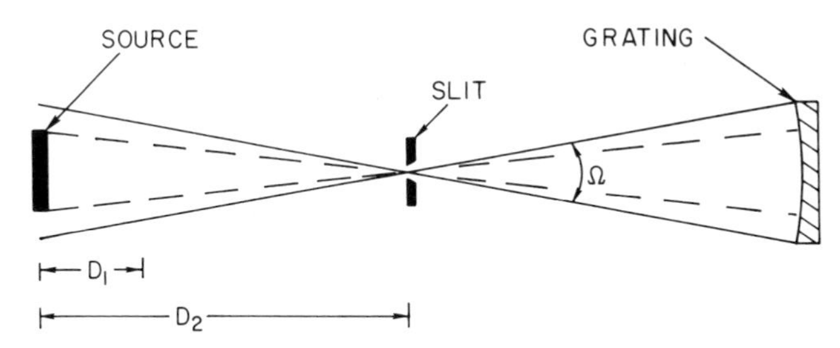

FIG. 5.25. The optical layout used to calibrate a spectrometer with a spectral irradiance standard source. The source distance D_2 is large enough to insure that the emitting area of the source falls within the angular aperture Ω of the spectrometer. D_1 is the distance from the source for which the spectral irradiance of the source is given.

length scanning is accomplished by stepping either the exit slit or the grating, the number of steps corresponding to the FWHM intensity of a monochromatic emission line will establish $\Delta\lambda_i$ as discussed in Section 5.4,A. When the calibrated spectrometer is used to measure line spectra, the count rate of the line N_0 can be determined from the peak reading of the line profile or from the integrated counts obtained by summing the counts of the line and dividing the sum by the steps corresponding to the FWHM intensity. For blended or broad lines, the integrated count rate is used to determine N_0. This technique can also be used to determine N_0 produced by a continuous source of radiation. However, to determine the flux emitted from a continuum source, a simpler method would be to sum the counts over fixed wavelength intervals during calibration with the continuum standard source. The sum of the counts for the wavelength intervals as a function of wavelength can then be applied to obtain the flux emitted from the source to be measured in the same wavelength intervals.

Standard sources of spectral radiance are also used for radiometric calibrations. For these sources, the spectral radiance of the lamp is given in units of photons/cm^2 sec sr Å. The lamps are calibrated for an emitting area of uniform brightness. With these sources, the emitting area is focused on the entrance slit of the spectrometer being calibrated, as shown in Fig. 5.26. The area of the source that is viewed by the spectrometer is determined by the optical parameters of the components used to illuminate the spectrometer aperture. The divergence of the beam entering the aperture is limited by a diaphragm to insure that the grating is not overfilled. The position and area of the diaphragm, therefore, determine the solid angle of the cone of radiation from the standard source that enters the aperture of the spectrometer. The rate of photons per increment of wave-

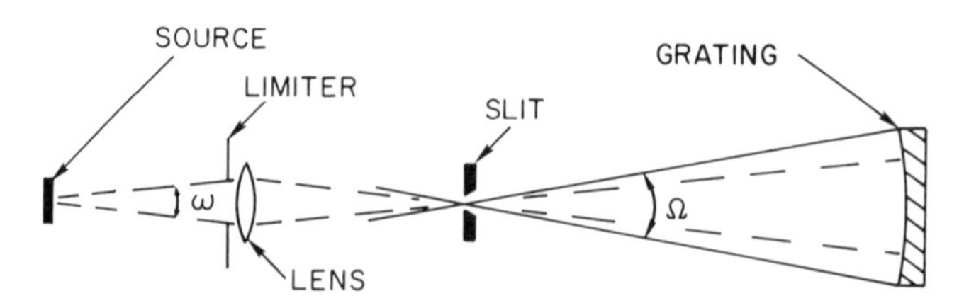

FIG. 5.26. The optical arrangement used to calibrate a spectrometer with a spectral radiance standard source. The lens focuses part of the emitting area of the standard lamp on the spectrometer entrance slit. The limiter restricts the divergence of the beam entering the spectrometer to a value less than the angular aperture Ω of the spectrometer. ω is the solid angle of the photon beam that enters the aperture of the spectrometer. The area of the source that is viewed by the spectrometer must also be determined from the parameters of the optical components.

length entering the entrance slit having an area A can then be determined and used to calibrate the spectrometer.

The branching ratio method is also an effective technique for obtaining radiometric calibrations of spectrometers. This method consists of measuring the counting rates of groups of atomic spectral lines or molecular lines and bands emitted from common upper levels for which the transition probabilities are known. The radiation source is viewed directly without predispersion. If absorption of the line or band emitted by the source is negligible, the emitted intensity in photons/sec is given by

$$I_{nl} = N_n A_{nl}, \tag{5.14}$$

where N_n is the population of the upper level n and A_{nl} the transition probability. The intensity ratio of any pair of emission lines from a common upper level n to lower levels m and l, as shown in Fig. 5.27, having transition probabilities A_{nm} and A_{nl} is independent of the upper level population and given by

$$I_{nl}/I_{nm} = A_{nl}/A_{nm}. \tag{5.15}$$

Therefore, when the transition probabilities are known, the relative intensities of the line pairs can be determined and from them a relative calibration at the wavelengths of the line pairs obtained. To apply the branching ratio techniques to radiometric calibrations below ~1200 Å, atomic emission lines are customarily used. Emission lines from hydrogenlike ions are ideally suited for this method, because their probabilities can be calculated precisely. However, when hydrogenlike ions are used, it is important that the closely spaced sublevels of the upper level be populated according to their statistical weights. These sublevels will be statistically populated in a reasonably dense plasma where frequent collisions of the excited ions occur during the lifetime of the excited levels. If statistical populations of the sublevels are achieved, the average transition prob-

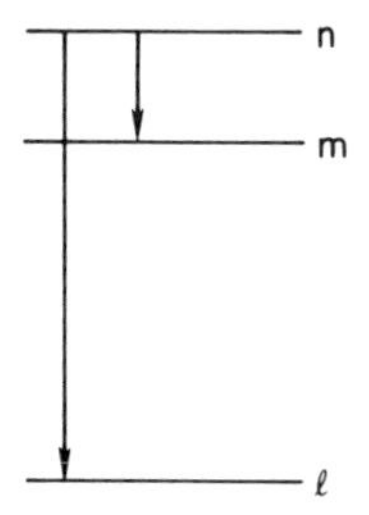

FIG. 5.27. Illustration of line pairs originating from a common upper level n that can be used to apply the branching ratio method to obtain radiometric calibrations.

abilities for the transition from the upper level to the lower levels can be used to determine the intensity ratio of the line pair. The average values of the transition probabilities for hydrogen are tabulated by Wiese *et al.* (1966). The average values of transition probabilities for other hydrogenlike ions can also be determined from this tabulation.

When lines of other species of atoms and ions are used, the accuracy of the relative calibration will be determined by the accuracy of the transition probabilities. Although the number of lines available for use at wavelengths below ~1200 Å is limited, this method can be used to obtain radiometric calibrations, as has been demonstrated by Griffin and McWhirter (1962) and Hinnov and Hofmann (1963). To establish an absolute calibration with the branching ratio method, the absolute intensity of one of the lines must be known. Typically the line pair is selected so that one of the lines falls in the far ultraviolet region and one in the visible or near ultraviolet region, where it is relatively simple to obtain a radiometric calibration of the spectrometer with standard sources. As an example, the branching ratio method has been applied to the Lyman β line at 1026 Å and the Balmer H_α line at 6563 Å, both of which originate from the $n = 3$ level of atomic hydrogen. A measurement of the absolute intensity of the visible line at 6563 Å establishes the absolute intensity of the UV line at 1026 Å provided that the $n = 3$ level is statistically populated and the lines are optically thin. There are many other pairs of lines of atomic species for which the branching ratio method has been used, e.g., line pairs from neutral helium (Van Eck *et al.*, 1963) and from singly ionized helium and the lithiumlike ions C(IV), N(V), O(VI), and Ne(VIII) (Boland *et al.*, 1968; Hinnov, 1976). The wavelengths of several emission lines that have been used for the branching ratio method are given in Table 5.I.

The branching ratio technique is particularly well suited for obtaining photometric calibrations in the wavelength region from ~1100 to 2500 Å, which contains many closely spaced molecular lines and bands that satisfy the criteria for the application of the branching ratio method. Aarts and DeHeer (1968) applied the branching ratio method to radiometric calibrations between 1500 and 2600 Å by comparing the intensities of vibrational transitions from the same upper levels of the fourth positive system of CO and the first negative systems of CO^+, which are tabulated in their paper. Becker *et al.* (1971) have used relative intensity measurements in band progressions of the Lyman system of H_2 and HD to obtain a relative calibration of a spectrometer in the region 1100–1650 Å. Their calibrations were extended to 1820 Å by using band progressions of the Lyman–Birge–Hopfield (LBH) system of nitrogen. In their paper, Becker *et al.* give the wavelengths and transition probabilities of many lines and bands of H_2, HD, and N_2 that can be used to obtain calibrations between

TABLE 5.I

SOME EMISSION LINES USED IN THE BRANCHING RATIO CALIBRATION[a]

Transition[b]		Wavelength (Å)	g(upper)[c]	$A(10^8\ sec^{-1})$[d]
H	1–3 (Ly β)	1025.72	18	0.575
	2–3 (H α)	6562.80	18	0.441
HeI	$1s^2$–1s3p	537.03	3	5.66
	1s2s–1s3p	5015.68	3	0.134
HeI	$1s^2$–1s4p	522.21	3	2.46
	1s2s–1s4p	3964.73	3	0.072
HeII	1–3	256.32	18	8.92
	2–3	1640.4	18	7.06
HeII	1–4	243.03	32	2.04
	2–4	1215.09	32	1.35
	3–4	4685.68	32	1.44
HeII	1–5	237.33	50	0.66
	3–5	3203.15	50	0.35
C(IV)	2s–3p	312.43	6	45.6
	3s–3p($P_{3/2}$)	5801.51	4	0.32
N(V)	2s–3p	209.28	6	120
	3s–3p($P_{3/2}$)	4603.83	4	0.42
O(VI)	2s-3p	150.10	6	259
	3s-3p($P_{3/2}$)	3811.35	4	0.51
Ne(VIII)	2s-3p	88.10	6	853
	3s-3p	2860.1	6	0.70

[a] Data from Wiese *et al.* (1966).
[b] Principal quantum number given for H and HeII.
[c] g is the statistical weight of upper level.
[d] Average transition probabilities of upper level n given for H and HeII.

~1100 and 1820 Å with estimated errors of about 5% in the relative calibration. Mumma and Zipf (1971) have also used the LBH bands of N_2 to obtain a relative calibration of a spectrometer in the wavelength region 1275–1850 Å and the CO fourth positive system to extend the calibration to 2500 Å.

A proportional counter can also be used to obtain radiometric calibrations at wavelengths shorter than ~120 Å. When a proportional counter is used, the characteristic radiation emitted from an x-ray source is viewed directly without predispersion. The discriminator of the pulse counting apparatus is adjusted to count all pulses of the distribution of anode voltage pulses produced by a particular emission band radiated by the source. Since the mean voltage pulse height of the anode distribution is proportional to the photon energy of the emission band, the discriminator window can be set to record only those pulses originating from a particular wavelength. However, when more than one emission band is produced by the source, the wavelength separation of the emission bands must be large enough to insure that the pulse distributions do not overlap. Because the width of the pulse height distribution narrows with decreasing wavelength (see Eq. 5.10 and Fig. 5.21), the ability of the counter to resolve different emission bands increases with decreasing wavelength. As discussed by Manson (1972), the pulse height resolution of a proportional counter is sufficient to resolve the characteristic x-ray emission bands of nitrogen, carbon, boron, and berylium that are distributed in wavelength between ~32 and 113 Å. These bands were used by Manson to calibrate a rocket spectrometer in the wavelength region 31–130 Å. Mack *et al.* (1976) have used some of these bands and also bands of silicon and aluminum to measure the quantum efficiencies of photomultipliers between 44 and 170 Å. The production and identification of these x-ray emission bands are discussed below in Section 5.5,C.

When the transmittance of the proportional counter window is known, the integrated counts of the pulse height distribution produced by a particular emission band will give the absolute photon rate of the band radiation incident on the counter window, since each anode pulse corresponds to the absorption of one photon within the active volume of the counter. The proportional counter can be used, therefore, as a standard detector to calibrate a spectrometer with the same procedures as those discussed in Section 5.4,A for the case when the spectral width of the incident radiation illuminating the aperture of the spectrometer is wider than the instrumental linewidth of the spectrometer.

5.5. Ultraviolet Light Sources

There are a variety of light sources available to produce both continuum and emission line radiation throughout the ultraviolet region of the spectrum. An excellent survey of pulsed and dc discharge light sources is given by Samson (1967). Since 1967, several additional sources have been developed that have significantly increased the wavelength range of radiation available for radiometric calibrations. Several relatively simple light sources will be described here that have been used to obtain radiometric calibrations in the wavelength region from ~30 to 2500 Å. For wavelengths above 2500 Å, commercially available quartz halogen and tungsten ribbon filament standard lamps are widely used for radiometric calibrations, as discussed in Section 5.4,B. The sources described here are limited to steady state sources of radiation produced with dc gas-discharge and x-ray sources. These sources, in contrast to pulsed sources, are electrically quiet and emit radiation continuously. They are, therefore, ideally suited for radiometric calibrations of photon counting systems, when the calibration is carried out using dispersed radiation and transfer standard detectors.

A. Capillary Discharge Source

Capillary discharge sources have been used for many years to produce line radiation from neutral atoms and band spectra of molecules. Figure 5.28 shows one of these sources described by Newburgh *et al.* (1962), which is rugged and easily constructed. As indicated in the figure caption, the source consists of a high-purity 1100 aluminum alloy cathode having fins for air cooling. The discharge is confined within a water-jacketed capillary that has an inner diameter of 5 mm. The outer diameter of the water jacket is 25 mm. The entire capillary assembly is constructed by quartz tubing and has an overall length of ~10 cm. The capillary is joined to the cathode and to the water-cooled entrance flange of a monochromator by compressing neoprene O rings between spacers with threaded brass collars. Gas enters the source through glass or plastic tubing connected to the cathode with a Swagelock or similar vacuum fitting. The gas flows through the capillary and through the slit of a differential pumping unit. The entrance slit of the monochromator (not shown) is mounted parallel to the differential pumping slit, so that the discharge is viewed through both slits along the axis of the capillary. A differential pumping slit may not be necessary when the capillary gas pressure is less

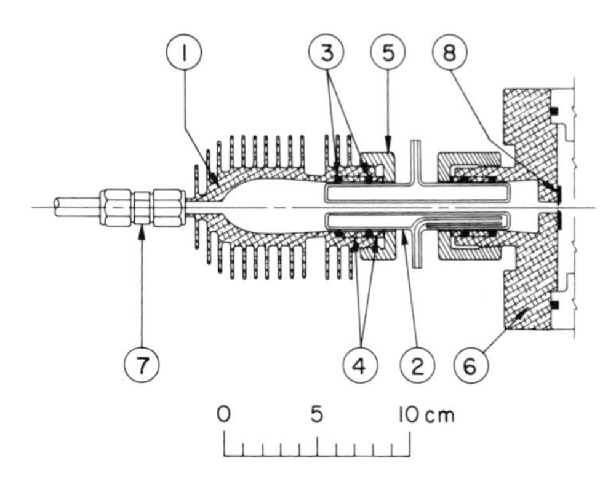

FIG. 5.28. Capillary discharge source: air-cooled aluminum cathode (1), water-jacketed quartz capillary (2), O ring seals (3), aluminum spacers (4), brass collar to compress O rings (5), monochromator entrance flange (6), Swagelok vacuum fitting (7), and slit to differential pumping unit (8). [Courtesy of Newburgh *et al.* (1962, p. 734).]

than ~1 Torr and the spectrometer to be calibrated is contained in a separately pumped vacuum chamber that is isolated from the monochromator chamber except for the area of the spectrometer entrance slit (see Fig. 5.23).

To operate the lamp, the gas flow is controlled with a variable leak valve to maintain a gas pressure of ~1 Torr in the capillary. As shown in Fig. 5.29, the cathode of the source is connected in series with a ballast resistor to a high-voltage power supply. The cathode is maintained at negative high voltage and the anode flange at ground potential. A power supply having a voltage range extending to 3 kV and a current range to 500 mA can be used with a 2–3 kΩ ballast resistor. When operated with commercial grade gases at pressures of ~1 Torr, the capillary source will produce intense radiation from the resonance lines of H, He, N, O, Ne, Ar, and Kr. Additional resonance series lines of these neutral atoms are also produced although their intensities are lower than those of the resonance lines. If a trace of hydrogen is added to argon, the resonance series lines of hydrogen become the dominant hydrogen lines of the spectrum. Since the optimum discharge pressure will vary with different gases, the pressure should be adjusted to achieve maximum line intensities and stable discharge operation. Plots of the spectra of several gases produced with a capillary source are given by Samson (1967).

When hydrogen is used, the molecular hydrogen spectrum is produced, which consists of a sharp band structure extending from ~900 to 1600 Å (Samson, 1967). When a mixture of CO_2 and He is used, the capillary source produces radiation from many closely spaced emission bands of the fourth positive system of CO distributed throughout the UV region from 1280 to 2800 Å (Krupenie, 1966). These emission bands can be used to obtain radiometric calibrations when the procedures described in Section 5.4,A for obtaining calibrations with transfer standard detectors and emission band radiation are used. The capillary discharge can also be operated in a different mode to produce the molecular helium continuum in the 600–900 Å region (Newburgh *et al.*, 1962). This mode of operation

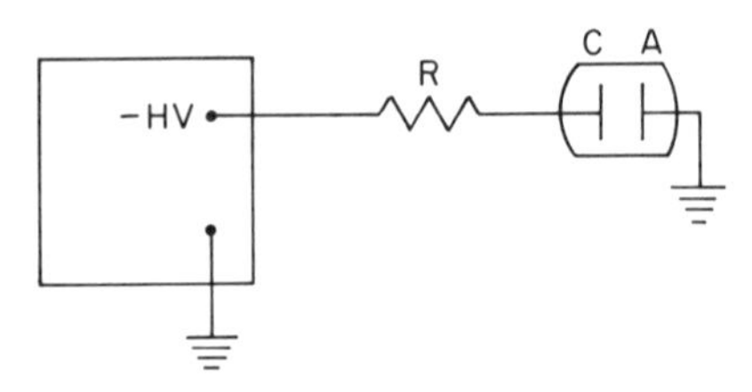

FIG. 5.29. Schematic of electrical circuit for operation of capillary source. R is the ballast resistor; C, the cathode, and A, the anode.

necessitates high-purity He at pressures well above 100 Torr, and therefore requires differential pumping between the source and entrance slit of the monochromator. Although a dc discharge can be used to produce the He_2 continuum, a pulsed discharge gives a continuum of higher intensity at lower pressures. However, the operation of the source in a pulsed mode generally is not compatible with photon counting techniques.

B. Hollow-Cathode Source

The spectral lines of neutral atoms produced with a capillary discharge source fall predominantly at wavelengths longer than ~500 Å. The spectral lines of the ion species, however, fall in general at shorter wavelengths than those of the neutral atoms. A hollow-cathode discharge source is an efficient source for producing the ion emission lines as well as the neutral atomic lines. This source can be used to produce emission line radiation at wavelengths extending down to ~150 Å.

The hollow-cathode source described here is based on a source described by Newburgh *et al.* (1962) and Newburgh (1963) that has been modified for use with a water-cooled cathode and anode. A brief description of the source has been given by Manson (1973). The water-cooled cathode was first introduced to permit operation of the source at high power levels to test ultraviolet spectrometers for satellite instrumentation (Bedo and Hinteregger, 1965). The source is stable at low pressures and the discharge easily started at voltages of between 2 and 3 kV. Paresce *et al.* (1971) also describe a continuous discharge source that is similar in geometry and operation to the source described here, but is substantially different in mechanical design.

The source is shown in cross section in Fig. 5.30. Both the cathode and the anode are constructed of high-purity 1100 aluminum alloy. The inner diameter of the cathode is 1.75 cm, the outer diameter 5 cm, and the length of the cathode from the opening to the surface of the cathode flange 20 cm. Water enters and leaves the cathode jacket through the cathode flange, which has a thickness of ~2 cm. The water-cooled anode has an inner diameter of 6.25 cm, an outer diameter of 9 cm, and an overall length of 11 cm. The cathode structure is fastened to a commercially available Pyrex pipe which has end grooves to accept either O rings or standard Teflon type T gaskets. The mechanical connection between the cathode flange and Pyrex pipe is made with a flange and a molded fiber insert manufactured for use with the pipe. Identical fixtures are used to connect the pipe to the anode flange, as can be seen in the figure. The assembled source can then be bolted to the entrance flange of the monochromator and sealed with an O ring.

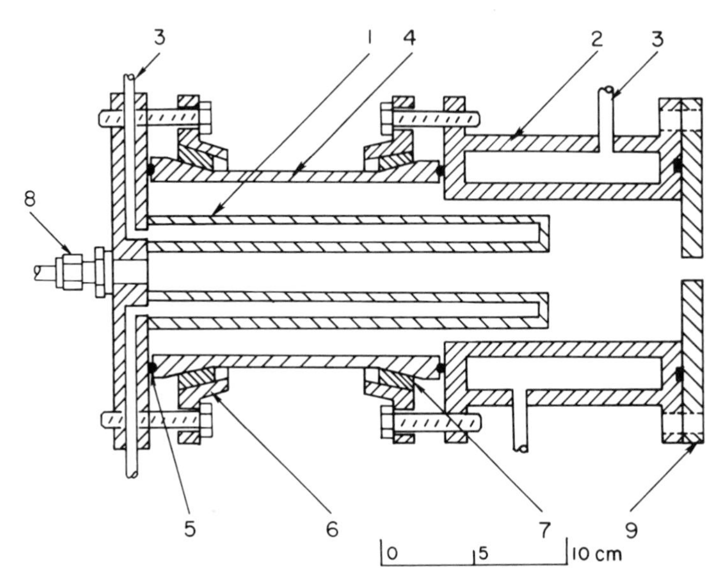

FIG. 5.30. Hollow cathode source: water cooled cathode (1), water-cooled anode (2), water connections (3), Pyrex 3-in. glass pipe (4), O ring seals (5), aluminum flanges (6), molded fiber inserts (7), gas inlet tubing (8), and monochromator entrance flange (9).

The operation of the hollow-cathode source is similar to that of the capillary in that the gas flow is controlled with a variable leak valve, and the gas passes through the source and then through the entrance slit of the monochromator. The same power supply, ballast resistor, and electrical connections as those used for the capillary source can be used with the hollow cathode source. Commercial grade gas is also suitable for source operation. The gas flow and voltage should be adjusted until the discharge strikes within the cathode. When the source is operating properly, a bright discharge will be confined within the cathode, and only a very faint discharge will be visible in the region outside of the cathode. After this mode of operation has been achieved, the pressure and voltage can be varied to obtain maximum intensity of the emission lines. By adjusting the pressure and voltage, many emission lines of single and doubly ionized species having wavelengths between ~150 and 1200 Å can be produced with sufficient intensity to obtain radiometric calibration. Paresce *et al.* (1971) give plots of the spectra of He, Ne, Ar, and H when these gases are used with their source. When the source of Fig. 5.30 is operated with a mixture of 90% neon and 10% helium, emission lines of NeIII and IV between about 135 and 285 Å are produced (Manson, 1973). Manson gives a plot of the neon spectrum in this wavelength region measured with a GM counter having a VYNS window and a VYNS prefilter mounted between the hollow-cathode source and grating to protect the grating from contamination from the source. Because the window of the GM counter suppresses the intense neon lines between 300 and 800 Å, the signal-to-background ratio of the spectrum between 135 and 285 Å is high enough to provide emission line radiation for calibrations in this wavelength region.

After prolonged use, an insulating layer builds up on the inner surface of an aluminum cathode, so that the source becomes unstable and difficult to start. When this occurs, it is necessary to disassemble the source and remachine the inner cathode surface. The use of other cathode materials, e.g., stainless steel, will reduce this problem and increase the source stability. The water-cooled cathode of the source shown in Fig. 5.30 can be replaced with a cathode that is simpler to construct and is suitable for use when high power levels are not required and when the source is not operated within a vacuum chamber. This cathode was first used with a hollow-cathode source by Newburgh (1963). As shown in Fig. 5.31, the cathode is constructed of seamless tantalum tubing, approximately 20 cm long, 1.25-cm outer diameter, and 0.5 mm thick. The tantalum tubing is surrounded by Vycor tubing having inner and outer diameters of 1.45 and 1.75 cm, respectively, to protect the tantalum from positive ion bombardment. Both tubes are supported by an aluminum holder, which is attached to the cathode flange. This cathode structure is interchangeable with the water-cooled cathode structure shown in Fig. 5.30. The identical fixture is used to fasten the cathode assembly to the Pyrex pipe. The operation of the hollow-cathode source with this cathode structure is also stable over a wider range of gas pressure and voltage.

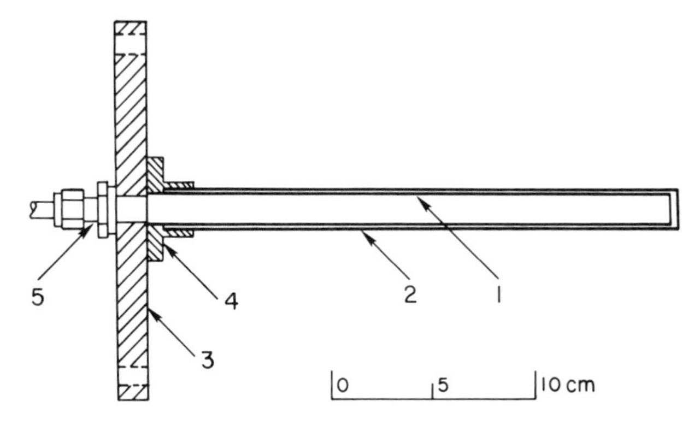

FIG. 5.31. Alternate cathode for hollow-cathode source: tantalum cathode (1), Vycor tube (2), aluminum flange (3), aluminum cathode support (4), and gas inlet tubing (5). [Courtesy of Newburgh (1963, p. 864).]

C. X-Ray Source

X-ray sources are widely used to extend the wavelength range of radiometric calibrations to wavelengths below ~150 Å. The K, L and M band radiation from several elements, e.g., nitrogen, carbon, molybdenum, boron, beryllium, and aluminum, are easily excited with simple sources and provide radiation in the wavelength range 31–180 Å. Caruso and Neupert (1965a,b) describe a particularly simple source consisting of an elliptically shaped anode and a V-shaped filament cathode, which they used to excite the K emission bands of C and B at 44 and 67 Å, respectively. Figure 5.32 illustrates a source developed by Manson (unpublished) that has been used frequently at the Air Force Geophysics Laboratory to obtain radiometric calibrations of solar spectrometers between 31 and 180 Å (Manson, 1972; Heroux *et al.,* 1972). This source is commercially available from the J. E. Manson Company, Concord, Massachusetts. The source, shown in Fig. 5.32, consists of a solid stainless steel cylindrical anode having a diameter of 10 mm and a length of 12 mm. The anode is fastened to a ceramic feed-through insulator, which in turn is fastened to an anode flange. The flange is fastened with screws to the aluminum body of the source so that the anode can be detached and various compounds deposited on the front surface of the anode. The aluminum body of the source has a cylindrical opening with a diameter of 12 mm. The axis of the anode coincides with the axis of the cylindrical opening. A Phillips EM-75 electron microscope filament is also aligned with this axis so that the tip of the V-shaped filament is located about

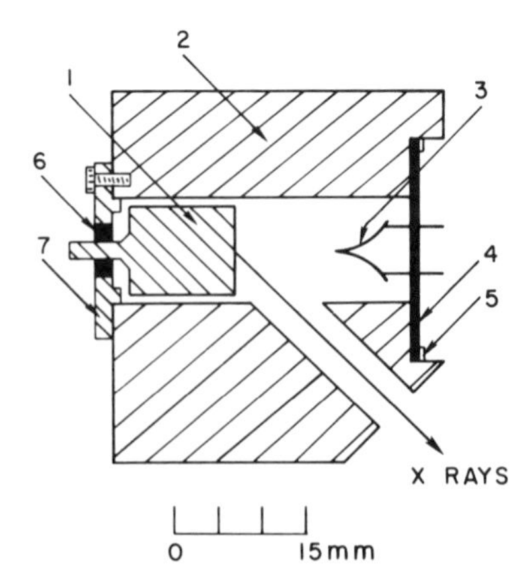

FIG. 5.32. X-ray source: stainless steel anode (1), aluminum body (2), filament (3), Teflon support (4), retaining ring (5), ceramic insulator (6), and anode support flange (7). [Courtesy of J. E. Manson (unpublished).]

10 mm from the front surface of the anode. The filament is mounted in a circular Teflon support that is fastened to the body of the source with a retaining ring. The radiation is viewed at 45° from the cathode normal through a cylindrical opening, 6 mm in diameter.

As shown in Fig. 5.33, the anode of the source is connected to positive high voltage through a resistor. The filament is heated with a step-down transformer, and the filament voltage varied with a VARIAC connected to the primary winding of the transformer. The filament current is typically in the range 4–5 A. The emission current to the anode can then be adjusted with the potentiometer in the filament circuit. An anode voltage of ~3 kV and an emission current of 0.2–0.3 μA is usually sufficient to produce x-ray band emissions of sufficient intensity for calibrations between 30 and 180 Å.

Various materials can be applied to the face of the anode to produce their characteristic emission bands. A slurry of the powder made with a few drops of water or alcohol can be brushed on the anode, and, after drying, the anode is fastened to the body of the source. This procedure can be used with powdered BN, B, Mo, Be, and Al to produce their characteristic emission bands. Aquadag can be painted directly on the anode face to produce the C–K emission band. Some emission bands that have been produced with this source are given in Table 5.II. Examples of the flux produced with the source are given in Table 5.III.

Examples of the x-ray emission bands produced with the source are given in Fig. 5.34, where the K emission bands of B and Be are plotted.

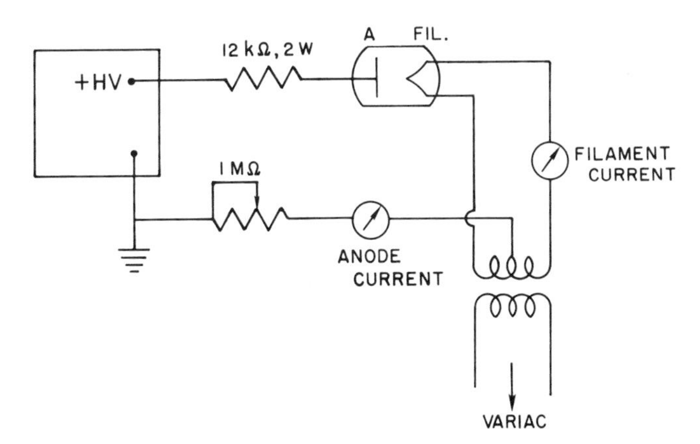

FIG. 5.33. Schematic of electrical circuit for operation of x-ray source. The filament current is controlled with a VARIAC. The anode current is controlled with a potentiometer connected to the center tap of the filament transformer. [Courtesy of J. E. Manson (unpublished).]

TABLE 5.II

SEVERAL X-RAY EMISSION BANDS PRODUCED WITH THE MANSON X-RAY SOURCE

Band	Wavelength (Å)
O–K	23.7
N–K	31.6
C–K	44.0
Mo–$M_V N_{III}$	64.2
B–K	67
Be–K	113
Al–L	180

TABLE 5.III

FLUX PRODUCED WITH X-RAY SOURCE[a]

Element	λ(Å)	Anode voltage (V)	Beam current (μA)	Flux (photons/sec sr)
C	44	1500	200	4×10^{10}
B	67	1500	60	7×10^{9}
		4000	200	9×10^{10}
Be	113	3000	200	3×10^{10}

[a] Courtesy of J. E. Manson, unpublished.

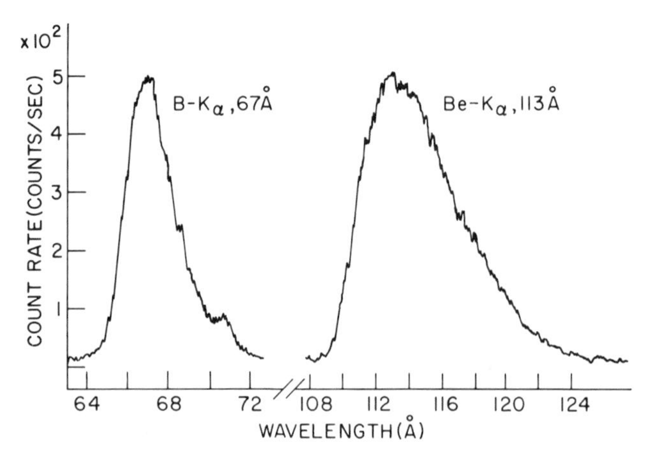

FIG. 5.34. The B–K and Be–K emission bands produced with the Manson x-ray source. The data were obtained with the source mounted on a grazing incidence spectrometer. A GM counter having a VYNS window was used as the detector.

The anode voltage and current used to produce these bands were 3 kV and 0.3 μA, respectively. The data were obtained with a GM counter mounted behind the exit slit of a grazing-incidence spectrometer. The GM counter, shown in Fig. 5.18, had a VYNS window and was operated with a mixture of 96% argon and 4% isobutene at a pressure of 200 Torr. As is evident in the figure, the widths of the emission bands vary with different elements. The Al–L band, which peaks near 180 Å, is a particularly broad band. In addition, contamination of the aluminum-coated anode during source operation often causes a slow decrease in the intensity of the band radiation. For these reasons, the use of a hollow-cathode source operated with a gas mixture of 90% neon and 10% helium rather than the x-ray source is preferable for obtaining radiometric calibrations near 180 Å.

An x-ray source described by Henke and Tester (1974) is also used frequently to excite x-ray emission bands in the approximate wavelength region 10–150 Å. Because of its construction, the anode of this source does not view the filament directly, and therefore anode contamination caused by material evaporated from the filament reaching the anode is significantly reduced. This source can be operated at high power levels for long periods of time and can be used to excite many characteristic x-ray bands, which are given in their paper.

REFERENCES

AARTS, J. F. M., and DEHEER, F. J. (1968). *J. Opt. Soc. Am.* **58,** 1666.

ANDRESEN, R. D., and PAGE, D. E. (1971). *Rev. Sci. Instrum.* **42,** 371.

BECKER, K. H., FINK, E. W., and ALLISON, A. C. (1971). *J. Opt. Soc. Am.* **61,** 495.

BEDO, D. E., and HINTEREGGER, H. E. (1965). *Jpn. J. Appl. Phys.* **4,** 473.

BOLAND, B. C., JONES, T. J. L., and MCWHIRTER, R. W. P. (1968). Calibration Methods in the Ultraviolet and X-Ray Regions of the Spectrum, ESRO Rep. No. SP-33, p. 59. Euorpean Space Research Organization, Paris.

BRIDGES, J. M., and OTT, W. R. (1977). *Appl. Opt.* **16,** 367.

BUCKLEY, J. L. (1971). *Appl. Opt.* **10,** 1114.

CAIRNS, R. B., and SAMSON, J. A. R. (1966). *J. Opt. Soc. Am.* **56,** 1568.

CANFIELD, L. R., JOHNSTON, R. G., and MADDEN, R. P. (1973). *Appl. Opt.* **12,** 1611.

CANN, M. W. P. (1969). *Appl. Opt.* **8,** 1645.

CARUSO, A. J., and NEUPERT, W. M. (1965a). *Appl. Opt.* **4,** 247.

CARUSO, A. J., and NEUPERT, W. M. (1965b). *Rev. Sci. Instrum.* **36,** 554.

DUCKETT, S. W., and METZGER, P. H. (1965). *Phys. Rev.* **137,** A953.

EDERER, D. L., and TOMBOULIAN, D. H. (1964). *Appl. Opt.* **3,** 1073.

EDERER, D. L., SALOMAN, E. B., EBNER, S. C., and MADDEN, R. P. (1975). *J. Res. Nat. Bur. Std. A Phys. Chem.* **79A,** 761.

EVANS, D. S. (1965). *Rev. Sci. Instrum.* **36,** 375.

FASTIE, W. G., and KERR, D. E. (1975). *Appl. Opt.* **14,** 2133.

GRIFFIN, W. G., and MCWHIRTER, R. W. P. (1962). *In Conf. Opt. Instrum. Tech.* (K. J. Habell, ed.), p. 14. Chapman and Hall, London.

HENKE, B. L., and TESTER, M. A. (1974). *Adv. X-Ray Anal.* **18,** 76.
HENKE, B. L., WHITE, R., and LUNDBERG, B. (1957). *J. Appl. Phys.* **28,** 98.
HEROUX, L. (1968). *Appl. Opt.* **7,** 2351.
HEROUX, L., and HINTEREGGER, H. E. (1960). *Rev. Sci. Instrum.* **31,** 280.
HEROUX, L., and HINTEREGGER, H. E. (1962). *Appl. Opt.* **1,** 701.
HEROUX, L., MANSON, J. E., HINTEREGGER, H. E., and MCMAHON, W. J. (1965). *J. Opt. Soc. Am.* **55,** 103.
HEROUX, L., MCMAHON, W. J., and HINTEREGGER, H. E. (1966). *Appl. Opt.* **5,** 1338.
HEROUX, L., NEWBURGH, R. G., MCMAHON, W. J., and HINTEREGGER, H. E. (1968). *Appl. Opt.* **7,** 37.
HEROUX, L., COHEN, M., and MALINOVSKY, M. (1972). *Solar Phys.* **23,** 369.
HINNOV, E. (1976). *Phys. Rev. A* **14,** 1533.
HINNOV, E., and HOFMANN, F. (1963). *J. Opt. Soc. Am.* **53,** 1259.
HUDSON, R. D., and KIEFFER, L. J. (1971). *In* "Atomic Data," Vol. 2, 205–262. Academic Press, New York.
KRUPENIE, P. H. (1966). "The Band Spectrum of Carbon Monoxide," NSRDS-NBS 5. U.S. Government Printing Office, Washington, D.C.
LAMPTON, M., and PRIMBSCH, J. H. (1971). *Rev. Sci. Instrum.* **42,** 731.
LAPSON, L. B., and TIMOTHY, J. G. (1973). *Appl. Opt.* **12,** 388.
LAPSON, L. B., and TIMOTHY, J. G. (1976). *Appl. Opt.* **15,** 1218.
LOMBARD, F. J., and MARTIN, F. (1961). *Rev. Sci. Instrum.* **32,** 200.
LUKIRSKII, A. P., RUMSH, M. A., and SMIRNOV, L. A. (1960a). *Opt. Spectrosc. (USSR)* **9,** 265.
LUKIRSKII, A. P., RUMSH, M. A., and KARPOVITCH, I. A. (1960b). *Opt. Spectrosc. (USSR)* **9,** 343.
MACK, J. E., PARESCE, F., and BOWYER, S. (1976). *Appl. Opt.* **15,** 861.
MADDEN, R. P., EDERER, D. L., and CODLING, K. (1967). *Appl. Opt.* **6,** 31.
MANSON, J. E. (1967). *Ap. J.* **147,** 703.
MANSON, J. E. (1972). *Solar Phys.* **27,** 107.
MANSON, J. E. (1973). *Appl. Opt.* **12,** 1394.
MUMMA, M. J., and ZIPF, E. C. (1971). *J. Opt. Soc. Am.* **61,** 83.
NEWBURGH, R. G. (1963). *Appl. Opt.* **2,** 864.
NEWBURGH, R. G., HEROUX, L., and HINTEREGGER, H. E. (1962). *Appl. Opt.* **1,** 733.
OTT, W. R., FIEFFE-PREVOST, P., and WIESE, W. L. (1973). *Appl. Opt.* **12,** 1618.
PARESCE, F., KUMAR, S., and BOWYER, C. S. (1971). *Appl. Opt.* **10,** 1904.
PITZ, E. (1969). *Appl. Opt.* **8,** 255.
PRESCOTT, J. R. (1966). *Nucl. Instrum. Methods* **39,** 173.
RUMSH, M. A., LUKIRSKII, A. P., and SHCHEMELEV, V. N. (1960). *Dokl. Akad. Nauk USSR* **135,** 55 [*English trans.: Sov. Phys.-Dokl.* **5,** 1231].
SALOMAN, E. B., and EDERER, D. L. (1975). *Appl. Opt.* **14,** 1029.
SAMSON, J. A. R. (1967). "Techniques of Vacuum Ultraviolet Spectroscopy." Wiley, New York.
TIMOTHY, J. G., and BYBEE, R. L. (1978). *Rev. Sci. Instrum.* **49,** 1192.
TIMOTHY, J. G., and LAPSON, L. B. (1974). *Appl. Opt.* **13,** 1417.
TIMOTHY, A. F., TIMOTHY, J. G., and WILLMORE, A. P. (1967). *Appl. Opt.* **6,** 1319.
VAN ECK, J., DEHEER, F. J., and KISTEMAKER, J. (1963). *Phys. Rev.* **130,** 656.
WIESE, W. L., SMITH, M. W., and GLENNON, B. M. (1966). "Atomic Transition Probabilities," Vol. 1. U.S. Government Printing Office, Washington, D.C.
YOUNG, A. T., and SCHILD, R. E. (1971). *Appl. Opt.* **10,** 1668.

Index

E

F

G

H

I

J

K

L

M

N

O

P

T

U

V

W

Z